Introduction to Embedded Systems

Introduction to Embedded Systems

A Cyber-Physical Systems Approach

Second Edition

Edward Ashford Lee and Sanjit Arunkumar Seshia

The MIT Press
Cambridge, Massachusetts
London, England

This book was set in Times Roman using LaTeX.

Library of Congress Cataloging-in-Publication Data:

Names: Lee, Edward A., 1957- author. | Seshia, Sanjit A., author.
Title: Introduction to embedded systems : a cyber-physical systems approach / Edward A. Lee and Sanjit A. Seshia.
Description: Second edition. | Cambridge, MA : MIT Press, [2017] | Includes bibliographical references and index.
Identifiers: LCCN 2016039490 | ISBN 9780262533812 (pbk. : alk. paper)
Subjects: LCSH: Embedded computer systems.
Classification: LCC TK7895.E42 L445 2017 | DDC 006.2/2--dc23 LC record available at https://lccn.loc.gov/2016039490

This book is dedicated to our families.

Contents

Preface

What This Book Is About

The most visible use of computers and software is processing information for human consumption. We use them to write books (like this one), search for information on the web, communicate via email, and keep track of financial data. The vast majority of computers in use, however, are much less visible. They run the engine, brakes, seatbelts, airbag, and audio system in your car. They digitally encode your voice and construct a radio signal to send it from your cell phone to a base station. They control your microwave oven, refrigerator, and dishwasher. They run printers ranging from desktop inkjet printers to large industrial high-volume printers. They command robots on a factory floor, power generation in a power plant, processes in a chemical plant, and traffic lights in a city. They search for microbes in biological samples, construct images of the inside of a human body, and measure vital signs. They process radio signals from space looking for supernovae and for extraterrestrial intelligence. They bring toys to life, enabling them to react to human touch and to sounds. They control aircraft and trains. These less visible computers are called **embedded systems**, and the software they run is called **embedded software**.

Despite this widespread prevalence of embedded systems, computer science has, throughout its relatively short history, focused primarily on information processing. Only recently have embedded systems received much attention from researchers. And only recently has the community recognized that the engineering techniques required to design and analyze these systems are distinct. Although embedded systems have been in use since the 1970s, for most of their history they were seen simply as small computers. The principal engineering problem was understood to be one of coping with limited resources (limited processing power, limited energy sources, small memories, etc.). As such, the engineering challenge was to optimize the designs. Since all designs benefit from optimization, the discipline was not distinct from anything else in computer science. It just had to be more aggressive about applying the same optimization techniques.

Recently, the community has come to understand that the principal challenges in embedded systems stem from their interaction with physical processes, and not from their limited resources. The term cyber-physical systems (CPS) was coined by Helen Gill at the National Science Foundation in the U.S. to refer to the integration of computation with physical processes. In CPS, embedded computers and networks monitor and control the physical processes, usually with feedback loops where physical processes affect computations and vice versa. The design of such systems, therefore, requires understanding the joint dynamics of computers, software, networks, and physical processes. It is this study of *joint* dynamics that sets this discipline apart.

When studying CPS, certain key problems emerge that are rare in so-called general-purpose computing. For example, in general-purpose software, the time it takes to perform a task is an issue of *performance*, not *correctness*. It is not incorrect to take longer to perform a task. It is merely less convenient and therefore less valuable. In CPS, the time it takes to perform a task may be critical to correct functioning of the system. In the physical world, as opposed to the cyber world, the passage of time is inexorable.

In CPS, moreover, many things happen at once. Physical processes are compositions of many things going on at once, unlike software processes, which are deeply rooted in sequential steps. Abelson and Sussman (1996) describe computer science as "procedural epistemology," knowledge through procedure. In the physical world, by contrast, processes are rarely procedural. Physical processes are compositions of many parallel processes. Measuring and controlling the dynamics of these processes by orchestrating actions that influence the processes are the main tasks of embedded systems. Consequently, concurrency is intrinsic in CPS. Many of the technical challenges in designing and analyzing embedded software stem from the need to bridge an inherently sequential semantics with an intrinsically concurrent physical world.

Why We Wrote This Book

The mechanisms by which software interacts with the physical world are changing rapidly. Today, the trend is towards "smart" sensors and actuators, which carry microprocessors, network interfaces, and software that enables remote access to the sensor data and remote activation of the actuator. Called variously the Internet of Things (IoT), Industry 4.0, the Industrial Internet, Machine-to-Machine (M2M), the Internet of Everything, the Smarter Planet, TSensors (Trillion Sensors), or The Fog (like The Cloud, but closer to the ground), the vision is of a technology that deeply connects our physical world with our information world. In the IoT world, the interfaces between these worlds are inspired by and derived from information technology, particularly web technology.

IoT interfaces are convenient, but not yet suitable for tight interactions between the two worlds, particularly for real-time control and safety-critical systems. Tight interactions still require technically intricate, low-level design. Embedded software designers are forced to struggle with interrupt controllers, memory architectures, assembly-level programming (to exploit specialized instructions or to precisely control timing), device driver design, network interfaces, and scheduling strategies, rather than focusing on specifying desired behavior.

The sheer mass and complexity of these technologies (at both the high level and the low level) tempts us to focus an introductory course on mastering them. But a better introductory course would focus on how to model and design the joint dynamics of software, networks, and physical processes. Such a course would present the technologies only as today's (rather primitive) means of accomplishing those joint dynamics. This book is our attempt at a textbook for such a course.

Most texts on embedded systems focus on the collection of technologies needed to get computers to interact with physical systems (Barr and Massa, 2006; Berger, 2002; Burns and Wellings, 2001; Kamal, 2008; Noergaard, 2005; Parab et al., 2007; Simon, 2006; Valvano, 2007; Wolf, 2000). Others focus on adaptations of computer-science techniques (like programming languages, operating systems, networking, etc.) to deal with technical problems in embedded systems (Buttazzo, 2005a; Edwards, 2000; Pottie and Kaiser, 2005). While these implementation technologies are (today) necessary for system designers to get embedded systems working, they do not form the intellectual core of the discipline. The intellectual core is instead in models and abstractions that conjoin computation and physical dynamics.

A few textbooks offer efforts in this direction. Jantsch (2003) focuses on concurrent models of computation, Marwedel (2011) focuses on models of software and hardware behavior, and Sriram and Bhattacharyya (2009) focus on dataflow models of signal processing behav- p.142
ior and their mapping onto programmable DSPs. Alur (2015) focuses on formal modeling, specification, and verification of cyber-physical systems. These are excellent textbooks that cover certain topics in depth. Models of concurrency (such as dataflow) and abstract models of software (such as Statecharts) provide a better starting point than imperative programming languages (like C), interrupts and threads, and architectural annoyances that a designer must work around (like caches). These texts, however, do not address all the needs of an introductory course. They are either too specialized or too advanced or both. This book is our attempt to provide an introductory text that follows the spirit of focusing on models and their relationship to realizations of systems.

The major theme of this book is on models and their relationship to realizations of sys- p.11 tems. The models we study are primarily about dynamics, the evolution of a system state in time. We do not address structural models, which represent static information about the construction of a system, although these too are important to embedded system design.

Working with models has a major advantage. Models can have formal properties. We can p.56 say definitive things about models. For example, we can assert that a model is determinate, meaning that given the same inputs it will always produce the same outputs. No such absolute assertion is possible with any physical realization of a system. If our model is p.361 a good abstraction of the physical system (here, "good abstraction" means that it omits only inessential details), then the definitive assertion about the model gives us confidence in the physical realization of the system. Such confidence is hugely valuable, particularly for embedded systems where malfunctions can threaten human lives. Studying models of systems gives us insight into how those systems will behave in the physical world.

Our focus is on the interplay of software and hardware with the physical environment in which they operate. This requires explicit modeling of the temporal dynamics of software and networks and explicit specification of concurrency properties intrinsic to the application. The fact that the implementation technologies have not yet caught up with this perspective should not cause us to teach the wrong engineering approach. We should teach design and modeling as it should be, and enrich this with a *critical* presentation of how it is. Embedded systems technologies today, therefore, should not be presented dispassionately as a collection of facts and tricks, as they are in many of the above cited books, but rather as stepping stones towards a sound design practice. The focus should be on what that sound design practice is, and on how today's technologies both impede and achieve it.

Stankovic et al. (2005) support this view, stating that "existing technology for RTES [real-time embedded systems] design does not effectively support the development of reliable and robust embedded systems." They cite a need to "raise the level of programming abstraction." We argue that raising the level of abstraction is insufficient. We also have to fundamentally change the abstractions that are used. Timing properties of software, for example, cannot be effectively introduced at higher levels of abstraction if they are entirely absent from the lower levels of abstraction on which these are built.

We require robust and predictable designs with repeatable temporal dynamics (Lee, 2009a). We must do this by building abstractions that appropriately reflect the realities of cyber-physical systems. The result will be CPS designs that can be much more sophisticated, including more adaptive control logic, evolvability over time, and improved safety and reliability, all without suffering from the brittleness of today's designs, where small changes have big consequences.

In addition to dealing with temporal dynamics, CPS designs invariably face challenging concurrency issues. Because software is so deeply rooted in sequential abstractions, concurrency mechanisms such as interrupts and multitasking, using semaphores and mutual exclusion, loom large. We therefore devote considerable effort in this book to developing a critical understanding of threads, message passing, deadlock avoidance, race conditions, and data determinism.

Note about This Edition

This is the second edition of the textbook. In addition to several bug fixes and improvements to presentation and wording, it includes two new chapters. Chapter 7 covers sensors and actuators with an emphasis on modeling. Chapter 17 covers the basics of security and privacy for embedded systems.

What Is Missing

Even with the new additions, this version of the book is not complete. It is arguable, in fact, that complete coverage of embedded systems in the context of CPS is impossible. Specific topics that we cover in the undergraduate Embedded Systems course at Berkeley (see http://LeeSeshia.org) and hope to include in future versions of this book include networking, fault tolerance, simulation techniques, control theory, and hardware/software codesign.

How to Use This Book

This book is divided into three major parts, focused on modeling, design, and analysis, as shown in Figure 1. The three parts of the book are relatively independent of one another and are largely meant to be read concurrently. A systematic reading of the text can be accomplished in eight segments, shown with dashed outlines. Most segments include two chapters, so complete coverage of the text is possible in a 15 week semester, allowing two weeks for most modules.

The appendices provide background material that is well covered in other textbooks, but which can be quite helpful in reading this text. Appendix A reviews the notation of sets and functions. This notation enables a higher level of precision than is common in the study of embedded systems. Appendix B reviews basic results in the theory of computability and complexity. This facilitates a deeper understanding of the challenges in modeling and analysis of systems. Note that Appendix B relies on the formalism of state machines covered in Chapter 3, and hence should be read after reading Chapter 3.

In recognition of recent advances in technology that are fundamentally changing the technical publishing industry, this book is published in a non-traditional way. At least the present

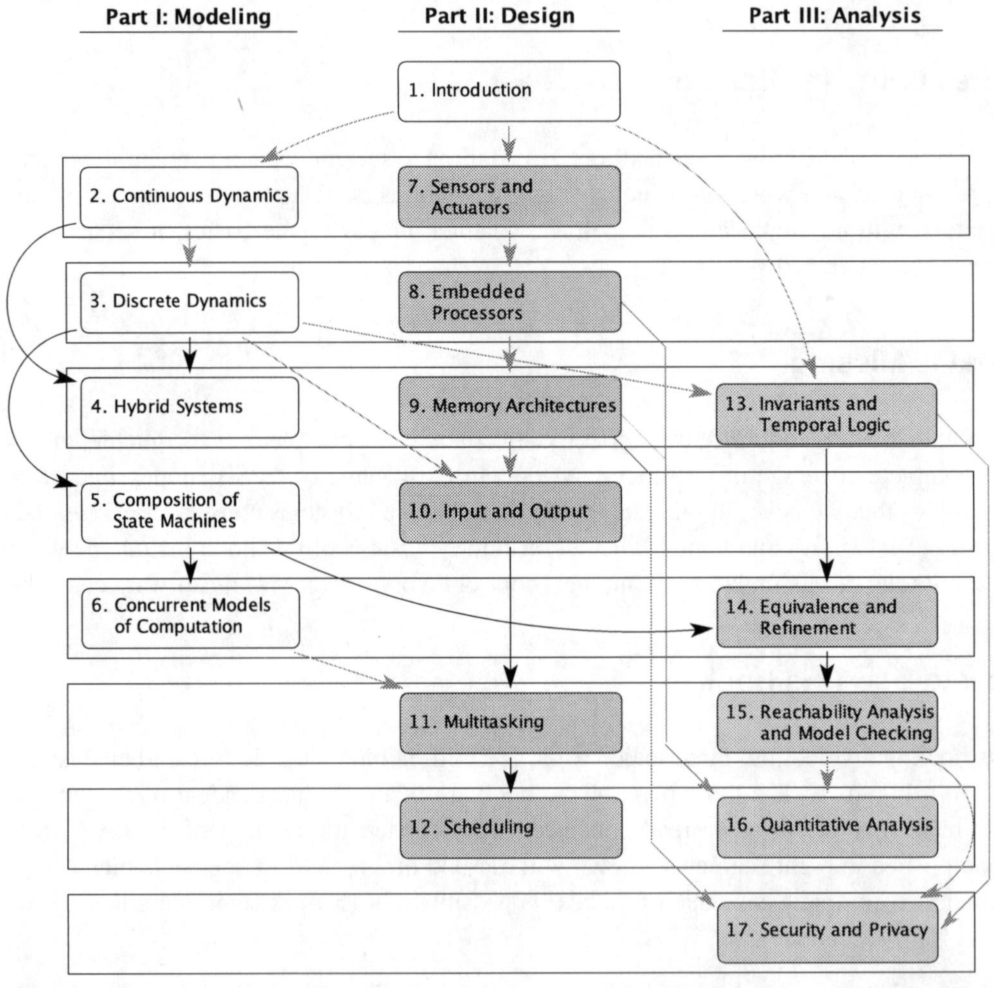

Figure 1: Map of the book with strong and weak dependencies between chapters. Strong dependencies between chapters are shown with arrows in black. Weak dependencies are shown in grey. When there is a weak dependency from chapter i to chapter j, then j may mostly be read without reading i, at most requiring skipping some examples or specialized analysis techniques.

version is available free in the form of PDF file designed specifically for reading on tablet computers. It can be obtained from the website http://LeeSeshia.org. The layout is optimized for medium-sized screens, particularly laptop computers and the iPad and other tablets. Extensive use of hyperlinks and color enhance the online reading experience.

We attempted to adapt the book to e-book formats, which, in theory, enable reading on various sized screens, attempting to take best advantage of the available screen. However, like HTML documents, e-book formats use a reflow technology, where page layout is recomputed on the fly. The results are highly dependent on the screen size and prove ludicrous on many screens and suboptimal on all. As a consequence, we have opted for controlling the layout, and we do not recommend attempting to read the book on an smartphone.

Although the electronic form is convenient, we recognize that there is real value in a tangible manifestation on paper, something you can thumb through, something that can live on a bookshelf to remind you of its existence. This edition is published by MIT Press, who has assured us that they will keep the book affordable.

Two disadvantages of print media compared to electronic media are the lack of hyperlinks and the lack of text search. We have attempted to compensate for those limitations by providing page number references in the margins whenever a term is used that is defined elsewhere. The term that is defined elsewhere is underlined with a discrete light gray line. In addition, we have provided an extensive index, with more than 2,000 entries.

There are typographic conventions worth noting. When a term is being defined, it will appear in **bold face**, and the corresponding index entry will also be in bold face. Hyperlinks are shown in blue in the electronic version. The notation used in diagrams, such as those for finite-state machines, is intended to be familiar, but not to conform with any particular programming or modeling language.

Reporting Errors

If you find errors or typos in this book, or if you have suggestions for improvements or other comments, please send email to:

authors@leeseshia.org

Please include the version number of the book, whether it is the electronic or the hard-copy distribution, and the relevant page numbers. Thank you!

Intended Audience

This book is intended for students at the advanced undergraduate level or introductory graduate level, and for practicing engineers and computer scientists who wish to understand the engineering principles of embedded systems. We assume that the reader has some exposure to machine structures (e.g., should know what an ALU is), computer programming (we use C throughout the text), basic discrete mathematics and algorithms, and at least an appreciation for signals and systems (what it means to sample a continuous-time signal, for example).

Acknowledgments

The authors gratefully acknowledge contributions and helpful suggestions from Murat Arcak, Dai Bui, Janette Cardoso, Gage Eads, Stephen Edwards, Suhaib Fahmy, Shanna-Shaye Forbes, Daniel Holcomb, Jeff C. Jensen, Garvit Juniwal, Hokeun Kim, Jonathan Kotker, Wenchao Li, Isaac Liu, Slobodan Matic, Mayeul Marcadella, Le Ngoc Minh, Christian Motika, Chris Myers, Steve Neuendorffer, David Olsen, Minxue Pan, Hiren Patel, Jan Reineke, Rhonda Righter, Alberto Sangiovanni-Vincentelli, Chris Shaver, Shih-Kai Su (together with students in CSE 522, lectured by Dr. Georgios E. Fainekos at Arizona State University), Stavros Tripakis, Pravin Varaiya, Reinhard von Hanxleden, Armin Wasicek, Kevin Weekly, Maarten Wiggers, Qi Zhu, and the students in UC Berkeley's EECS 149 class over the past years, particularly Ned Bass and Dan Lynch. The authors are especially grateful to Elaine Cheong, who carefully read most chapters and offered helpful editorial suggestions. We also acknowledge the bug fixes and suggestions sent in by several readers which has helped us improve the book since its initial publication. We give special thanks to our families for their patience and support, particularly to Helen, Katalina, and Rhonda (from Edward), and Amma, Appa, Ashwin, Bharathi, Shriya, and Viraj (from Sanjit).

This book is almost entirely constructed using open-source software. The typesetting is done using LaTeX, and many of the figures are created using Ptolemy II. See:

http://ptolemy.org

Further Reading

Many textbooks on embedded systems have appeared in recent years. These books approach the subject in surprisingly diverse ways, often reflecting the perspective of a more established discipline that has migrated into embedded systems, such as VLSI design, control systems, signal processing, robotics, real-time systems, or software engineering. Some of these books complement the present one nicely. We strongly recommend them to the reader who wishes to broaden his or her understanding of the subject.

Specifically, Patterson and Hennessy (1996), although not focused on embedded processors, is the canonical reference for computer architecture, and a must-read for anyone interested embedded processor architectures. Sriram and Bhattacharyya (2009) focus on signal processing applications, such as wireless communications and digital media, and give particularly thorough coverage to dataflow programming methodologies. Wolf p.142 (2000) gives an excellent overview of hardware design techniques and microprocessor architectures and their implications for embedded software design. Mishra and Dutt (2005) give a view of embedded architectures based on architecture description languages (ADLs). Oshana (2006) specializes in DSP processors from Texas Instruments, p.206 giving an overview of architectural approaches and a sense of assembly-level programming.

Focused more on software, Buttazzo (2005a) is an excellent overview of scheduling techniques for real-time software. Liu (2000) gives one of the best treatments yet of techniques for handling sporadic real-time events in software. Edwards (2000) gives a good overview of domain-specific higher-level programming languages used in some embedded system designs. Pottie and Kaiser (2005) give a good overview of networking technologies, particularly wireless, for embedded systems. Koopman (2010) focuses on design process for embedded software, including requirements management, project management, testing plans, and security plans. Alur (2015) provides an excellent, in-depth treatment of formal modeling and verification of cyber-physical systems.

No single textbook can comprehensively cover the breadth of technologies available to the embedded systems engineer. We have found useful information in many of the books that focus primarily on today's design techniques (Barr and Massa, 2006; Berger, 2002; Burns and Wellings, 2001; Gajski et al., 2009; Kamal, 2008; Noergaard, 2005; Parab et al., 2007; Simon, 2006; Schaumont, 2010; Vahid and Givargis, 2010).

Notes for Instructors

At Berkeley, we use this text for an advanced undergraduate course called *Introduction to Embedded Systems*. A great deal of material for lectures and labs can be found via the main web page for this text:

http://leeseshia.org

In addition, a solutions manual and other instructional material are available to qualified instructors at bona fide teaching institutions. See

http://chess.eecs.berkeley.edu/instructors/

or contact authors@leeseshia.org.

1 Introduction

Contents

A **cyber-physical system** (**CPS**) is an integration of computation with physical processes whose behavior is defined by *both* cyber and physical parts of the system. Embedded computers and networks monitor and control the physical processes, usually with feedback loops where physical processes affect computations and vice versa. As an intellectual challenge, CPS is about the *intersection*, not the union, of the physical and the cyber. It is not sufficient to separately understand the physical components and the computational components. We must instead understand their interaction.

In this chapter, we use a few CPS applications to outline the engineering principles of such systems and the processes by which they are designed.

1.1 Applications

CPS applications arguably have the potential to eclipse the 20th century information technology (IT) revolution. Consider the following examples.

Example 1.1: Heart surgery often requires stopping the heart, performing the surgery, and then restarting the heart. Such surgery is extremely risky and carries many detrimental side effects. A number of research teams have been working on an alternative where a surgeon can operate on a beating heart rather than stopping the heart. There are two key ideas that make this possible. First, surgical tools can be robotically controlled so that they move with the motion of the heart (Kremen, 2008). A surgeon can therefore use a tool to apply constant pressure to a point on the heart while the heart continues to beat. Second, a stereoscopic video system can present to the surgeon a video illusion of a still heart (Rice, 2008). To the surgeon, it looks as if the heart has been stopped, while in reality, the heart continues to beat. To realize such a surgical system requires extensive modeling of the heart, the tools, the computational hardware, and the software. It requires careful design of the software that ensures precise timing and safe fallback behaviors to handle malfunctions. And it requires detailed analysis of the models and the designs to provide high confidence.

Example 1.2: Consider a city where traffic lights and cars cooperate to ensure efficient flow of traffic. In particular, imagine never having to stop at a red light unless there is actual cross traffic. Such a system could be realized with expensive infrastructure that detects cars on the road. But a better approach might be to have the cars themselves cooperate. They track their position and communicate to cooperatively use shared resources such as intersections. Making such a system reliable, of course, is essential to its viability. Failures could be disastrous.

Example 1.3: Imagine an airplane that refuses to crash. While preventing all possible causes of a crash is not possible, a well-designed flight control system can prevent certain causes. The systems that do this are good examples of cyber-physical systems.

In traditional aircraft, a pilot controls the aircraft through mechanical and hydraulic linkages between controls in the cockpit and movable surfaces on the wings and tail of the aircraft. In a **fly-by-wire** aircraft, the pilot commands are mediated by a flight computer and sent electronically over a network to actuators in the wings and tail. Fly-by-wire aircraft are much lighter than traditional aircraft, and therefore more fuel efficient. They have also proven to be more reliable. Virtually all new aircraft designs are fly-by-wire systems.

In a fly-by-wire aircraft, since a computer mediates the commands from the pilot, the computer can modify the commands. Many modern flight control systems modify pilot commands in certain circumstances. For example, commercial airplanes made by Airbus use a technique called **flight envelope protection** to prevent an airplane from going outside its safe operating range. They can prevent a pilot from causing a stall, for example.

The concept of flight envelope protection could be extended to help prevent certain other causes of crashes. For example, the **soft walls** system proposed by Lee (2001), if implemented, would track the location of the aircraft on which it is installed and prevent it from flying into obstacles such as mountains and buildings. In Lee's proposal, as an aircraft approaches the boundary of an obstacle, the fly-by-wire flight control system creates a virtual pushing force that forces the aircraft away. The pilot feels as if the aircraft has hit a soft wall that diverts it. There are many challenges, both technical and non-technical, to designing and deploying such a system. See Lee (2003) for a discussion of some of these issues.

Although the soft walls system of the previous example is rather futuristic, there are modest versions in automotive safety that have been deployed or are in advanced stages of research and development. For example, many cars today detect inadvertent lane changes and warn the driver. Consider the much more challenging problem of automatically correcting the driver's actions. This is clearly much harder than just warning the driver. How can you ensure that the system will react and take over only when needed, and only exactly to the extent to which intervention is needed?

It is easy to imagine many other applications, such as systems that assist the elderly; telesurgery systems that allow a surgeon to perform an operation at a remote location; and home appliances that cooperate to smooth demand for electricity on the power grid.

Moreover, it is easy to envision using CPS to improve many existing systems, such as robotic manufacturing systems; electric power generation and distribution; process control in chemical factories; distributed computer games; transportation of manufactured goods; heating, cooling, and lighting in buildings; people movers such as elevators; and bridges that monitor their own state of health. The impact of such improvements on safety, energy consumption, and the economy is potentially enormous.

Many of the above examples will be deployed using a structure like that sketched in Figure 1.1. There are three main parts in this sketch. First, the **physical plant** is the "physical" part of a cyber-physical system. It is simply that part of the system that is not realized with computers or digital networks. It can include mechanical parts, biological or chemical processes, or human operators. Second, there are one or more computational **platforms**, which consist of sensors, actuators, one or more computers, and (possibly) one or more operating systems. Third, there is a **network fabric**, which provides the mechanisms for the computers to communicate. Together, the platforms and the network fabric form the "cyber" part of the cyber-physical system.

Figure 1.1 shows two networked platforms each with its own sensors and/or actuators. The action taken by the actuators affects the data provided by the sensors through the physical plant. In the figure, Platform 2 controls the physical plant via Actuator 1. It measures the processes in the physical plant using Sensor 2. The box labeled Computation 2 implements a **control law**, which determines based on the sensor data what commands to issue to the actuator. Such a loop is called a **feedback control** loop. Platform 1 makes additional measurements using Sensor 1, and sends messages to Platform 2 via the network fabric. Computation 3 realizes an additional control law, which is merged with that of Computation 2, possibly preempting it.

Example 1.4: Consider a high-speed printing press for a print-on-demand service. This might be structured similarly to Figure 1.1, but with many more platforms, sensors, and actuators. The actuators may control motors that drive paper through the press and ink onto the paper. The control laws may include a strategy for compensating for paper stretch, which will typically depend on the type of paper, the temperature, and the humidity. A networked structure like that in Figure 1.1 might be used to induce rapid shutdown to prevent damage to the equipment in case of paper jams. Such shutdowns need to be tightly orchestrated across the entire system to prevent disasters. Similar situations are found in high-end instrumentation systems and in energy production and distribution (Eidson et al., 2009).

About the Term "Cyber-Physical Systems"

The term "cyber-physical systems" emerged in 2006, coined by Helen Gill at the National Science Foundation in the US. We may be tempted to associate the term "**cyberspace**" with CPS, but the roots of the term CPS are older and deeper. It would be more accurate to view the terms "cyberspace" and "cyber-physical systems" as stemming from the same root, "**cybernetics**," rather than viewing one as being derived from the other.

The term "cybernetics" was coined by Norbert Wiener (Wiener, 1948), an American mathematician who had a huge impact on the development of control systems theory. During World War II, Wiener pioneered technology for the automatic aiming and firing of anti-aircraft guns. Although the mechanisms he used did not involve digital computers, the principles involved are similar to those used today in a huge variety of computer-based feedback control systems. Wiener derived the term from the Greek κυβερνητης (kybernetes), meaning helmsman, governor, pilot, or rudder. The metaphor is apt for control systems. Wiener described his vision of cybernetics as the conjunction of control and communication. His notion of control was deeply rooted in closed-loop feedback, where the control logic is driven by measurements of physical processes, and in turn drives the physical processes. Even though Wiener did not use digital computers, the control logic is effectively a computation, and therefore cybernetics is the conjunction of physical processes, computation, and communication. Wiener could not have anticipated the powerful effects of digital computation and networks. The fact that the term "cyber-physical systems" may be ambiguously interpreted as the conjunction of cyberspace with physical processes, therefore, helps to underscore the enormous impact that CPS will have. CPS leverages an information technology that far outstrips even the wildest dreams of Wiener's era.

p.31

The term CPS relates to the currently popular terms Internet of Things (IoT), Industry 4.0, the Industrial Internet, Machine-to-Machine (M2M), the Internet of Everything, TSensors (trillion sensors), and the Fog (like the Cloud, but closer to the ground). All of these reflect a vision of a technology that deeply connects our physical world with our information world. In our view, the term CPS is more foundational and durable than all of these, because it does not directly reference either implementation approaches (e.g., the "Internet" in IoT) nor particular applications (e.g., "Industry" in Industry 4.0). It focuses instead on the fundamental intellectual problem of conjoining the engineering traditions of the cyber and the physical worlds.

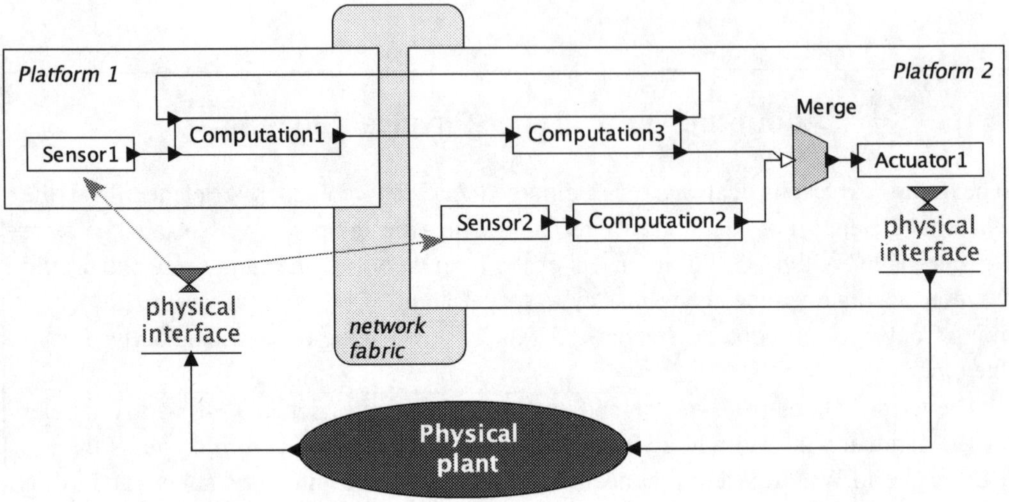

Figure 1.1: Example structure of a cyber-physical system.

1.2 Motivating Example

In this section, we describe a motivating example of a cyber-physical system. Our goal is to use this example to illustrate the importance of the breadth of topics covered in this text. The specific application is the Stanford testbed of autonomous rotorcraft for multi agent control (**STARMAC**), developed by Claire Tomlin and colleagues as a cooperative effort at Stanford and Berkeley (Hoffmann et al., 2004). The STARMAC is a small **quadrotor** aircraft; it is shown in flight in Figure 1.2. Its primary purpose is to serve as a testbed for experimenting with multi-vehicle autonomous control techniques. The objective is to be able to have multiple vehicles cooperate on a common task.

There are considerable challenges in making such a system work. First, controlling the vehicle is not trivial. The main actuators are the four rotors, which produce a variable amount of downward thrust. By balancing the thrust from the four rotors, the vehicle can take off, land, turn, and even flip in the air. How do we determine what thrust to apply? Sophisticated control algorithms are required.

Second, the weight of the vehicle is a major consideration. The heavier it is, the more stored energy it needs to carry, which of course makes it even heavier. The heavier it is, the more thrust it needs to fly, which implies bigger and more powerful motors and rotors. The design crosses a major threshold when the vehicle is heavy enough that the rotors become

Figure 1.2: The STARMAC quadrotor aircraft in flight (reproduced with permission).

dangerous to humans. Even with a relatively light vehicle, safety is a considerable concern, and the system needs to be designed with fault handling.

Third, the vehicle needs to operate in a context, interacting with its environment. It might, for example, be under the continuous control of a watchful human who operates it by remote control. Or it might be expected to operate autonomously, to take off, perform some mission, return, and land. Autonomous operation is enormously complex and challenging because it cannot benefit from the watchful human. Autonomous operation demands more sophisticated sensors. The vehicle needs to keep track of where it is (it needs to perform **localization**). It needs to sense obstacles, and it needs to know where the ground is. With good design, it is even possible for such vehicles to autonomously land on the pitching deck of a ship. The vehicle also needs to continuously monitor its own health, to detect malfunctions and react to them so as to contain the damage.

It is not hard to imagine many other applications that share features with the quadrotor problem. The problem of landing a quadrotor vehicle on the deck of a pitching ship is similar to the problem of operating on a beating heart (see Example 1.1). It requires detailed modeling of the dynamics of the environment (the ship, the heart), and a clear understanding of the interaction between the dynamics of the embedded system (the quadrotor, the robot) and its environment.

The rest of this chapter will explain the various parts of this book, using the quadrotor example to illustrate how the various parts contribute to the design of such a system.

1.3 The Design Process

The goal of this book is to understand how to go about designing and implementing cyber-physical systems. Figure 1.3 shows the three major parts of the process, **modeling**, **design**, and **analysis**. Modeling is the process of gaining a deeper understanding of a system through imitation. Models imitate the system and reflect properties of the system. Models specify **what** a system does. Design is the structured creation of artifacts. It specifies **how** a system does what it does. Analysis is the process of gaining a deeper understanding of a system through dissection. It specifies **why** a system does what it does (or fails to do what a model says it should do).

As suggested in Figure 1.3, these three parts of the process overlap, and the design process iteratively moves among the three parts. Normally, the process will begin with modeling, where the goal is to understand the problem and to develop solution strategies.

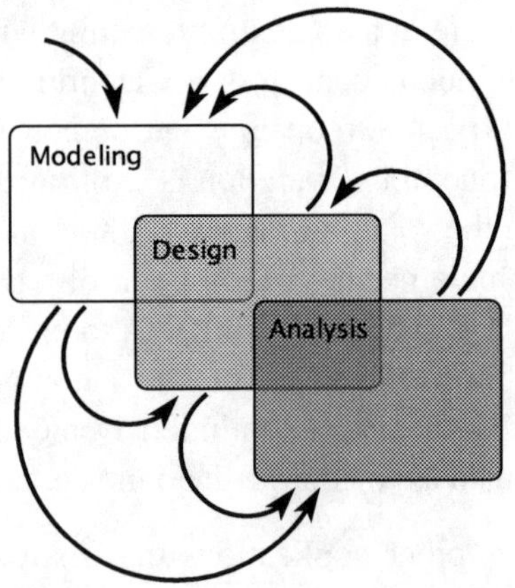

Figure 1.3: Creating embedded systems requires an iterative process of modeling, design, and analysis.

Example 1.5: For the quadrotor problem of Section 1.2, we might begin by constructing models that translate commands from a human to move vertically or laterally into commands to the four motors to produce thrust. A model will reveal that if the thrust is not the same on the four rotors, then the vehicle will tilt and move laterally.

Such a model might use techniques like those in Chapter 2 (Continuous Dynamics), constructing differential equations to describe the dynamics of the vehicle. It p.19
would then use techniques like those in Chapter 3 (Discrete Dynamics) to build state machines that model the modes of operation such as takeoff, landing, hov- p.47
ering, and lateral flight. It could then use the techniques of Chapter 4 (Hybrid Systems) to blend these two types of models, creating hybrid system models of p.75
the system to study the transitions between modes of operation. The techniques of Chapters 5 (Composition of State Machines) and 6 (Concurrent Models of Computation) would then provide mechanisms for composing models of multiple vehicles, models of the interactions between a vehicle and its environment, and models of the interactions of components within a vehicle.

The process may progress quickly to the design phase, where we begin selecting components and putting them together (motors, batteries, sensors, microprocessors, memory systems, operating systems, wireless networks, etc.). An initial prototype may reveal flaws in the models, causing a return to the modeling phase and revision of the models.

Example 1.6: The hardware architecture of the first generation STARMAC quadrotor is shown in Figure 1.4. At the left and bottom of the figure are a number p.6
of sensors used by the vehicle to determine where it is (localization) and what is p.7
around it. In the middle are three boxes showing three distinct microprocessors. The Robostix is an Atmel AVR 8-bit microcontroller that runs with no operating p.204
system and performs the low-level control algorithms to keep the craft flying. The other two processors perform higher-level tasks with the help of an operating system. Both processors include wireless links that can be used by cooperating vehicles and ground controllers.

Chapter 7 (Sensors and Actuators) considers sensors and actuators, including the IMU and p.190
rangers shown in Figure 1.4. Chapter 8 (Embedded Processors) considers processor ar-

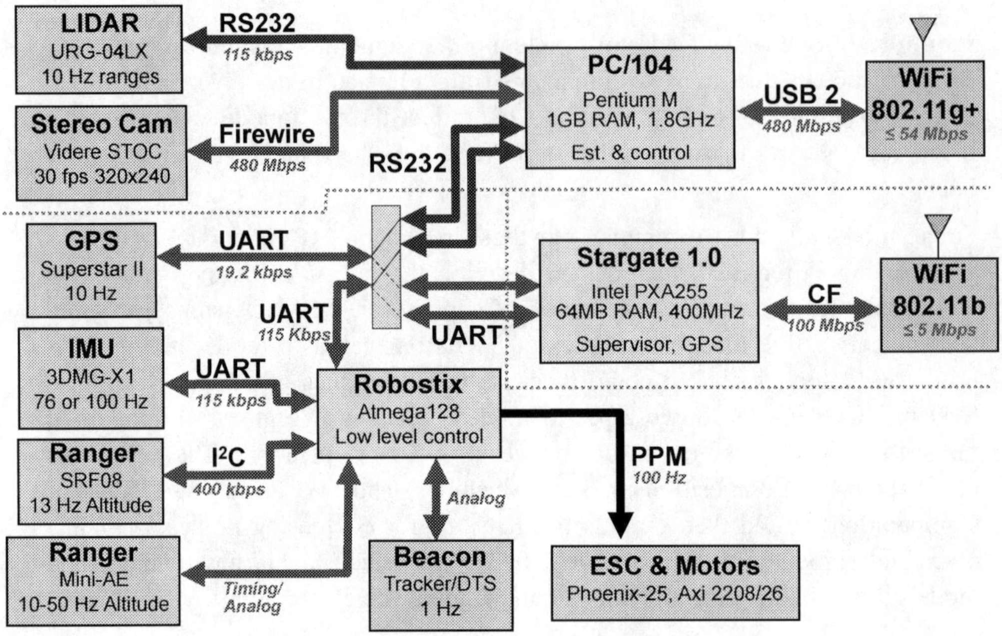

Figure 1.4: The STARMAC architecture (reproduced with permission).

chitectures, offering some basis for comparing the relative advantages of one architecture or another. Chapter 9 (Memory Architectures) considers the design of memory systems, emphasizing the impact that they can have on overall system behavior. Chapter 10 (Input and Output) considers the interfacing of processors with sensors and actuators. Chapters 11 (Multitasking) and 12 (Scheduling) focus on software architecture, with particular emphasis on how to orchestrate multiple real-time tasks.

In a healthy design process, analysis figures prominently early in the process. Analysis will be applied to the models and to the designs. The models may be analyzed for safety conditions, p.344 for example to ensure an invariant that asserts that if the vehicle is within one meter of the ground, then its vertical speed is no greater than 0.1 meter/sec. The designs may be analyzed for the timing behavior of software, for example to determine how long it takes the system to respond to an emergency shutdown command. Certain analysis problems will involve details of both models and designs. For the quadrotor example, it is important to understand how the system will behave if network connectivity is lost and it becomes impossible to communicate with the vehicle. How can the vehicle detect that communication has been lost? This will require accurate modeling of the network and the software.

Example 1.7: For the quadrotor problem, we use the techniques of Chapter 13 (Invariants and Temporal Logic) to specify key safety requirements for operation of the vehicles. We would then use the techniques of Chapters 14 (Equivalence and Refinement) and 15 (Reachability Analysis and Model Checking) to verify that these safety properties are satisfied by implementations of the software. The techniques of Chapter 16 (Quantitative Analysis) would be used to determine whether real-time constraints are met by the software. Finally, the techniques of Chapter 17 would be used to ensure that malicious parties cannot take control of the quadrotor and that any confidential data it may be gathering is not leaked to an adversary.

Corresponding to a design process structured as in Figure 1.3, this book is divided into three major parts, focused on modeling, design, and analysis (see Figure 1 on page xviii). We now describe the approach taken in the three parts.

1.3.1 Modeling

The modeling part of the book, which is the first part, focuses on models of dynamic behavior. It begins with a light coverage of the big subject of modeling of physical dynamics in Chapter 2, specifically focusing on continuous dynamics in time. It then talks about discrete dynamics in Chapter 3, using state machines as the principal formalism. It then combines the two, continuous and discrete dynamics, with a discussion of hybrid systems in Chapter 4. Chapter 5 (Composition of State Machines) focuses on concurrent composition of state machines, emphasizing that the semantics of composition is a critical issue with which designers must grapple. Chapter 6 (Concurrent Models of Computation) gives an overview of concurrent models of computation, including many of those used in design tools that practitioners frequently leverage, such as Simulink and LabVIEW.

In the modeling part of the book, we define a **system** to be simply a combination of parts that is considered as a whole. A **physical system** is one realized in matter, in contrast to a conceptual or **logical system** such as software and algorithms. The **dynamics** of a system is its evolution in time: how its state changes. A **model** of a physical system is a description of certain aspects of the system that is intended to yield insight into properties of the system. In this text, models have mathematical properties that enable systematic analysis. The model imitates properties of the system, and hence yields insight into that system.

A model is itself a system. It is important to avoid confusing a model and the system that it models. These are two distinct artifacts. A model of a system is said to have high **fidelity** if it accurately describes properties of the system. It is said to **abstract** the system if it omits details. Models of physical systems inevitably *do* omit details, so they are always abstractions of the system. A major goal of this text is to develop an understanding of how to use models, of how to leverage their strengths and respect their weaknesses.

p.1 A cyber-physical system (CPS) is a system composed of physical subsystems together with computing and networking. Models of cyber-physical systems normally include all three parts. The models will typically need to represent both dynamics and **static properties** (those that do not change during the operation of the system). It is important to note that a model of a cyber-physical system need not have both discrete and continuous parts. It is possible for a purely discrete (or purely continuous) model to have high fidelity for the properties of interest.

Each of the modeling techniques described in this part of the book is an enormous subject, much bigger than one chapter, or even one book. In fact, such models are the focus of many branches of engineering, physics, chemistry, and biology. Our approach is aimed at engineers. We assume some background in mathematical modeling of dynamics (calculus courses that give some examples from physics are sufficient), and then focus on how to compose diverse models. This will form the core of the cyber-physical system problem, since joint modeling of the cyber side, which is logical and conceptual, with the physical side, which is embodied in matter, is the core of the problem. We therefore make no attempt to be comprehensive, but rather pick a few modeling techniques that are widely used by engineers and well understood, review them, and then compose them to form a cyber-physical whole.

1.3.2 Design

The second part of the book has a very different flavor, reflecting the intrinsic heterogeneity of the subject. This part focuses on the design of embedded systems, with emphasis on the role they play *within* a CPS. Chapter 7 (Sensors and Actuators) considers sensors and actuators, with emphasis on how to model them so that their role in overall system dynamics is understood. Chapter 8 (Embedded Processors) discusses processor architectures, with emphasis on specialized properties most suited to embedded systems. Chapter 9 (Memory Architectures) describes memory architectures, including abstractions such as memory models in programming languages, physical properties such as memory technologies, and architectural properties such as memory hierarchy (caches, scratchpads, etc.). The emphasis is on how memory architecture affects dynamics. Chapter 10 (Input and Output) is about the interface between the software world and the physical world. It discusses input/output mechanisms in software and computer architectures, and the digital/analog interface, including sampling. Chapter 11 (Multitasking) introduces the notions that underlie operating systems, with particular emphasis on multitasking. The emphasis is on the pitfalls of using low-level mechanisms such as threads, with a hope of convincing the reader that there is real value in using the modeling techniques covered in the first part of the book. Those modeling techniques help designers build confidence in system designs. Chapter 12 (Scheduling) introduces real-time scheduling, covering many of the classic results in the area.

In all chapters in the design part, we particularly focus on the mechanisms that provide concurrency and control over timing, because these issues loom large in the design of cyber-physical systems. When deployed in a product, embedded processors typically have a dedicated function. They control an automotive engine or measure ice thickness in the Arctic. They are not asked to perform arbitrary functions with user-defined software. Consequently, the processors, memory architectures, I/O mechanisms, and operating systems can be more specialized. Making them more specialized can bring enormous benefits. For example, they may consume far less energy, and consequently be usable with small batteries for long periods of time. Or they may include specialized hardware to perform operations that would be costly to perform on general-purpose hardware, such as image analysis. Our goal in this part is to enable the reader to *critically* evaluate the numerous available technology offerings.

One of the goals in this part of the book is to teach students to implement systems while *thinking across traditional abstraction layers* — e.g., hardware *and* software, computation *and* physical processes. While such cross-layer thinking is valuable in implementing systems in general, it is particularly essential in embedded systems given their heterogeneous nature. For example, a programmer implementing a control algorithm expressed in terms of real-valued quantities must have a solid understanding of computer arithmetic (e.g., of fixed-point numbers) in order to create a reliable implementation. Similarly, an implementor of automotive software that must satisfy real-time constraints must be aware of processor features – such as pipelines and caches – that can affect the execution time of tasks and hence the real-time behavior of the system. Likewise, an implementor of interrupt-driven or multi-threaded software must understand the atomic operations provided by the underlying software-hardware platform and use appropriate synchronization constructs to ensure correctness. Rather than doing an exhaustive survey of different implementation methods and platforms, this part of the book seeks to give the reader an appreciation for such cross-layer topics, and uses homework exercises to facilitate a deeper understanding of them.

p.224
p.236
p.264

1.3.3 Analysis

Every system must be designed to meet certain requirements. For embedded systems, which are often intended for use in safety-critical, everyday applications, it is essential to certify that the system meets its requirements. Such system requirements are also called **properties** or **specifications**. The need for specifications is aptly captured by the following quotation, paraphrased from Young et al. (1985):

> "A design without specifications cannot be right or wrong, it can only be surprising!"

The analysis part of the book focuses on precise specifications of properties, on techniques for comparing specifications, and on techniques for analyzing specifications and the resulting designs. Reflecting the emphasis on dynamics in the text, Chapter 13 (Invariants and Temporal Logic) focuses on temporal logics, which provide precise descriptions of dynamic properties of systems. These descriptions are treated as models. Chapter 14 (Equivalence p.361 and Refinement) focuses on the relationships between models. Is one model an abstraction of another? Is it equivalent in some sense? Specifically, that chapter introduces type systems as a way of comparing static properties of models, and language containment and p.374 simulation relations as a way of comparing dynamic properties of models. Chapter 15 (Reachability Analysis and Model Checking) focuses on techniques for analyzing the large number of possible dynamic behaviors that a model may exhibit, with particular emphasis on model checking as a technique for exploring such behaviors. Chapter 16 (Quantitative Analysis) is about analyzing quantitative properties of embedded software, such as finding bounds on resources consumed by programs. It focuses particularly on execution time analysis, with some introduction to other quantitative properties such as energy and memory usage. Chapter 17 (Security and Privacy) introduces the basics of security and privacy for embedded systems design, including cryptographic primitives, protocol security, software security, secure information flow, side channels, and sensor security.

In present engineering practice, it is common to have system requirements stated in a natural language such as English. It is important to precisely state requirements to avoid ambiguities inherent in natural languages. The goal of this part of the book is to help replace descriptive techniques with *formal* ones, which we believe are less error prone.

Importantly, formal specifications also enable the use of automatic techniques for p.385 formal verification of both models and implementations. The analysis part of the book introduces readers to the basics of formal verification, including notions of equivalence and refinement checking, as well as reachability analysis and model checking. In discussing these verification methods, we attempt to give users of verification tools an appreciation of what is "under the hood" so that they may derive the most benefit from them. This *user's view* is supported by examples discussing, for example, how model checking can be applied to find subtle errors in concurrent software, or how reachability analysis can be used in computing a control strategy for a robot to achieve a particular task.

1.4 Summary

Cyber-physical systems are heterogeneous blends by nature. They combine computation, communication, and physical dynamics. They are harder to model, harder to design, and harder to analyze than homogeneous systems. This chapter gives an overview of the engineering principles addressed in this book for modeling, designing, and analyzing such systems.

Part I

Modeling Dynamic Behaviors

This part of this text studies modeling of embedded systems, with emphasis on joint modeling of software and physical dynamics. We begin in Chapter 2 with a discussion of established techniques for modeling the dynamics of physical systems, with emphasis on their continuous behaviors. In Chapter 3, we discuss techniques for modeling discrete behaviors, which reflect better the behavior of software. In Chapter 4, we bring these two classes of models together and show how discrete and continuous behaviors are jointly modeled by hybrid systems. Chapters 5 and 6 are devoted to reconciling the inherently concurrent nature of the physical world with the inherently sequential world of software. Chapter 5 shows how state machine models, which are fundamentally sequential, can be composed concurrently. That chapter specifically introduces the notion of synchronous composition. Chapter 6 shows that synchronous composition is but one of the ways to achieve concurrent composition.

p.8
p.11

2 Continuous Dynamics

Contents

This chapter reviews a few of the many modeling techniques for studying dynamics of a physical system. We begin by studying mechanical parts that move (this problem is known as **classical mechanics**). The techniques used to study the dynamics of such parts extend broadly to many other physical systems, including circuits, chemical processes, and biological processes. But mechanical parts are easiest for most people to visualize, so they make our example concrete. Motion of mechanical parts can often be modeled using **differential equations**, or equivalently, **integral equations**. Such models really only work well for "smooth" motion (a concept that we can make more precise using notions of linearity, time invariance, and continuity). For motions that are not smooth, such as those modeling collisions of mechanical parts, we can use modal models that represent distinct modes of operation with abrupt (conceptually instantaneous) transitions between modes. Collisions

p.11

of mechanical objects can be usefully modeled as discrete, instantaneous events. The problem of jointly modeling smooth motion and such discrete events is known as hybrid systems modeling and is studied in Chapter 4. Such combinations of discrete and continuous behaviors bring us one step closer to joint modeling of cyber and physical processes.

We begin with simple equations of motion, which provide a model of a system in the form of **ordinary differential equations (ODEs)**. We then show how these ODEs can be represented in actor models, which include the class of models in popular modeling languages such as LabVIEW (from National Instruments) and Simulink (from The MathWorks, Inc.). We then consider properties of such models such as linearity, time invariance, and stability, and consider consequences of these properties when manipulating models. We develop a simple example of a feedback control system that stabilizes an unstable system. Controllers for such systems are often realized using software, so such systems can serve as a canonical example of a cyber-physical system. The properties of the overall system emerge from properties of the cyber and physical parts.

2.1 Newtonian Mechanics

In this section, we give a brief working review of some principles of classical mechanics. This is intended to be just enough to be able to construct interesting models, but is by no means comprehensive. The interested reader is referred to many excellent texts on classical mechanics, including Goldstein (1980); Landau and Lifshitz (1976); Marion and Thornton (1995).

Motion in space of physical objects can be represented with **six degrees of freedom**, illustrated in Figure 2.1. Three of these represent position in three dimensional space, and three represent orientation in space. We assume three axes, x, y, and z, where by convention x is drawn increasing to the right, y is drawn increasing upwards, and z is drawn increasing out of the page. **Roll** θ_x is an angle of rotation around the x axis, where by convention an angle of 0 radians represents horizontally flat along the z axis (i.e., the angle is given relative to the z axis). **Yaw** θ_y is the rotation around the y axis, where by convention 0 radians represents pointing directly to the right (i.e., the angle is given relative to the x axis). **Pitch** θ_z is rotation around the z axis, where by convention 0 radians represents pointing horizontally (i.e., the angle is given relative to the x axis).

The position of an object in space, therefore, is represented by six functions of the form $f\colon \mathbb{R} \to \mathbb{R}$, where the domain represents time and the codomain represents either distance

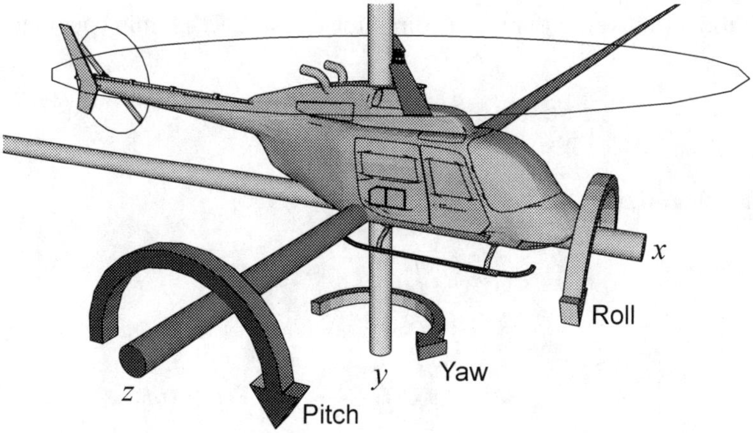

Figure 2.1: Modeling position with six degrees of freedom requires including pitch, roll, and yaw, in addition to position.

along an axis or angle relative to an axis.[1] Functions of this form are known as **continuous-time signals.**[2] These are often collected into vector-valued functions $\mathbf{x}\colon \mathbb{R} \to \mathbb{R}^3$ and $\theta\colon \mathbb{R} \to \mathbb{R}^3$, where $\mathbf{x}$ represents position, and θ represents orientation.

Changes in position or orientation are governed by **Newton's second law**, relating force with acceleration. Acceleration is the second derivative of position. Our first equation handles the position information,

$$\mathbf{F}(t) = M\ddot{\mathbf{x}}(t), \tag{2.1}$$

where $\mathbf{F}$ is the force vector in three directions, M is the mass of the object, and $\ddot{\mathbf{x}}$ is the second derivative of $\mathbf{x}$ with respect to time (i.e., the acceleration). Velocity is the integral of acceleration, given by

$$\forall\, t > 0, \quad \dot{\mathbf{x}}(t) \quad = \quad \dot{\mathbf{x}}(0) + \int_0^t \ddot{\mathbf{x}}(\tau) d\tau$$

[1] If the notation is unfamiliar, see Appendix A.

[2] The domain of a continuous-time signal may be restricted to a connected subset of $\mathbb{R}$, such as $\mathbb{R}_+$, the non-negative reals, or $[0, 1]$, the interval between zero and one, inclusive. The codomain may be an arbitrary set, though when representing physical quantities, real numbers are most useful.

where $\dot{\mathbf{x}}(0)$ is the initial velocity in three directions. Using (2.1), this becomes

$$\forall\, t > 0, \quad \dot{\mathbf{x}}(t) = \dot{\mathbf{x}}(0) + \frac{1}{M}\int_0^t \mathbf{F}(\tau)d\tau,$$

Position is the integral of velocity,

$$\begin{aligned} \mathbf{x}(t) &= \mathbf{x}(0) + \int_0^t \dot{\mathbf{x}}(\tau)d\tau \\ &= \mathbf{x}(0) + t\dot{\mathbf{x}}(0) + \frac{1}{M}\int_0^t\int_0^\tau \mathbf{F}(\alpha)d\alpha d\tau, \end{aligned}$$

where $\mathbf{x}(0)$ is the initial position. Using these equations, if you know the initial position and initial velocity of an object and the forces on the object in all three directions as a function of time, you can determine the acceleration, velocity, and position of the object at any time.

The versions of these equations of motion that affect orientation use **torque**, the rotational version of force. It is again a three-element vector as a function of time, representing the net rotational force on an object. It can be related to angular velocity in a manner similar to equation (2.1),

$$\mathbf{T}(t) = \frac{d}{dt}\left(\mathbf{I}(t)\dot{\theta}(t)\right), \tag{2.2}$$

where $\mathbf{T}$ is the torque vector in three axes and $\mathbf{I}(t)$ is the **moment of inertia tensor** of the object. The moment of inertia is a 3×3 matrix that depends on the geometry and orientation of the object. Intuitively, it represents the reluctance that an object has to spin around any axis as a function of its orientation along the three axes. If the object is spherical, for example, this reluctance is the same around all axes, so it reduces to a constant scalar I (or equivalently, to a diagonal matrix $\mathbf{I}$ with equal diagonal elements I). The equation then looks much more like (2.1),

$$\mathbf{T}(t) = I\ddot{\theta}(t). \tag{2.3}$$

To be explicit about the three dimensions, we might write (2.2) as

$$\begin{bmatrix} T_x(t) \\ T_y(t) \\ T_z(t) \end{bmatrix} = \frac{d}{dt}\left(\begin{bmatrix} I_{xx}(t) & I_{xy}(t) & I_{xz}(t) \\ I_{yx}(t) & I_{yy}(t) & I_{yz}(t) \\ I_{zx}(t) & I_{zy}(t) & I_{zz}(t) \end{bmatrix}\begin{bmatrix} \dot{\theta}_x(t) \\ \dot{\theta}_y(t) \\ \dot{\theta}_z(t) \end{bmatrix}\right).$$

Here, for example, $T_y(t)$ is the net torque around the y axis (which would cause changes in yaw), $I_{yx}(t)$ is the inertia that determines how acceleration around the x axis is related to torque around the y axis.

Rotational velocity is the integral of acceleration,

$$\dot{\theta}(t) = \dot{\theta}(0) + \int_0^t \ddot{\theta}(\tau)d\tau,$$

where $\dot{\theta}(0)$ is the initial rotational velocity in three axes. For a spherical object, using (2.3), this becomes

$$\dot{\theta}(t) = \dot{\theta}(0) + \frac{1}{I}\int_0^t \mathbf{T}(\tau)d\tau.$$

Orientation is the integral of rotational velocity,

$$\begin{aligned} \theta(t) &= \theta(0) + \int_0^t \dot{\theta}(\tau)d\tau \\ &= \theta(0) + t\dot{\theta}(0) + \frac{1}{I}\int_0^t \int_0^\tau \mathbf{T}(\alpha)d\alpha d\tau \end{aligned}$$

where $\theta(0)$ is the initial orientation. Using these equations, if you know the initial orientation and initial rotational velocity of an object and the torques on the object in all three axes as a function of time, you can determine the rotational acceleration, velocity, and orientation of the object at any time.

Often, as we have done for a spherical object, we can simplify by reducing the number of dimensions that are considered. In general, such a simplification is called a **model-order reduction**. For example, if an object is a moving vehicle on a flat surface, there may be little reason to consider the y axis movement or the pitch or roll of the object.

Example 2.1: Consider a simple control problem that admits such reduction of dimensionality. A helicopter has two rotors, one above, which provides lift, and one on the tail. Without the rotor on the tail, the body of the helicopter would spin. The rotor on the tail counteracts that spin. Specifically, the force produced by the tail rotor must counter the torque produced by the main rotor. Here we consider this role of the tail rotor independently from all other motion of the helicopter.

A simplified model of the helicopter is shown in Figure 2.2. Here, we assume that the helicopter position is fixed at the origin, so there is no need to consider equations describing position. Moreover, we assume that the helicopter remains vertical, so

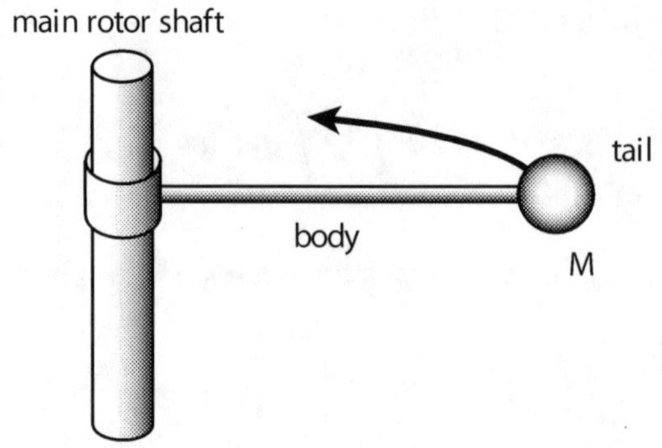

Figure 2.2: Simplified model of a helicopter.

pitch and roll are fixed at zero. These assumptions are not as unrealistic as they may seem since we can define the coordinate system to be fixed to the helicopter.

With these assumptions, the **moment of inertia** reduces to a scalar that represents a torque that resists changes in yaw. The changes in yaw will be due to **Newton's third law**, the **action-reaction law**, which states that every action has an equal and opposite reaction. This will tend to cause the helicopter to rotate in the opposite direction from the rotor rotation. The tail rotor has the job of countering that torque to keep the body of the helicopter from spinning.

We model the simplified helicopter by a system that takes as input a continuous-time signal T_y, the torque around the y axis (which causes changes in yaw). This torque is the sum of the torque caused by the main rotor and that caused by the tail rotor. When these are perfectly balanced, that sum is zero. The output of our system will be the angular velocity $\dot{\theta}_y$ around the y axis. The dimensionally-reduced version of (2.2) can be written as

$$\ddot{\theta}_y(t) = T_y(t)/I_{yy}.$$

Integrating both sides, we get the output $\dot{\theta}$ as a function of the input T_y,

$$\dot{\theta}_y(t) = \dot{\theta}_y(0) + \frac{1}{I_{yy}} \int_0^t T_y(\tau)d\tau. \tag{2.4}$$

The critical observation about this example is that if we were to choose to model the helicopter by, say, letting $\mathbf{x}\colon \mathbb{R} \to \mathbb{R}^3$ represent the absolute position in space of the tail of the helicopter, we would end up with a far more complicated model. Designing the control system would also be much more difficult.

2.2 Actor Models

In the previous section, a model of a physical system is given by a differential or an integral equation that relates input signals (force or torque) to output signals (position, orientation, velocity, or rotational velocity). Such a physical system can be viewed as a component in a larger system. In particular, a **continuous-time system** (one that operates on continuous-time signals) may be modeled by a box with an input **port** and an output port as follows:

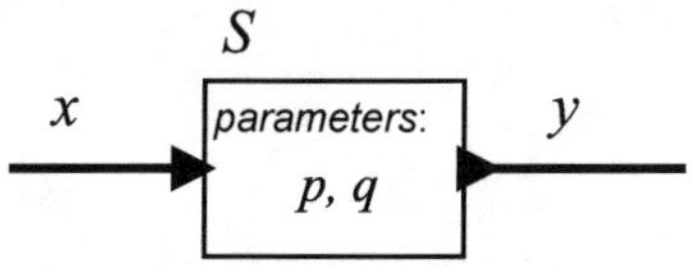

where the input signal x and the output signal y are functions of the form

$$x\colon \mathbb{R} \to \mathbb{R}, \quad y\colon \mathbb{R} \to \mathbb{R}.$$

Here the domain represents **time** and the codomain represents the value of the signal at a particular time. The domain $\mathbb{R}$ may be replaced by $\mathbb{R}_+$, the non-negative reals, if we wish to explicitly model a system that comes into existence and starts operating at a particular point in time.

The model of the system is a function of the form

$$S\colon X \to Y, \tag{2.5}$$

where $X = Y = \mathbb{R}^{\mathbb{R}}$, the set of functions that map the reals into the reals, like x and y above.[3] The function S may depend on parameters of the system, in which case the parameters may be optionally shown in the box, and may be optionally included in the function notation. For example, in the above figure, if there are parameters p and q, we might write the system function as $S_{p,q}$ or even $S(p,q)$, keeping in mind that both notations represent functions of

[3] As explained in Appendix A, the notation $\mathbb{R}^{\mathbb{R}}$ (which can also be written $(\mathbb{R} \to \mathbb{R})$) represents the set of all functions with domain $\mathbb{R}$ and codomain $\mathbb{R}$.

the form in 2.5. A box like that above, where the inputs are functions and the outputs are functions, is called an **actor**.

Example 2.2: The actor model for the helicopter of example 2.1 can be depicted as follows:

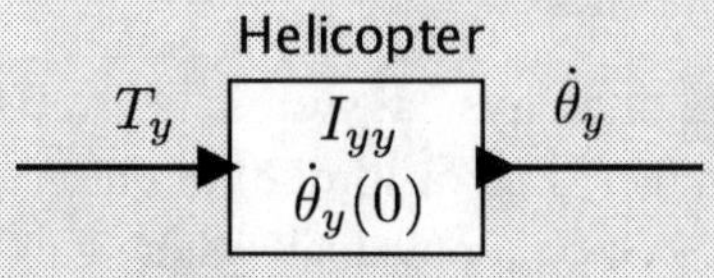

The input and output are both continuous-time functions. The parameters of the actor are the initial angular velocity $\dot{\theta}_y(0)$ and the moment of inertia I_{yy}. The function of the actor is defined by (2.4).

Actor models are composable. In particular, given two actors S_1 and S_2, we can form a **cascade composition** as follows:

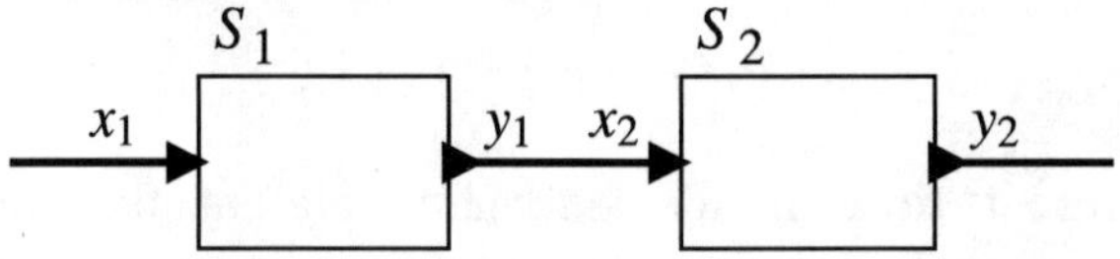

In the diagram, the "wire" between the output of S_1 and the input of S_2 means precisely that $y_1 = x_2$, or more pedantically,

$$\forall\, t \in \mathbb{R}, \quad y_1(t) = x_2(t).$$

Example 2.3: The actor model for the helicopter can be represented as a cascade composition of two actors as follows:

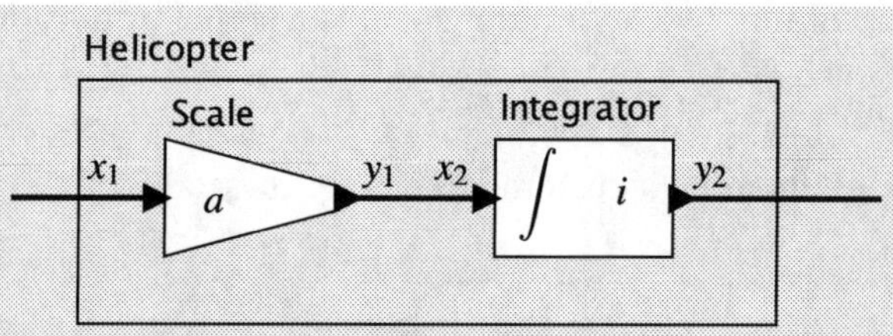

The left actor represents a Scale actor parameterized by the constant a defined by

$$\forall\, t \in \mathbb{R}, \quad y_1(t) = a x_1(t). \tag{2.6}$$

More compactly, we can write $y_1 = a x_1$, where it is understood that the product of a scalar a and a function x_1 is interpreted as in (2.6). The right actor represents an integrator parameterized by the initial value i defined by

$$\forall\, t \in \mathbb{R}, \quad y_2(t) = i + \int_0^t x_2(\tau) d\tau.$$

If we give the parameter values $a = 1/I_{yy}$ and $i = \dot{\theta}_y(0)$, we see that this system represents (2.4) where the input $x_1 = T_y$ is torque and the output $y_2 = \dot{\theta}_y$ is angular velocity.

In the above figure, we have customized the **icons**, which are the boxes representing the actors. These particular actors (scaler and integrator) are particularly useful building blocks for building up models of physical dynamics, so assigning them recognizable visual notations is useful.

We can have actors that have multiple input signals and/or multiple output signals. These are represented similarly, as in the following example, which has two input signals and one output signal:

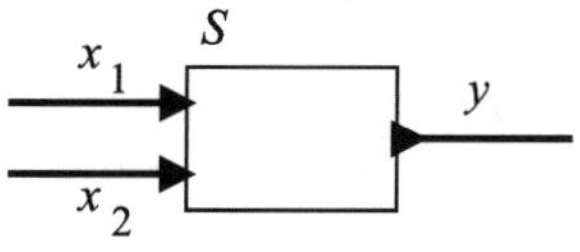

A particularly useful building block with this form is a signal **adder**, defined by

$$\forall\, t \in \mathbb{R}, \quad y(t) = x_1(t) + x_2(t).$$

This will often be represented by a custom icon as follows:

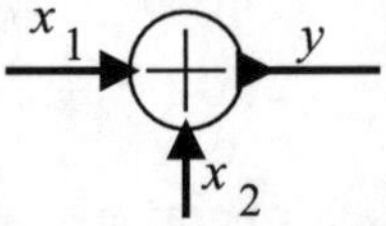

Sometimes, one of the inputs will be subtracted rather than added, in which case the icon is further customized with minus sign near that input, as below:

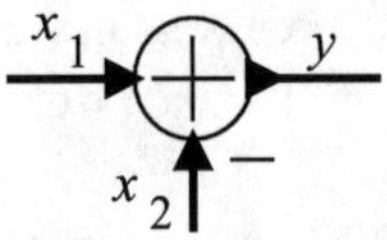

This actor represents a function $S\colon (\mathbb{R} \to \mathbb{R})^2 \to (\mathbb{R} \to \mathbb{R})$ given by

$$\forall\, t \in \mathbb{R},\ \forall\, x_1, x_2 \in (\mathbb{R} \to \mathbb{R}), \quad (S(x_1, x_2))(t) = y(t) = x_1(t) - x_2(t).$$

Notice the careful notation. $S(x_1, x_2)$ is a function in $\mathbb{R}^{\mathbb{R}}$. Hence, it can be evaluated at a $t \in \mathbb{R}$.

In the rest of this chapter, we will not make a distinction between a system and its actor model, unless the distinction is essential to the argument. We will assume that the actor model captures everything of interest about the system. This is an admittedly bold assumption. Generally the properties of the actor model are only approximate descriptions of the actual system.

2.3 Properties of Systems

In this section, we consider a number of properties that actors and the systems they compose may have, including causality, memorylessness, linearity, time invariance, and stability.

2.3.1 Causal Systems

Intuitively, a system is **causal** if its output depends only on current and past inputs. Making this notion precise is a bit tricky, however. We do this by first giving a notation for "current and past inputs." Consider a continuous-time signal $x\colon \mathbb{R} \to A$, for some set A. Let $x|_{t\le\tau}$ represent a function called the **restriction in time** that is only defined for times $t \le \tau$, and where it is defined, $x|_{t\le\tau}(t) = x(t)$. Hence if x is an input to a system, then $x|_{t\le\tau}$ is the "current and past inputs" at time τ.

Consider a continuous-time system $S\colon X \to Y$, where $X = A^{\mathbb{R}}$ and $Y = B^{\mathbb{R}}$ for some sets A and B. This system is causal if for all $x_1, x_2 \in X$ and $\tau \in \mathbb{R}$,

$$x_1|_{t\le\tau} = x_2|_{t\le\tau} \Rightarrow S(x_1)|_{t\le\tau} = S(x_2)|_{t\le\tau}$$

That is, the system is causal if for two possible inputs x_1 and x_2 that are identical up to (and including) time τ, the outputs are identical up to (and including) time τ. All systems we have considered so far are causal.

A system is **strictly causal** if for all $x_1, x_2 \in X$ and $\tau \in \mathbb{R}$,

$$x_1|_{t<\tau} = x_2|_{t<\tau} \Rightarrow S(x_1)|_{t\le\tau} = S(x_2)|_{t\le\tau}$$

That is, the system is strictly causal if for two possible inputs x_1 and x_2 that are identical up to (and *not* including) time τ, the outputs are identical up to (and including) time τ. The output at time t of a strictly causal system does not depend on its input at time t. It only depends on past inputs. A strictly causal system, of course, is also causal. The Integrator actor is strictly causal. The adder is not strictly causal, but it is causal. Strictly causal actors are useful for constructing feedback systems. p.31

2.3.2 Memoryless Systems

Intuitively, a system has memory if the output depends not only on the current inputs, but also on past inputs (or future inputs, if the system is not causal). Consider a continuous-time system $S\colon X \to Y$, where $X = A^{\mathbb{R}}$ and $Y = B^{\mathbb{R}}$ for some sets A and B. Formally, this system is **memoryless** if there exists a function $f\colon A \to B$ such that for all $x \in X$,

$$(S(x))(t) = f(x(t))$$

for all $t \in \mathbb{R}$. That is, the output $(S(x))(t)$ at time t depends only on the input $x(t)$ at time t.

The Integrator considered above is not memoryless, but the adder is. Exercise 2 shows that if a system is strictly causal and memoryless then its output is constant for all inputs.

2.3.3 Linearity and Time Invariance

Systems that are linear and time invariant (LTI) have particularly nice mathematical properties. Much of the theory of control systems depends on these properties. These properties form the main body of courses on signals and systems, and are beyond the scope of this text. But we will occasionally exploit simple versions of the properties, so it is useful to determine when a system is LTI.

A system $S\colon X \to Y$, where X and Y are sets of signals, is linear if it satisfies the **superposition** property:

$$\forall\, x_1, x_2 \in X \text{ and } \forall\, a, b \in \mathbb{R}, \quad S(ax_1 + bx_2) = aS(x_1) + bS(x_2).$$

It is easy to see that the helicopter system defined in Example 2.1 is linear if and only if the initial angular velocity $\dot{\theta}_y(0) = 0$ (see Exercise 3).

More generally, it is easy to see that an integrator as defined in Example 2.3 is linear if and only if the initial value $i = 0$, that the Scale actor is always linear, and that the cascade of any two linear actors is linear. We can trivially extend the definition of linearity to actors with more than one input or output signal and then determine that the adder is also linear.

To define time invariance, we first define a specialized continuous-time actor called a **delay**. Let $D_\tau\colon X \to Y$, where X and Y are sets of continuous-time signals, be defined by

$$\forall\, x \in X \text{ and } \forall\, t \in \mathbb{R}, \quad (D_\tau(x))(t) = x(t - \tau). \tag{2.7}$$

Here, τ is a parameter of the delay actor. A system $S\colon X \to Y$ is time invariant if

$$\forall\, x \in X \text{ and } \forall\, \tau \in \mathbb{R}, \quad S(D_\tau(x)) = D_\tau(S(x)).$$

The helicopter system defined in Example 2.1 and (2.4) is not time invariant. A minor variant, however, is time invariant:

$$\dot{\theta}_y(t) = \frac{1}{I_{yy}} \int_{-\infty}^{t} T_y(\tau)\, d\tau.$$

This version does not allow for an initial angular rotation.

A **linear time-invariant system** (**LTI**) is a system that is both linear and time invariant. A major objective in modeling physical dynamics is to choose an LTI model whenever possible. If a reasonable approximation results in an LTI model, it is worth making this approximation. It is not always easy to determine whether the approximation is reasonable, or to find models for which the approximation is reasonable. It is often easy to construct models that are more complicated than they need to be (see Exercise 4).

2.3.4 Stability

A system is said to be **bounded-input bounded-output stable** (**BIBO stable** or just **stable**) if the output signal is bounded for all input signals that are bounded.

Consider a continuous-time system with input w and output v. The input is bounded if there is a real number $A < \infty$ such that $|w(t)| \leq A$ for all $t \in \mathbb{R}$. The output is bounded if there is a real number $B < \infty$ such that $|v(t)| \leq B$ for all $t \in \mathbb{R}$. The system is stable if for any input bounded by some A, there is some bound B on the output.

Example 2.4: It is now easy to see that the helicopter system developed in Example 2.1 is unstable. Let the input be $T_y = u$, where u is the **unit step**, given by

$$\forall\, t \in \mathbb{R}, \quad u(t) = \begin{cases} 0, & t < 0 \\ 1, & t \geq 0 \end{cases} . \tag{2.8}$$

This means that prior to time zero, there is no torque applied to the system, and starting at time zero, we apply a torque of unit magnitude. This input is clearly bounded. It never exceeds one in magnitude. However, the output grows without bound.

In practice, a helicopter uses a feedback system to determine how much torque to apply at the tail rotor to keep the body of the helicopter straight. We study how to do that next.

2.4 Feedback Control

A system with **feedback** has directed cycles, where an output from an actor is fed back to affect an input of the same actor. An example of such a system is shown in Figure 2.3. Most control systems use feedback. They make measurements of an **error** (e in the figure), which is a discrepancy between desired behavior (ψ in the figure) and actual behavior ($\dot{\theta}_y$ in the figure), and use that measurement to correct the behavior. The error measurement is feedback, and the corresponding correction signal (T_y in the figure) should compensate to reduce future error. Note that the correction signal normally can only affect *future* errors, so a feedback system must normally include at least one strictly causal actor (the Helicopter in the figure) in every directed cycle. p.29

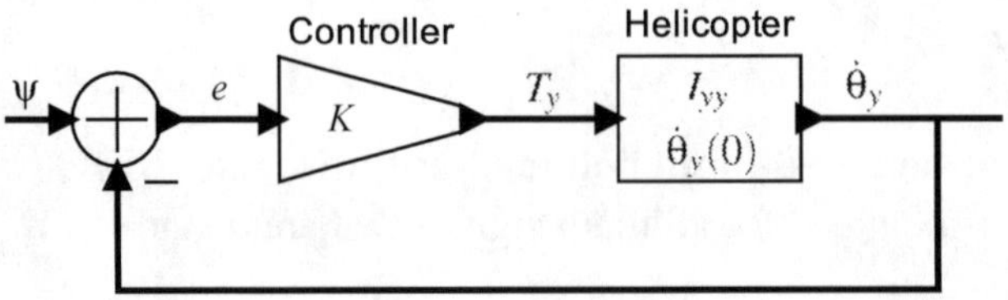

Figure 2.3: Proportional control system that stabilizes the helicopter.

Feedback control is a sophisticated topic, easily occupying multiple texts and complete courses. Here, we only barely touch on the subject, just enough to motivate the interactions between software and physical systems. Feedback control systems are often implemented using embedded software, and the overall physical dynamics is a composition of the software and physical dynamics. More detail can be found in Chapters 12-14 of Lee and Varaiya (2011).

Example 2.5: Recall that the helicopter model of Example 2.1 is not stable. We can stabilize it with a simple feedback control system, as shown in Figure 2.3. The input ψ to this system is a continuous-time system specifying the desired angular velocity. The **error signal** e represents the difference between the actual and the desired angular velocity. In the figure, the controller simply scales the error signal by a constant K, providing a control input to the helicopter. We use (2.4) to write

$$\dot{\theta}_y(t) = \dot{\theta}_y(0) + \frac{1}{I_{yy}} \int_0^t T_y(\tau) d\tau \tag{2.9}$$

$$= \dot{\theta}_y(0) + \frac{K}{I_{yy}} \int_0^t (\psi(\tau) - \dot{\theta}_y(\tau)) d\tau, \tag{2.10}$$

where we have used the facts (from the figure),

$$e(t) = \psi(t) - \dot{\theta}_y(t), \quad \text{and}$$

$$T_y(t) = Ke(t).$$

Equation (2.10) has $\dot{\theta}_y(t)$ on both sides, and therefore is not trivial to solve. The easiest solution technique uses Laplace transforms (see Lee and Varaiya (2011)

Chapter 14). However, for our purposes here, we can use a more brute-force technique from calculus. To make this as simple as possible, we assume that $\psi(t) = 0$ for all t; i.e., we wish to control the helicopter simply to keep it from rotating at all. The desired angular velocity is zero. In this case, (2.10) simplifies to

$$\dot{\theta}_y(t) = \dot{\theta}_y(0) - \frac{K}{I_{yy}} \int_0^t \dot{\theta}_y(\tau) d\tau. \tag{2.11}$$

Using the fact from calculus that, for $t \geq 0$,

$$\int_0^t a e^{a\tau} d\tau = e^{at} u(t) - 1,$$

where u is given by (2.8), we can infer that the solution to (2.11) is

$$\dot{\theta}_y(t) = \dot{\theta}_y(0) e^{-Kt/I_{yy}} u(t). \tag{2.12}$$

(Note that although it is easy to verify that this solution is correct, deriving the solution is not so easy. For this purpose, Laplace transforms provide a far better mechanism.)

We can see from (2.12) that the angular velocity approaches the desired angular velocity (zero) as t gets large as long as K is positive. For larger K, it will approach more quickly. For negative K, the system is unstable, and angular velocity will grow without bound.

The previous example illustrates a **proportional control** feedback loop. It is called this because the control signal is proportional to the error. We assumed a desired signal of zero. It is equally easy to assume that the helicopter is initially at rest (the angular velocity is zero) and then determine the behavior for a particular non-zero desired signal, as we do in the following example.

Example 2.6: Assume that the helicopter is **initially at rest**, meaning that

$$\dot{\theta}(0) = 0,$$

and that the desired signal is

$$\psi(t) = au(t)$$

for some constant a. That is, we wish to control the helicopter to get it to rotate at a fixed rate.

We use (2.4) to write

$$\begin{aligned}
\dot{\theta}_y(t) &= \frac{1}{I_{yy}} \int_0^t T_y(\tau) d\tau \\
&= \frac{K}{I_{yy}} \int_0^t (\psi(\tau) - \dot{\theta}_y(\tau)) d\tau \\
&= \frac{K}{I_{yy}} \int_0^t a d\tau - \frac{K}{I_{yy}} \int_0^t \dot{\theta}_y(\tau) d\tau \\
&= \frac{Kat}{I_{yy}} - \frac{K}{I_{yy}} \int_0^t \dot{\theta}_y(\tau) d\tau.
\end{aligned}$$

Using the same (black magic) technique of inferring and then verifying the solution, we can see that the solution is

$$\dot{\theta}_y(t) = au(t)(1 - e^{-Kt/I_{yy}}). \tag{2.13}$$

Again, the angular velocity approaches the desired angular velocity as t gets large as long as K is positive. For larger K, it will approach more quickly. For negative K, the system is unstable, and angular velocity will grow without bound.

Note that the first term in the above solution is exactly the desired angular velocity. The second term is an error called the **tracking error**, that for this example asymptotically approaches zero.

The above example is somewhat unrealistic because we cannot independently control the *net* torque of the helicopter. In particular, the net torque T_y is the sum of the torque T_t due to the top rotor and the torque T_r due to the tail rotor,

$$\forall\, t \in \mathbb{R}, \quad T_y(t) = T_t(t) + T_r(t)\ .$$

T_t will be determined by the rotation required to maintain or achieve a desired altitude, quite independent of the rotation of the helicopter. Thus, we will actually need to design a control system that controls T_r and stabilizes the helicopter for any T_t (or, more precisely, any T_t within operating parameters). In the next example, we study how this changes the performance of the control system.

Example 2.7: In Figure 2.4(a), we have modified the helicopter model so that it has two inputs, T_t and T_r, the torque due to the top rotor and tail rotor respectively. The feedback control system is now controlling only T_r, and T_t is treated as an external (uncontrolled) input signal. How well will this control system behave?

Again, a full treatment of the subject is beyond the scope of this text, but we will study a specific example. Suppose that the torque due to the top rotor is given by

$$T_t = bu(t)$$

for some constant b. That is, at time zero, the top rotor starts spinning a constant velocity, and then holds that velocity. Suppose further that the helicopter is initially at rest. We can use the results of Example 2.6 to find the behavior of the system.

First, we transform the model into the equivalent model shown in Figure 2.4(b). This transformation simply relies on the algebraic fact that for any real numbers a_1, a_2, K,

$$Ka_1 + a_2 = K(a_1 + a_2/K).$$

We further transform the model to get the equivalent model shown in Figure 2.4(c), which has used the fact that addition is commutative. In Figure 2.4(c), we see that the portion of the model enclosed in the box is exactly the same as the control system analyzed in Example 2.6, shown in Figure 2.3. Thus, the same analysis as in Example 2.6 still applies. Suppose that desired angular rotation is

$$\psi(t) = 0.$$

Then the input to the original control system will be

$$x(t) = \psi(t) + T_t(t)/K = (b/K)u(t).$$

From (2.13), we see that the solution is

$$\dot{\theta}_y(t) = (b/K)u(t)(1 - e^{-Kt/I_{yy}}). \qquad (2.14)$$

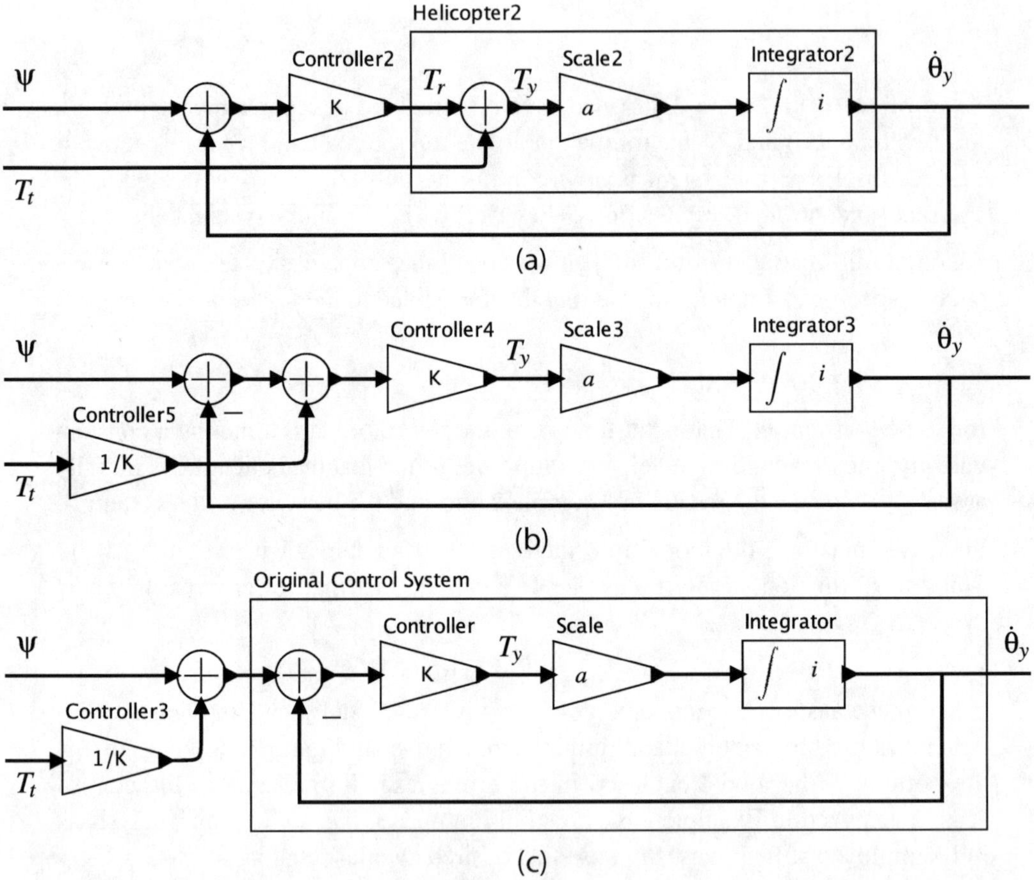

Figure 2.4: (a) Helicopter model with separately controlled torques for the top and tail rotors. (b) Transformation to an equivalent model (assuming $K > 0$). (c) Further transformation to an equivalent model that we can use to understand the behavior of the controller.

The desired angular rotation is zero, but the control system asymptotically approaches a non-zero angular rotation of b/K. This tracking error can be made arbitrarily small by increasing the control system feedback gain K, but with this controller design, it cannot be made to go to zero. An alternative controller design that yields an asymptotic tracking error of zero is studied in Exercise 7.

2.5 Summary

This chapter has described two distinct modeling techniques that describe physical dynamics. The first is ordinary differential equations, a venerable toolkit for engineers, and the second is actor models, a newer technique driven by software modeling and simulation tools. The two are closely related. This chapter has emphasized the relationship between these models, and the relationship of those models to the systems being modeled. These relationships, however, are quite a deep subject that we have barely touched upon. Our objective is to focus the attention of the reader on the fact that we may use multiple models for a system, and that models are distinct from the systems being modeled. The fidelity p.12 of a model (how well it approximates the system being modeled) is a strong factor in the success or failure of any engineering effort.

Exercises

1. A **tuning fork**, shown in Figure 2.5, consists of a metal finger (called a **tine**) that is displaced by striking it with a hammer. After being displaced, it vibrates. If the tine has no friction, it will vibrate forever. We can denote the displacement of the tine after being struck at time zero as a function $y\colon \mathbb{R}_+ \to \mathbb{R}$. If we assume that the initial displacement introduced by the hammer is one unit, then using our knowledge of physics we can determine that for all $t \in \mathbb{R}_+$, the displacement satisfies the differential equation

$$\ddot{y}(t) = -\omega_0^2 y(t)$$

where ω_0^2 is a constant that depends on the mass and stiffness of the tine, and where $\ddot{y}(t)$ denotes the second derivative with respect to time of y. It is easy to verify that y given by

$$\forall\, t \in \mathbb{R}_+, \quad y(t) = \cos(\omega_0 t)$$

is a solution to the differential equation (just take its second derivative). Thus, the displacement of the tuning fork is sinusoidal. If we choose materials for the tuning fork so that $\omega_0 = 2\pi \times 440$ radians/second, then the tuning fork will produce the tone of A-440 on the musical scale.

(a) Is $y(t) = \cos(\omega_0 t)$ the only solution? If not, give some others.

(b) Assuming the solution is $y(t) = \cos(\omega_0 t)$, what is the initial displacement?

(c) Construct a model of the tuning fork that produces y as an output using generic actors like Integrator, adder, scaler, or similarly simple actors. Treat the initial displacement as a parameter. Carefully label your diagram.

2. Show that if a system $S\colon A^{\mathbb{R}} \to B^{\mathbb{R}}$ is strictly causal and memoryless then its output is constant. Constant means that the output $(S(x))(t)$ at time t does not depend on t.

3. This exercise studies linearity.

(a) Show that the helicopter model defined in Example 2.1 is linear if and only if the initial angular velocity $\dot{\theta}_y(0) = 0$.

(b) Show that the cascade of any two linear actors is linear.

(c) Augment the definition of linearity so that it applies to actors with two input signals and one output signal. Show that the adder actor is linear.

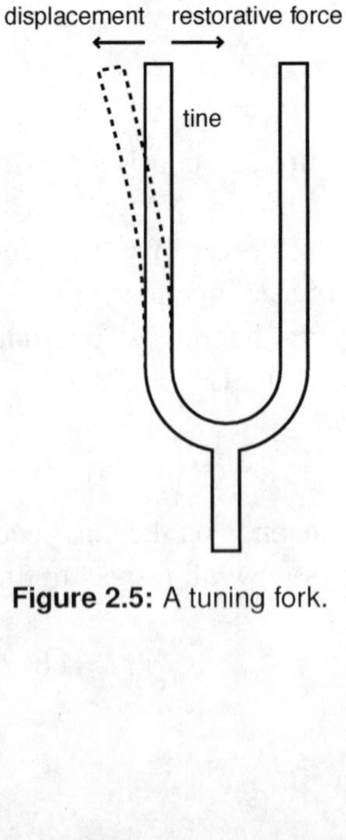

Figure 2.5: A tuning fork.

4. Consider the helicopter of Example 2.1, but with a slightly different definition of the input and output. Suppose that, as in the example, the input is $T_y\colon \mathbb{R} \to \mathbb{R}$, as in the example, but the output is the position of the tail relative to the main rotor shaft. Specifically, let the x-y plane be the plane orthogonal to the rotor shaft, and let the position of the tail at time t be given by a tuple $((x(t), y(t))$. Is this model LTI? Is it BIBO stable?

5. Consider a rotating robot where you can control the angular velocity around a fixed axis.

   (a) Model this as a system where the input is angular velocity $\dot{\theta}$ and the output is angle θ. Give your model as an equation relating the input and output as functions of time.

   (b) Is this model BIBO stable?

   (c) Design a proportional controller to set the robot onto a desired angle. That is, assume that the initial angle is $\theta(0) = 0$, and let the desired angle be $\psi(t) = au(t)$, where u is the unit step function. Find the actual angle as a function of time and the proportional controller feedback gain K. What is your output at $t = 0$? What does it approach as t gets large? p.31

6. A DC motor produces a torque that is proportional to the current through the windings of the motor. Neglecting friction, the net torque on the motor, therefore, is this torque minus the torque applied by whatever load is connected to the motor. Newton's second law (the rotational version) gives p.22 p.21

$$k_T i(t) - x(t) = I \frac{d}{dt}\omega(t), \tag{2.15}$$

where k_T is the motor torque constant, $i(t)$ is the current at time t, $x(t)$ is the torque applied by the load at time t, I is the moment of inertia of the motor, and $\omega(t)$ is the angular velocity of the motor. p.195 p.24

   (a) Assuming the motor is initially at rest, rewrite (2.15) as an integral equation. p.33

   (b) Assuming that both x and i are inputs and ω is an output, construct an actor model (a block diagram) that models this motor. You should use only primitive actors such as integrators and basic arithmetic actors such as scale and adder.

   (c) In reality, the input to a DC motor is not a current, but is rather a voltage. If we assume that the inductance of the motor windings is negligible, then the relationship between voltage and current is given by

$$v(t) = Ri(t) + k_b\omega(t),$$

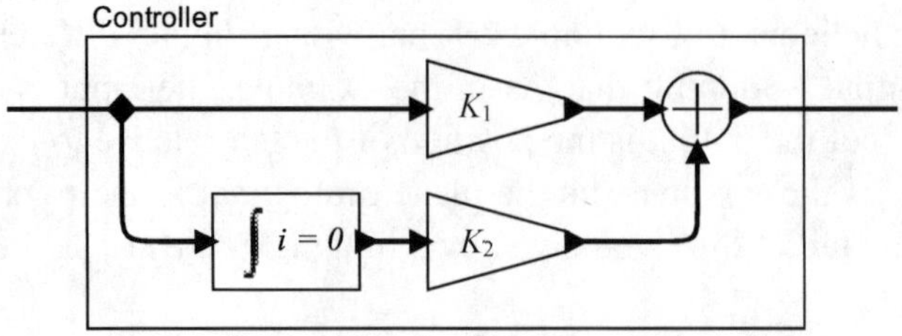

Figure 2.6: A PI controller for the helicopter.

where R is the resistance of the motor windings and k_b is a constant called the motor back electromagnetic force constant. The second term appears because a rotating motor also functions as an electrical generator, where the voltage generated is proportional to the angular velocity.

Modify your actor model so that the inputs are v and x rather than i and x.

7. (a) Using your favorite continuous-time modeling software (such as LabVIEW, Simulink, or Ptolemy II), construct a model of the helicopter control system shown in Figure 2.4. Choose some reasonable parameters and plot the actual angular velocity as a function of time, assuming that the desired angular velocity is zero, $\psi(t) = 0$, and that the top-rotor torque is non-zero, $T_t(t) = bu(t)$. Give your plot for several values of K and discuss how the behavior varies with K.

   (b) Modify the model of part (a) to replace the Controller of Figure 2.4 (the simple scale-by-K actor) with the alternative controller shown in Figure 2.6. This alternative controller is called a **proportional-integrator (PI) controller**. It has two parameter K_1 and K_2. Experiment with the values of these parameters, give some plots of the behavior with the same inputs as in part (a), and discuss the behavior of this controller in contrast to the one of part (a).

3 Discrete Dynamics

Contents

Models of embedded systems include both **discrete** and **continuous** components. Loosely speaking, continuous components evolve smoothly, while discrete components evolve abruptly. The previous chapter considered continuous components, and showed that the physical dynamics of the system can often be modeled with ordinary differential or integral

equations, or equivalently with actor models that mirror these equations. Discrete compo-
p.20 nents, on the other hand, are not conveniently modeled by ODEs. In this chapter, we study how state machines can be used to model discrete dynamics. In the next chapter, we will show how these state machines can be combined with models of continuous dynamics to get hybrid system models.

3.1 Discrete Systems

A **discrete system** operates in a sequence of discrete steps and is said to have **discrete dynamics**. Some systems are inherently discrete.

> **Example 3.1:** Consider a system that counts the number of cars that enter and leave a parking garage in order to keep track of how many cars are in the garage at any time. It could be modeled as shown in Figure 3.1. We ignore for now how to design the sensors that detect the entry or departure of cars. We simply assume that the ArrivalDetector actor produces an event when a car arrives, and the DepartureDetector actor produces an event when a car departs. The Counter actor keeps a running count, starting from an initial value i. Each time the count changes, it produces an output event that updates a display.

In the above example, each entry or departure is modeled as a **discrete event**. A discrete event occurs at an instant of time rather than over time. The Counter actor in Figure 3.1 is analogous to the Integrator actor used in the previous chapter, shown here in Figure 3.2.

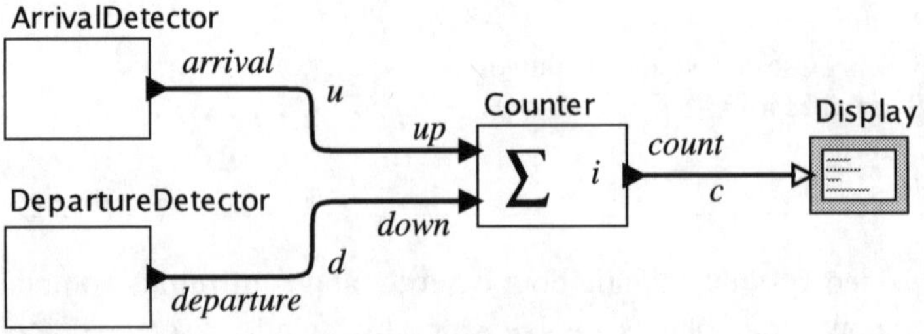

Figure 3.1: Model of a system that keeps track of the number of cars in a parking garage.

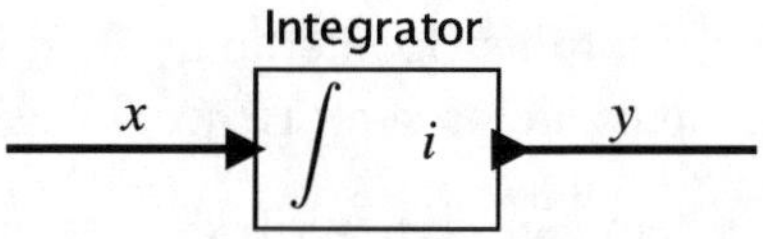

Figure 3.2: Icon for the Integrator actor used in the previous chapter.

Like the Counter actor, the Integrator accumulates input values. However, it does so very differently. The input of an Integrator is a function of the form $x\colon \mathbb{R} \to \mathbb{R}$ or $x\colon \mathbb{R}_+ \to \mathbb{R}$, a continuous-time signal. The signal u going into the *up* input port of the Counter, on the other hand, is a function of the form p.21

$$u\colon \mathbb{R} \to \{absent, present\}.$$

This means that at any time $t \in \mathbb{R}$, the input $u(t)$ is either *absent*, meaning that there is no event at that time, or *present*, meaning that there is. A signal of this form is known as a **pure signal**. It carries no value, but instead provides all its information by being either present or absent at any given time. The signal d in Figure 3.1 is also a pure signal.

Assume our Counter operates as follows. When an event is present at the *up* input port, it increments its count and produces on the output the new value of the count. When an event is present at the *down* input, it decrements its count and produces on the output the new value of the count.[1] At all other times (when both inputs are absent), it produces no output (the *count* output is absent). Hence, the signal c in Figure 3.1 can be modeled by a function of the form

$$c\colon \mathbb{R} \to \{absent\} \cup \mathbb{Z}\ .$$

(See Appendix A for notation.) This signal is not pure, but like u and d, it is either absent or present. Unlike u and d, when it is present, it has a value (an integer).

Assume further that the inputs are absent most of the time, or more technically, that the inputs are discrete (see the sidebar on page 44). Then the Counter reacts in sequence to each of a sequence of input events. This is very different from the Integrator, which reacts continuously to a continuum of inputs.

The input to the Counter is a pair of discrete signals that at certain times have an event (are present), and at other times have no event (are absent). The output also is a discrete signal that, when an input is present, has a value that is a natural number, and at other times is

[1] It would be wise to design this system with a fault handler that does something reasonable if the count drops below zero, but we ignore this for now.

absent.[2] Clearly, there is no need for this Counter to do anything when the input is absent. It only needs to operate when inputs are present. Hence, it has discrete dynamics.

The dynamics of a discrete system can be described as a sequence of steps that we call **reactions**, each of which we assume to be instantaneous. Reactions of a discrete system are triggered by the environment in which the discrete system operates. In the case of the example of Figure 3.1, reactions of the Counter actor are triggered when one or more input events are present. That is, in this example, reactions are **event triggered**. When both inputs to the Counter are absent, no reaction occurs.

A particular reaction will observe the values of the inputs at a particular time t and calculate output values for that same time t. Suppose an actor has input ports $P = \{p_1, \cdots, p_N\}$, where p_i is the name of the i-th input port. Assume further that for each input port $p \in P$,

2 As shown in Exercise 8, the fact that input signals are discrete does not necessarily imply that the output signal is discrete. However, for this application, there are physical limitations on the rates at which cars can arrive and depart that ensure that these signals are discrete. So it is safe to assume that they are discrete.

Probing Further: Discrete Signals

Discrete signals consist of a sequence of instantaneous events in time. Here, we make this intuitive concept precise.

Consider a signal of the form $e\colon \mathbb{R} \to \{absent\} \cup X$, where X is any set of values. This signal is a **discrete signal** if, intuitively, it is absent most of the time and we can count, in order, the times at which it is present (not absent). Each time it is present, we have a discrete event.

This ability to count the events in order is important. For example, if e is present at all rational numbers t, then we do not call this signal discrete. The times at which it is present cannot be counted in order. It is not, intuitively, a sequence of instantaneous events in time (it is a *set* of instantaneous events in time, but not a *sequence*).

To define this formally, let $T \subseteq \mathbb{R}$ be the set of times where e is present. Specifically,

$$T = \{t \in \mathbb{R} : e(t) \neq absent\}.$$

p.474 Then e is discrete if there exists a one-to-one function $f\colon T \to \mathbb{N}$ that is **order preserving**. Order preserving simply means that for all $t_1, t_2 \in T$ where $t_1 \leq t_2$, we have that $f(t_1) \leq f(t_2)$. The existence of such a one-to-one function ensures that we can count off the events *in temporal order*. Some properties of discrete signals are studied in Exercise 8.

a set V_p denotes the values that may be received on port p when the input is present. V_p is called the **type** of port p. At a reaction we treat each $p \in P$ as a variable that takes on a value $p \in V_p \cup \{absent\}$. A **valuation** of the inputs P is an assignment of a value in V_p to each variable $p \in P$ or an assertion that p is absent.

If port p receives a pure signal, then $V_p = \{present\}$, a singleton set (set with only one element). The only possible value when the signal is not absent is *present*. Hence, at a reaction, the variable p will have a value in the set $\{present, absent\}$. p.472

Probing Further: Modeling Actors as Functions

As in Section 2.2, the Integrator actor of Figure 3.2 can be modeled by a function of the form

$$I_i \colon \mathbb{R}^{\mathbb{R}_+} \to \mathbb{R}^{\mathbb{R}_+},$$

which can also be written

$$I_i \colon (\mathbb{R}_+ \to \mathbb{R}) \to (\mathbb{R}_+ \to \mathbb{R}).$$

(See Appendix A if the notation is unfamiliar.) In the figure,

$$y = I_i(x) \ ,$$

where i is the initial value of the integration and x and y are continuous-time signals. For example, if $i = 0$ and for all $t \in \mathbb{R}_+$, $x(t) = 1$, then

$$y(t) = i + \int_0^t x(\tau) d\tau = t \ .$$

Similarly, the Counter in Figure 3.1 can be modeled by a function of the form

$$C_i \colon (\mathbb{R}_+ \to \{absent, present\})^P \to (\mathbb{R}_+ \to \{absent\} \cup \mathbb{Z}),$$

where $\mathbb{Z}$ is the integers and P is the set of input ports, $P = \{up, down\}$. Recall that the notation A^B denotes the set of all functions from B to A. Hence, the input to the function C is a function whose domain is P that for each port $p \in P$ yields a function in $(\mathbb{R}_+ \to \{absent, present\})$. That latter function, in turn, for each time $t \in \mathbb{R}_+$ yields either *absent* or *present*. p.25

Example 3.2: For the garage counter, the set of input ports is $P = \{up, down\}$. Both receive pure signals, so the types are $V_{up} = V_{down} = \{present\}$. If a car is arriving at time t and none is departing, then at that reaction, $up = present$ and $down = absent$. If a car is arriving and another is departing at the same time, then $up = down = present$. If neither is true, then both are *absent*.

Outputs are similarly designated. Consider a discrete system with output ports $Q = \{q_1, \cdots, q_M\}$ with types $V_{q_1}, \cdots, V_{q_M}$. At each reaction, the system assigns a value $q \in V_q \cup \{absent\}$ to each $q \in Q$, producing a valuation of the outputs. In this chapter, we will assume that the output is *absent* at times t where a reaction does not occur. Thus, outputs of a discrete system are discrete signals. Chapter 4 describes systems whose outputs are not constrained to be discrete (see also box on page 57).

Example 3.3: The Counter actor of Figure 3.1 has one output port named *count*, so $Q = \{count\}$. Its type is $V_{count} = \mathbb{Z}$. At a reaction, *count* is assigned the count of cars in the garage.

3.2 The Notion of State

Intuitively, the **state** of a system is its condition at a particular point in time. In general, the state affects how the system reacts to inputs. Formally, we define the state to be an encoding of everything about the past that has an effect on the system's reaction to current or future inputs. The state is a summary of the past.

Consider the Integrator actor shown in Figure 3.2. This actor has state, which in this case happens to have the same value as the output at any time t. The state of the actor at a time t is the value of the integral of the input signal up to time t. In order to know how the subsystem will react to inputs at and beyond time t, we have to know what this value is at time t. We do not need to know anything more about the past inputs. Their effect on the future is entirely captured by the current value at t. The icon in Figure 3.2 includes i, an initial state value, which is needed to get things started at some starting time.

An Integrator operates in a time continuum. It integrates a continuous-time input signal, generating as output at each time the cumulative area under the curve given by the input plus the initial state. Its state at any given time is that accumulated area plus the initial state. The Counter actor in the previous section also has state, and that state is also an accumulation of past input values, but it operates discretely.

The state $y(t)$ of the Integrator at time t is a real number. Hence, we say that the **state space** of the Integrator is $States = \mathbb{R}$. For the Counter used in Figure 3.1, the state $s(t)$ at time t is an integer, so $States \subset \mathbb{Z}$. A practical parking garage has a finite and non-negative number M of spaces, so the state space for the Counter actor used in this way will be

$$States = \{0, 1, 2, \cdots, M\} .$$

(This assumes the garage does not let in more cars than there are spaces.) The state space for the Integrator is infinite (uncountably infinite, in fact). The state space for the garage counter is finite. Discrete models with finite state spaces are called finite-state machines (FSMs). There are powerful analysis techniques available for such models, so we consider them next.

3.3 Finite-State Machines

A **state machine** is a model of a system with discrete dynamics that at each reaction maps p.42
valuations of the inputs to valuations of the outputs, where the map may depend on its current state. A **finite-state machine** (**FSM**) is a state machine where the set *States* of possible states is finite.

If the number of states is reasonably small, then FSMs can be conveniently drawn using a graphical notation like that in Figure 3.3. Here, each state is represented by a bubble, so for

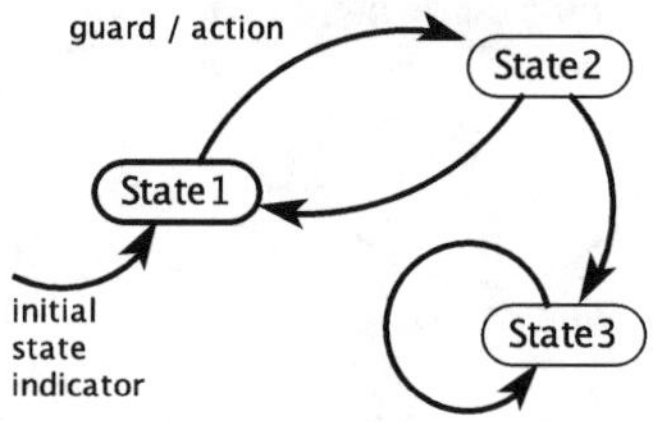

Figure 3.3: Visual notation for a finite state machine.

this diagram, the set of states is given by

$$States = \{\text{State1}, \text{State2}, \text{State3}\}.$$

At the beginning of each sequence of reactions, there is an **initial state**, State1, indicated in the diagram by a dangling arrow into it.

3.3.1 Transitions

Transitions between states govern the discrete dynamics of the state machine and the mapping of input valuations to output valuations. A transition is represented as a curved arrow, as shown in Figure 3.3, going from one state to another. A transition may also start and end at the same state, as illustrated with State3 in the figure. In this case, the transition is called a **self transition**.

In Figure 3.3, the transition from State1 to State2 is labeled with "guard / action." The **guard** determines whether the transition may be taken on a reaction. The **action** specifies what outputs are produced on each reaction.

A guard is a **predicate** (a boolean-valued expression) that evaluates to *true* when the transition should be taken, changing the state from that at the beginning of the transition to that at the end. When a guard evaluates to *true* we say that the transition is **enabled**. An action is an assignment of values (or *absent*) to the output ports. Any output port not mentioned in a transition that is taken is implicitly *absent*. If no action at all is given, then all outputs are implicitly *absent*.

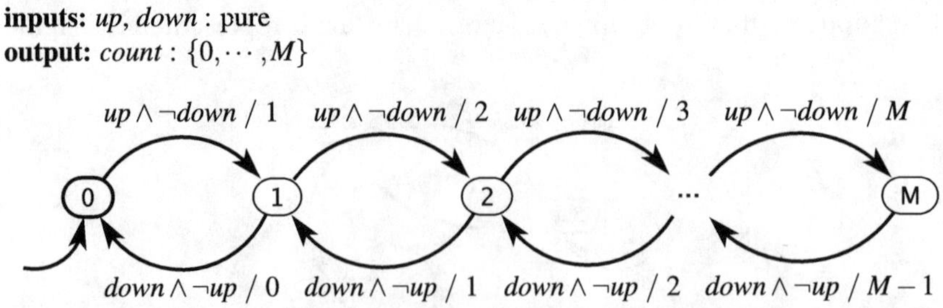

Figure 3.4: FSM model for the garage counter of Figure 3.1.

Example 3.4: Figure 3.4 shows an FSM model for the garage counter. The inputs and outputs are shown using the notation *name* : *type*. The set of states is $States = \{0, 1, 2, \cdots, M\}$. The transition from state 0 to 1 has a guard written as $up \land \neg down$. This is a predicate that evaluates to true when *up* is present and *down* is absent. If at a reaction the current state is 0 and this guard evaluates to true, then the transition will be taken and the next state will be 1. Moreover, the action indicates that the output should be assigned the value 1. The output port *count* is not explicitly named because there is only one output port, and hence there is no ambiguity.

If the guard expression on the transition from 0 to 1 had been simply *up*, then this could evaluate to true when *down* is also present, which would incorrectly count cars when a car was arriving at the same time that another was departing.

If p_1 and p_2 are pure inputs to a discrete system, then the following are examples of valid guards:

true	Transition is always enabled.
p_1	Transition is enabled if p_1 is *present*.
$\neg p_1$	Transition is enabled if p_1 is *absent*.
$p_1 \land p_2$	Transition is enabled if both p_1 and p_2 are *present*.
$p_1 \lor p_2$	Transition is enabled if either p_1 or p_2 is *present*.
$p_1 \land \neg p_2$	Transition is enabled if p_1 is *present* and p_2 is *absent*.

These are standard logical operators where *present* is taken as a synonym for *true* and *absent* as a synonym for *false*. The symbol $\neg$ represents logical **negation**. The operator $\land$ is logical **conjunction** (logical AND), and $\lor$ is logical **disjunction** (logical OR).

Suppose that in addition the discrete system has a third input port p_3 with type $V_{p_3} = \mathbb{N}$. Then the following are examples of valid guards:

p_3	Transition is enabled if p_3 is *present* (not *absent*).
$p_3 = 1$	Transition is enabled if p_3 is *present* and has value 1.
$p_3 = 1 \land p_1$	Transition is enabled if p_3 has value 1 and p_1 is *present*.
$p_3 > 5$	Transition is enabled if p_3 is *present* with value greater than 5.

Example 3.5: A major use of energy worldwide is in heating, ventilation, and air conditioning (**HVAC**) systems. Accurate models of temperature dynamics and

temperature control systems can significantly improve energy conservation. Such modeling begins with a modest **thermostat**, which regulates temperature to maintain a **setpoint**, or target temperature. The word "thermostat" comes from Greek words for "hot" and "to make stand."

Consider a thermostat modeled by an FSM with *States* = {heating, cooling} as shown in Figure 3.5. Suppose the setpoint is 20 degrees Celsius. If the heater is on, then the thermostat allows the temperature to rise past the setpoint to 22 degrees. If the heater is off, then it allows the temperature to drop past the setpoint to 18 degrees. This strategy is called hysteresis (see box on page 52). It avoids **chattering**, where the heater would turn on and off rapidly when the temperature is close to the setpoint temperature.

There is a single input *temperature* with type $\mathbb{R}$ and two pure outputs *heatOn* and *heatOff*. These outputs will be *present* only when a change in the status of the heater is needed (i.e., when it is on and needs to be turned off, or when it is off and needs to be turned on).

p.44 The FSM in Figure 3.5 could be event triggered, like the garage counter, in which case it will react whenever a *temperature* input is provided. Alternatively, it could be **time triggered**, meaning that it reacts at regular time intervals. The definition of the FSM does not change in these two cases. It is up to the environment in which an FSM operates when it should react.

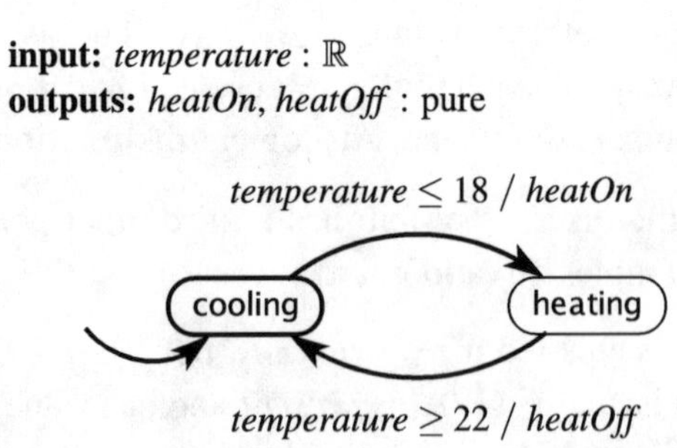

Figure 3.5: A model of a thermostat with hysteresis.

On a transition, the **action** (which is the portion after the slash) specifies the resulting valuation on the output ports when a transition is taken. If q_1 and q_2 are pure outputs and q_3 has type $\mathbb{N}$, then the following are examples of valid actions:

q_1	q_1 is present and q_2 and q_3 are *absent*.
q_1, q_2	q_1 and q_2 are both *present* and q_3 is *absent*.
$q_3 := 1$	q_1 and q_2 are *absent* and q_3 is *present* with value 1.
$q_3 := 1, q_1$	q_1 is *present*, q_2 is *absent*, and q_3 is *present* with value 1.
	(nothing) q_1, q_2, and q_3 are all *absent*.

Any output port that is not mentioned in a transition that is taken is implicitly *absent*. When assigning a value to an output port, we use the notation *name* := *value* to distinguish the **assignment** from a predicate, which would be written *name* = *value*. As in Figure 3.4, if p.48
there is only one output, then the assignment need not mention the port name.

3.3.2 When a Reaction Occurs

Nothing in the definition of a state machine constrains *when* it reacts. The environment determines when the machine reacts. Chapters 5 and 6 describe a variety of mechanisms and give a precise meaning to terms like event triggered and time triggered. For now, however, p.44
we just focus on what the machine does when it reacts.

When the environment determines that a state machine should react, the inputs will have a valuation. The state machine will assign a valuation to the output ports and (possibly) p.45
change to a new state. If no guard on any transition out of the current state evaluates to true, then the machine will remain in the same state.

It is possible for all inputs to be absent at a reaction. Even in this case, it may be possible for a guard to evaluate to true, in which case a transition is taken. If the input is absent and no guard on any transition out of the current state evaluates to true, then the machine will **stutter**. A **stuttering** reaction is one where the inputs and outputs are all absent and the machine does not change state. No progress is made and nothing changes.

Example 3.6: In Figure 3.4, if on any reaction both inputs are absent, then the machine will stutter. If we are in state 0 and the input *down* is *present*, then the guard on the only outgoing transition is false, and the machine remains in the same state. However, we do not call this a stuttering reaction because the inputs are not all *absent*.

Our informal description of the garage counter in Example 3.1 did not explicitly state what would happen if the count was at 0 and a car departed. A major advantage of FSM models is that they define all possible behaviors. The model in Figure 3.4 defines what happens in this circumstance. The count remains at 0. As a consequence, FSM models are amenable to formal checking, which determines whether the specified behaviors are in fact desirable behaviors. The informal specification cannot be subjected to such tests, or at least, not completely.

Probing Further: Hysteresis

The thermostat in Example 3.5 exhibits a particular form of state-dependent behavior
p.50 called **hysteresis**. Hysteresis is used to prevent chattering. A system with hysteresis has memory, but in addition has a useful property called **time-scale invariance**. In Example 3.5, the input signal as a function of time is a signal of the form

$$temperature\colon \mathbb{R} \to \{absent\} \cup \mathbb{R}\ .$$

Hence, *temperature*(t) is the temperature reading at time t, or *absent* if there is no temperature reading at that time. The output as a function of time has the form

$$heatOn, heatOff\colon \mathbb{R} \to \{absent, present\}\ .$$

Suppose that instead of *temperature* the input is given by

$$temperature'(t) = temperature(\alpha \cdot t)$$

for some $\alpha > 0$. If $\alpha > 1$, then the input varies faster in time, whereas if $\alpha < 1$ then the input varies more slowly, but in both cases, the input pattern is the same. Then for this FSM, the outputs $heatOn'$ and $heatOff'$ are given by

$$heatOn'(t) = heatOn(\alpha \cdot t) \qquad heatOff'(t) = heatOff(\alpha \cdot t)\ .$$

Time-scale invariance means that scaling the time axis at the input results in scaling the time axis at the output, so the absolute time scale is irrelevant.

An alternative implementation for the thermostat would use a single temperature threshold, but instead would require that the heater remain on or off for at least a minimum amount of time, regardless of the temperature. The consequences of this design choice are explored in Exercise 2.

Although it may seem that the model in Figure 3.4 does not define what happens if the state is 0 and *down* is *present*, it does so implicitly — the state remains unchanged and no output is generated. The reaction is not shown explicitly in the diagram. Sometimes it is useful to emphasize such reactions, in which case they can be shown explicitly. A convenient way to do this is using a **default transition**, shown in Figure 3.6. In that figure, the default transition is denoted with dashed lines and is labeled with "*true /* ". A default transition is enabled if no non-default transition is enabled and if its guard evaluates to true. In Figure 3.6, therefore, the default transition is enabled if $up \wedge \neg down$ evaluates to false, and when the default transition is taken the output is absent.

Default transitions provide a convenient notation, but they are not really necessary. Any default transition can be replaced by an ordinary transition with an appropriately chosen guard. For example, in Figure 3.6 we could use an ordinary transition with guard $\neg(up \wedge \neg down)$.

The use of both ordinary transitions and default transitions in a diagram can be thought of as a way of assigning priority to transitions. An ordinary transition has priority over a default transition. When both have guards that evaluate to true, the ordinary transition prevails. Some formalisms for state machines support more than two levels of priority. For example SyncCharts (André, 1996) associates with each transition an integer priority. This can make guard expressions simpler, at the expense of having to indicate priorities in the diagrams.

3.3.3 Update Functions

The graphical notation for FSMs defines a specific mathematical model of the dynamics of a state machine. A mathematical notation with the same meaning as the graphical notation

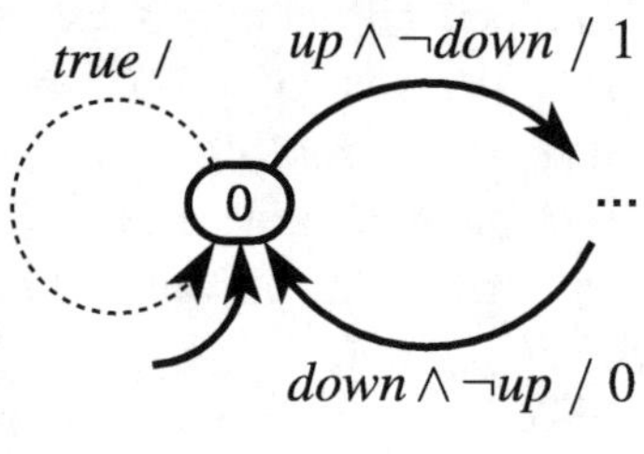

Figure 3.6: A default transition that need not be shown explicitly because it returns to the same state and produces no output.

sometimes proves convenient, particularly for large state machines where the graphical notation becomes cumbersome. In such a mathematical notation, a finite-state machine is a five-tuple

$$(States, Inputs, Outputs, update, initialState)$$

where

- p.46 *States* is a finite set of states;
- p.45 *Inputs* is a set of input valuations;
- *Outputs* is a set of output valuations;
- *update* : *States* $\times$ *Inputs* $\rightarrow$ *States* $\times$ *Outputs* is an **update function**, mapping a state and an input valuation to a *next* state and an output valuation;
- p.48 *initialState* is the initial state.

p.44 The FSM reacts in a sequence of reactions. At each reaction, the FSM has a *current state*, and the reaction may transition to a *next state*, which will be the current state of the next reaction. We can number these states starting with 0 for the initial state. Specifically, let $s\colon \mathbb{N} \rightarrow States$ be a function that gives the state of an FSM at reaction $n \in \mathbb{N}$. Initially, $s(0) = initialState$.

Let $x\colon \mathbb{N} \rightarrow Inputs$ and $y\colon \mathbb{N} \rightarrow Outputs$ denote that input and output valuations at each reaction. Hence, $x(0) \in Inputs$ is the first input valuation and $y(0) \in Outputs$ is the first

Software Tools Supporting FSMs

FSMs have been used in theoretical computer science and software engineering for quite some time (Hopcroft and Ullman, 1979). A number of software tools support design and analysis of FSMs. Statecharts (Harel, 1987), a notation for concurrent composition of hierarchical FSMs, has influenced many of these tools. One of the first tools supporting the Statecharts notation is STATEMATE (Harel et al., 1990), which subsequently evolved into Rational Rhapsody, sold by IBM. Many variants of Statecharts have arisen (von der Beeck, 1994), and some variant is now supported by nearly every software engineering tool that provides UML (unified modeling language) capabilities (Booch et al., 1998). SyncCharts (André, 1996) is a particularly nice variant in that p.104 it borrows the rigorous semantics of Esterel (Berry and Gonthier, 1992) for composition of concurrent FSMs. LabVIEW supports a variant of Statecharts that can operate within p.142 dataflow diagrams, and Simulink with its Stateflow extension supports a variant that can operate within continuous-time models.

output valuation. The dynamics of the state machine are given by the following equation:

$$(s(n+1), y(n)) = update(s(n), x(n)) \tag{3.1}$$

This gives the next state and output in terms of the current state and input. The *update* function encodes all the transitions, guards, and output specifications in an FSM. The term **transition function** is often used in place of update function.

The input and output valuations also have a natural mathematical form. Suppose an FSM has input ports $P = \{p_1, \cdots, p_N\}$, where each $p \in P$ has a corresponding type V_p. Then *Inputs* is a set of functions of the form

$$i \colon P \to V_{p_1} \cup \cdots \cup V_{p_N} \cup \{absent\} \ ,$$

where for each $p \in P$, $i(p) \in V_p \cup \{absent\}$ gives the value of port p. Thus, a function $i \in Inputs$ is a valuation of the input ports.

Example 3.7: The FSM in Figure 3.4 can be mathematically represented as follows:

$$\begin{aligned} States &= \{0, 1, \cdots, M\} \\ Inputs &= (\{up, down\} \to \{present, absent\}) \\ Outputs &= (\{count\} \to \{0, 1, \cdots, M, absent\}) \\ initialState &= 0 \end{aligned}$$

The update function is given by

$$update(s, i) = \begin{cases} (s+1, s+1) & \text{if } s < M \\ & \wedge\, i(up) = present \\ & \wedge\, i(down) = absent \\ (s-1, s-1) & \text{if } s > 0 \\ & \wedge\, i(up) = absent \\ & \wedge\, i(down) = present \\ (s, absent) & \text{otherwise} \end{cases} \tag{3.2}$$

for all $s \in States$ and $i \in Inputs$. Note that an output valuation $o \in Outputs$ is a function of the form $o \colon \{count\} \to \{0, 1, \cdots, M, absent\}$. In (3.2), the first alternative gives the output valuation as $o = s + 1$, which we take to mean the constant function that for all $q \in Q = \{count\}$ yields $o(q) = s + 1$. When there is more than one output port we will need to be more explicit about which output value is assigned to which output port. In such cases, we can use the same notation that we use for actions in the diagrams.

3.3.4 Determinacy and Receptiveness

The state machines presented in this section have two important properties:

Determinacy: A state machine is said to be **deterministic** if, for each state, there is at most one transition enabled by each input value. The formal definition of an FSM given above ensures that it is deterministic, since *update* is a function, not a one-to-many mapping. The graphical notation with guards on the transitions, however, has no such constraint. Such a state machine will be deterministic only if the guards leaving each state are non-overlapping. Note that a deterministic state machine is **determinate**, meaning that given the same inputs it will always produce the same outputs. However, not every determinate state machine is deterministic.

Receptiveness: A state machine is said to be **receptive** if, for each state, there is at least one transition possible on each input symbol. In other words, receptiveness ensures that a state machine is always ready to react to any input, and does not "get stuck" in any state. The formal definition of an FSM given above ensures that it is receptive,
p.472 since *update* is a function, not a partial function. It is defined for every possible state and input value. Moreover, in our graphical notation, since we have implicit
p.53 default transitions, we have ensured that all state machines specified in our graphical notation are also receptive.

It follows that if a state machine is both deterministic and receptive, for every state, there is *exactly* one transition possible on each input value.

3.4 Extended State Machines

The notation for FSMs becomes awkward when the number of states gets large. The garage counter of Figure 3.4 illustrates this point clearly. If M is large, the bubble-and-arc notation becomes unwieldy, which is why we resort to a less formal use of "..." in the figure.

An **extended state machine** solves this problem by augmenting the FSM model with variables that may be read and written as part of taking a transition between states.

Moore Machines and Mealy Machines

The state machines we describe in this chapter are known as **Mealy machines**, named after George H. Mealy, a Bell Labs engineer who published a description of these machines in 1955 (Mealy, 1955). Mealy machines are characterized by producing outputs when a transition is taken. An alternative, known as a **Moore machine**, produces outputs when the machine is in a state, rather than when a transition is taken. That is, the output is defined by the current state rather than by the current transition. Moore machines are named after Edward F. Moore, another Bell Labs engineer who described them in a 1956 paper (Moore, 1956).

The distinction between these machines is subtle but important. Both are discrete systems, and hence their operation consists of a sequence of discrete reactions. For a Moore machine, at each reaction, the output produced is defined by the current state (at the *start* of the reaction, not at the end). Thus, the output at the time of a reaction does not depend on the input at that same time. The input determines which transition is taken, but not what output is produced by the reaction. Hence, a Moore machine is strictly causal. p.29

A Moore machine version of the garage counter is shown in Figure 3.7. The outputs are shown in the state rather than on the transitions using a similar notation with a slash. Note, however, that this machine is *not* equivalent to the machine in Figure 3.4. To see that, suppose that on the first reaction, $up = present$ and $down = absent$. The output at that time will be 0 in Figure 3.7 and 1 in Figure 3.4. The output of the Moore machine represents the number of cars in the garage at the time of the arrival of a new car, not the number of cars after the arrival of the new car. Suppose instead that at the first reaction, $up = down = absent$. Then the output at that time is 0 in Figure 3.7 and *absent* in Figure 3.4. The Moore machine, when it reacts, always reports the output associated with the current state. The Mealy machine does not produce any output unless there is a transition explicitly denoting that output.

Any Moore machine may be converted to an equivalent Mealy machine. A Mealy machine may be converted to an almost equivalent Moore machine that differs only in that the output is produced on the *next* reaction rather than on the current one. We use Mealy machines because they tend to be more compact (requiring fewer states to represent the same functionality), and because it is convenient to be able to produce an output that instantaneously responds to the input.

inputs: *up*, *down*: pure
output: *count*: $\{0, \cdots, M\}$

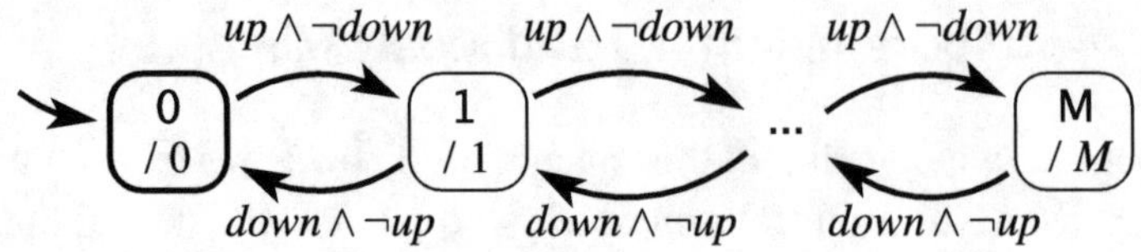

Figure 3.7: Moore machine for a system that keeps track of the number of cars in a parking garage. Note this machine is *not* equivalent to that in Figure 3.4.

Example 3.8: The garage counter of Figure 3.4 can be represented more compactly by the extended state machine in Figure 3.8.

That figure shows a variable c, declared explicitly at the upper left to make it clear that c is a variable and not an input or an output. The transition indicating the initial state initializes the value of this variable to zero.

The upper self-loop transition is then taken when the input *up* is present, the input *down* is absent, and the variable c is less than M. When this transition is taken, the state machine produces an output *count* with value $c+1$, and then the value of c is incremented by one.

variable: c: $\{0, \cdots, M\}$
inputs: *up*, *down*: pure
output: *count*: $\{0, \cdots, M\}$

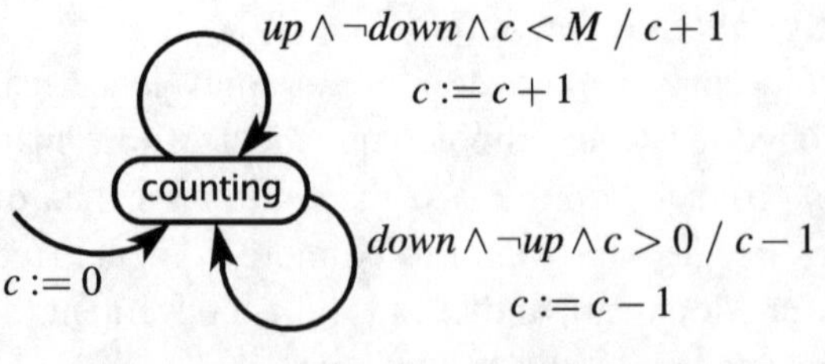

Figure 3.8: Extended state machine for the garage counter of Figure 3.4.

The lower self-loop transition is taken when the input *down* is present, the input *up* is absent, and the variable c is greater than zero. Upon taking the transition, the state machine produces an output with value $c - 1$, and then decrements the value of c.

Note that M is a parameter, not a variable. Specifically, it is assumed to be constant throughout execution.

The general notation for extended state machines is shown in Figure 3.9. This differs from the basic FSM notation of Figure 3.3 in three ways. First, variable declarations are shown explicitly to make it easy to determine whether an identifier in a guard or action refers to a variable or to an input or an output. Second, upon initialization, variables that have been declared may be initialized. The initial value will be shown on the transition that indicates the initial state. Third, transition annotations now have the form

guard / output action
set action(s)

The guard and output action are the same as for standard FSMs, except they may now refer to variables. The **set actions** are new. They specify assignments to variables that are made when the transition is taken. These assignments are made *after* the guard has been evaluated and the outputs have been produced. Thus, if the guard or output actions reference a variable, the value of the variable is that *before* the assignment in the set action. If there is more than one set action, then the assignments are made in sequence.

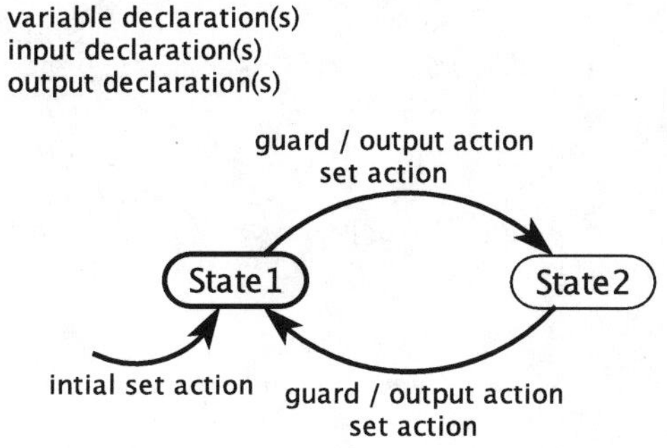

Figure 3.9: Notation for extended state machines.

Extended state machines can provide a convenient way to keep track of the passage of time.

Example 3.9: An extended state machine describing a traffic light at a pedestrian crosswalk is shown in Figure 3.10. This is a time triggered machine that assumes it will react once per second. It starts in the red state and counts 60 seconds with the help of the variable *count*. It then transitions to green, where it will remain until the pure input *pedestrian* is present. That input could be generated, for example, by a pedestrian pushing a button to request a walk light. When *pedestrian* is present, the machine transitions to yellow if it has been in state green for at least 60 seconds. Otherwise, it transitions to pending, where it stays for the remainder of the 60 second interval. This ensures that once the light goes green, it stays green for at least 60 seconds. At the end of 60 seconds, it will transition to yellow, where it will remain for 5 seconds before transitioning back to red.

p.50

The outputs produced by this machine are *sigG* to turn on the green light, *sigY* to change the light to yellow, and *sigR* to change the light to red.

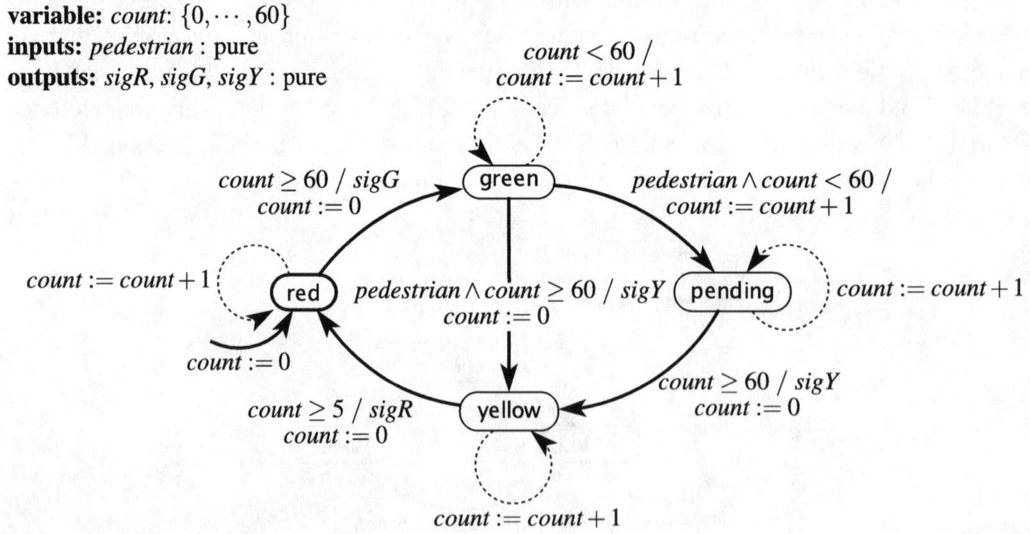

Figure 3.10: Extended state machine model of a traffic light controller that keeps track of the passage of time, assuming it reacts at regular intervals.

The state of an extended state machine includes not only the information about which discrete state (indicated by a bubble) the machine is in, but also what values any variables have. The number of possible states can therefore be quite large, or even infinite. If there are n discrete states (bubbles) and m variables each of which can have one of p possible values, then the size of the state space of the state machine is

$$|States| = np^m .$$

Example 3.10: The garage counter of Figure 3.8 has $n = 1$, $m = 1$, and $p = M + 1$, so the total number of states is $M + 1$.

Extended state machines may or may not be FSMs. In particular, it is not uncommon for p to be infinite. For example, a variable may have values in $\mathbb{N}$, the natural numbers, in which case, the number of states is infinite.

Example 3.11: If we modify the state machine of Figure 3.8 so that the guard on the upper transition is

$$up \wedge \neg down$$

instead of

$$up \wedge \neg down \wedge c < M$$

then the state machine is no longer an FSM.

Some state machines will have states that can never be reached, so the set of **reachable states** — comprising all states that can be reached from the initial state on some input sequence — may be smaller than the set of states.

Example 3.12: Although there are only four bubbles in Figure 3.10, the number of states is actually much larger. The *count* variable has 61 possible values and there are 4 bubbles, so the total number of combinations is $61 \times 4 = 244$. The size of the state space is therefore 244. However, not all of these states are reachable. In particular, while in the yellow state, the *count* variable will have only one of 6 values in $\{0, \cdots, 5\}$. The number of reachable states, therefore, is $61 \times 3 + 6 = 189$.

3.5 Nondeterminism

Most interesting state machines react to inputs and produce outputs. These inputs must come from somewhere, and the outputs must go somewhere. We refer to this "somewhere" as the **environment** of the state machine.

Example 3.13: The traffic light controller of Figure 3.10 has one pure input signal, *pedestrian*. This input is *present* when a pedestrian arrives at the crosswalk. The traffic light will remain green unless a pedestrian arrives. Some other subsystem is responsible for generating the *pedestrian* event, presumably in response to a pedestrian pushing a button to request a cross light. That other subsystem is part of the environment of the FSM in Figure 3.10.

A question becomes how to model the environment. In the traffic light example, we could construct a model of pedestrian flow in a city to serve this purpose, but this would likely be a very complicated model, and it is likely much more detailed than necessary. We want to ignore inessential details, and focus on the design of the traffic light. We can do this using a nondeterministic state machine.

Example 3.14: The FSM in Figure 3.11 models arrivals of pedestrians at a crosswalk with a traffic light controller like that in Figure 3.10. This FSM has three inputs, which are presumed to come from the outputs of Figure 3.10. Its single output, *pedestrian*, will provide the input for Figure 3.10.

The initial state is crossing. (Why? See Exercise 6.) When *sigG* is received, the FSM transitions to none. Both transitions from this state have guard *true*, indicat-

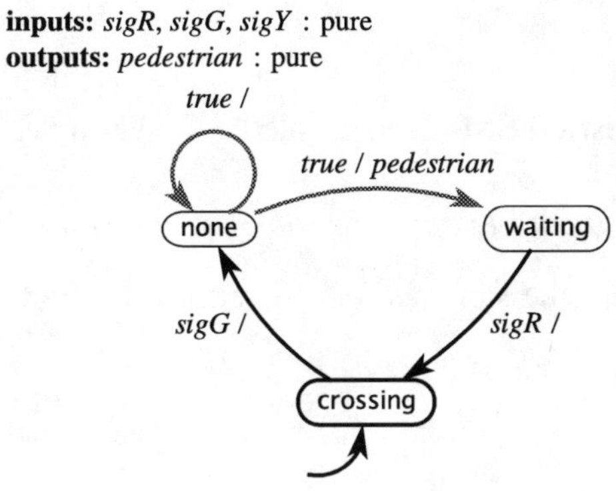

Figure 3.11: Nondeterministic model of pedestrians that arrive at a crosswalk.

ing that they are always enabled. Since both are enabled, this machine is nondeterministic. The FSM may stay in the same state and produce no output, or it may transition to waiting and produce pure output *pedestrian*.

The interaction between this machine and that of Figure 3.10 is surprisingly subtle. Variations on the design are considered in Exercise 6, and the composition of the two machines is studied in detail in Chapter 5.

If for any state of a state machine, there are two distinct transitions with guards that can evaluate to *true* in the same reaction, then the state machine is **nondeterministic**. In a diagram for such a state machine, the transitions that make the state machine nondeterministic may be colored red. In the example of Figure 3.11, the transitions exiting state none are the ones that make the state machine nondeterministic.

It is also possible to define state machines where there is more than one initial state. Such a state machine is also nondeterministic. An example is considered in Exercise 6.

In both cases, a nondeterministic FSM specifies a family of possible reactions rather than a single reaction. Operationally, all reactions in the family are possible. The nondeterministic FSM makes no statement at all about how *likely* the various reactions are. It is perfectly correct, for example, to always take the self loop in state none in Figure 3.11. A model that specifies likelihoods (in the form of probabilities) is a **stochastic model**, quite distinct from a nondeterministic model.

3.5.1 Formal Model

Formally, a **nondeterministic FSM** is represented as a five-tuple, similar to a deterministic FSM,

$$(States, Inputs, Outputs, possibleUpdates, initialStates)$$

The first three elements are the same as for a deterministic FSM, but the last two are not the same:

- *States* is a finite set of states;
- *Inputs* is a set of input valuations;
- *Outputs* is a set of output valuations;
- $possibleUpdates : States \times Inputs \to 2^{States \times Outputs}$ is an **update relation**, mapping a state and an input valuation to a *set of possible* (next state, output valuation) pairs;
- *initialStates* is a set of initial states.

The form of the function *possibleUpdates* indicates there can be more than one next state and/or output valuation given a current state and input valuation. The codomain is the powerset of $States \times Outputs$. We refer to the *possibleUpdates* function as an update *relation*, to emphasize this difference. The term **transition relation** is also often used in place of update relation.

To support the fact that there can be more than one initial state for a nondeterministic FSM, *initialStates* is a set rather than a single element of *States*.

Example 3.15: The FSM in Figure 3.11 can be formally represented as follows:

$$\begin{aligned} States &= \{\text{none}, \text{waiting}, \text{crossing}\} \\ Inputs &= (\{sigG, sigY, sigR\} \to \{present, absent\}) \\ Outputs &= (\{pedestrian\} \to \{present, absent\}) \\ initialStates &= \{\text{crossing}\} \end{aligned}$$

The update relation is given below:

$$possibleUpdates(s,i) = \begin{cases} \{(\text{none}, absent)\} & \\ \quad \text{if } s = \text{crossing} & \\ \quad \wedge\, i(sigG) = present & \\ \{(\text{none}, absent), (\text{waiting}, present)\} & \\ \quad \text{if } s = \text{none} & \\ \{(\text{crossing}, absent)\} & \\ \quad \text{if } s = \text{waiting} & \\ \quad \wedge\, i(sigR) = present & \\ \{(s, absent)\} \qquad \text{otherwise} & \end{cases} \tag{3.3}$$

for all $s \in States$ and $i \in Inputs$. Note that an output valuation $o \in Outputs$ is a function of the form $o\colon \{pedestrian\} \to \{present, absent\}$. In (3.3), the second alternative gives two possible outcomes, reflecting the nondeterminism of the machine.

3.5.2 Uses of Nondeterminism

While nondeterminism is an interesting mathematical concept in itself, it has two major uses in modeling embedded systems:

Environment Modeling: It is often useful to hide irrelevant details about how an environment operates, resulting in a nondeterministic FSM model. We have already seen one example of such environment modeling in Figure 3.11.

Specifications: System specifications impose requirements on some system features, while leaving other features unconstrained. Nondeterminism is a useful modeling technique in such settings as well. For example, consider a specification that the traffic light cycles through red, green, yellow, in that order, without regard for the timing between the outputs. The nondeterministic FSM in Figure 3.12 models this specification. The guard *true* on each transition indicates that the transition can be taken at any step. Technically, it means that each transition is enabled for any input valuation in *Inputs*.

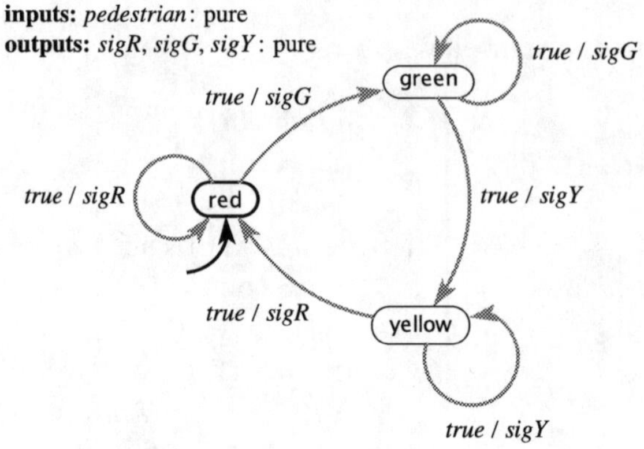

Figure 3.12: Nondeterministic FSM specifying order of signal lights, but not their timing. Notice that it ignores the *pedestrian* input.

3.6 Behaviors and Traces

p.42 p.44 An FSM has discrete dynamics. As we did in Section 3.3.3, we can abstract away the passage of time and consider only the *sequence* of reactions, without concern for when in time each reaction occurs. We do not need to talk explicitly about the amount of time that passes between reactions, since this is actually irrelevant to the behavior of an FSM.

p.45 Consider a port p of a state machine with type V_p. This port will have a sequence of values from the set $V_p \cup \{absent\}$, one value at each reaction. We can represent this sequence as a function of the form

$$s_p \colon \mathbb{N} \to V_p \cup \{absent\} .$$

This is the signal received on that port (if it is an input) or produced on that port (if it is an output).

A **behavior** of a state machine is an assignment of such a signal to each port such that the signal on any output port is the output sequence produced for the given input signals.

Example 3.16: The garage counter of Figure 3.4 has input port set $P = \{up, down\}$, with types $V_{up} = V_{down} = \{present\}$, and output port set $Q = \{count\}$

with type $V_{count} = \{0, \cdots, M\}$. An example of input sequences is

$$\begin{aligned} s_{up} &= (present, absent, present, absent, present, \cdots) \\ s_{down} &= (present, absent, absent, present, absent, \cdots) \end{aligned}$$

The corresponding output sequence is

$$s_{count} = (absent, absent, 1, 0, 1, \cdots) .$$

These three signals s_{up}, s_{down}, and s_{count} together are a behavior of the state machine. If we let

$$s'_{count} = (1, 2, 3, 4, 5, \cdots) ,$$

then s_{up}, s_{down}, and s'_{count} together *are not* a behavior of the state machine. The signal s'_{count} is not produced by reactions to those inputs.

Deterministic state machines have the property that there is exactly one behavior for each set of input sequences. That is, if you know the input sequences, then the output sequence is fully determined. That is, the machine is determinate. Such a machine can be viewed as a function that maps input sequences to output sequences. Nondeterministic state machines can have more than one behavior sharing the same input sequences, and hence cannot be viewed as a function mapping input sequences to output sequences. p.56

The set of all behaviors of a state machine M is called its **language**, written $L(M)$. Since our state machines are receptive, their languages always include all possible input sequences. p.56

A behavior may be more conveniently represented as a sequence of valuations called an **observable trace**. Let x_i represent the valuation of the input ports and y_i the valuation of the output ports at reaction i. Then an observable trace is a sequence p.45

$$((x_0, y_0), (x_1, y_1), (x_2, y_2), \cdots) .$$

An observable trace is really just another representation of a behavior.

It is often useful to be able to reason about the states that are traversed in a behavior. An **execution trace** includes the state trajectory, and may be written as a sequence

$$((x_0, s_0, y_0), (x_1, s_1, y_1), (x_2, s_2, y_2), \cdots) ,$$

where $s_0 = initialState$. This can be represented a bit more graphically as follows,

$$s_0 \xrightarrow{x_0/y_0} s_1 \xrightarrow{x_1/y_1} s_2 \xrightarrow{x_2/y_2} \cdots$$

This is an execution trace if for all $i \in \mathbb{N}$, $(s_{i+1}, y_i) = update(s_i, x_i)$ (for a deterministic machine), or $(s_{i+1}, y_i) \in possibleUpdates(s_i, x_i)$ (for a nondeterministic machine).

Example 3.17: Consider again the garage counter of Figure 3.4 with the same input sequences s_{up} and s_{down} from Example 3.16. The corresponding execution trace may be written

$$0 \xrightarrow{up \wedge down\ /} 0 \xrightarrow{\ /\ } 0 \xrightarrow{up\ /\ 1} 1 \xrightarrow{down\ /\ 0} 0 \xrightarrow{up\ /\ 1} \cdots$$

Here, we have used the same shorthand for valuations that is used on transitions in Section 3.3.1. For example, the label "*up* / 1" means that *up* is present, *down* is absent, and *count* has value 1. Any notation that clearly and unambiguously represents the input and output valuations is acceptable.

For a nondeterministic machine, it may be useful to represent all the possible traces that correspond to a particular input sequence, or even all the possible traces that result from all possible input sequences. This may be done using a **computation tree**.

Example 3.18: Consider the nondeterministic FSM in Figure 3.12. Figure 3.13 shows the computation tree for the first three reactions with any input sequence. Nodes in the tree are states and edges are labeled by the input and output valuations, where the notation *true* means any input valuation.

Traces and computation trees can be valuable for developing insight into the behaviors of a state machine and for verifying that undesirable behaviors are avoided.

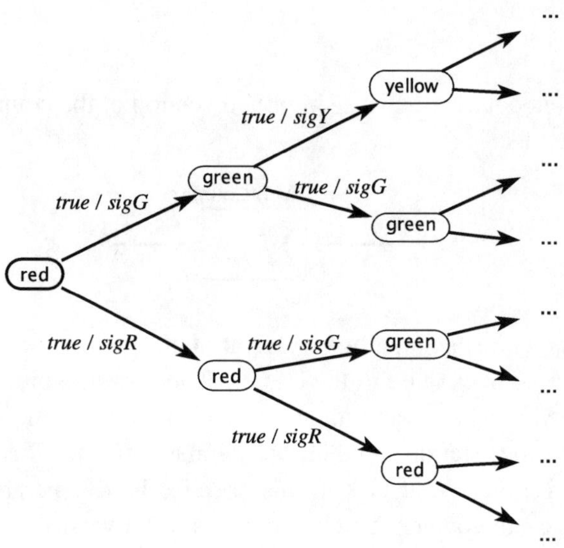

Figure 3.13: A computation tree for the FSM in Figure 3.12.

3.7 Summary

This chapter has given an introduction to the use of state machines to model systems with discrete dynamics. It gives a graphical notation that is suitable for finite state machines, and an extended state machine notation that can compactly represent large numbers of states. It also gives a mathematical model that uses sets and functions rather than visual notations. The mathematical notation can be useful to ensure precise interpretations of a model and to prove properties of a model. This chapter has also discussed nondeterminism, which can provide convenient abstractions that compactly represent families of behaviors.

Exercises

1. Consider an event counter that is a simplified version of the counter in Section 3.1. It has an icon like this:

This actor starts with state i and upon arrival of an event at the input, increments the state and sends the new value to the output. Thus, e is a pure signal, and c has the form $c\colon \mathbb{R} \to \{absent\} \cup \mathbb{N}$, assuming $i \in \mathbb{N}$. Suppose you are to use such an event counter in a weather station to count the number of times that a temperature rises above some threshold. Your task in this exercise is to generate a reasonable input signal e for the event counter. You will create several versions. For all versions, you will design a state machine whose input is a signal $\tau\colon \mathbb{R} \to \{absent\} \cup \mathbb{Z}$ that gives the current temperature (in degrees centigrade) once per hour. The output $e\colon \mathbb{R} \to \{absent, present\}$ will be a pure signal that goes to an event counter.

   (a) For the first version, your state machine should simply produce a *present* output whenever the input is *present* and greater than 38 degrees. Otherwise, the output should be absent.

   p.52 (b) For the second version, your state machine should have hysteresis. Specifically, it should produce a *present* output the first time the input is greater than 38 degrees, and subsequently, it should produce a *present* output anytime the input is greater than 38 degrees but has dropped below 36 degrees since the last time a *present* output was produced.

   (c) For the third version, your state machine should implement the same hysteresis as in part (b), but also produce a *present* output at most once per day.

2. Consider a variant of the thermostat of example 3.5. In this variant, there is only one temperature threshold, and to avoid chattering the thermostat simply leaves the heat on or off for at least a fixed amount of time. In the initial state, if the temperature is less than or equal to 20 degrees Celsius, it turns the heater on, and leaves it on for at least 30 seconds. After that, if the temperature is greater than 20 degrees, it turns the heater off and leaves it off for at least 2 minutes. It turns it on again only if the temperature is less than or equal to 20 degrees.

   (a) Design an FSM that behaves as described, assuming it reacts exactly once every 30 seconds.

(b) How many possible states does your thermostat have? Is this the smallest number of states possible?

(c) Does this model thermostat have the time-scale invariance property? p.52

3. Consider the following state machine:

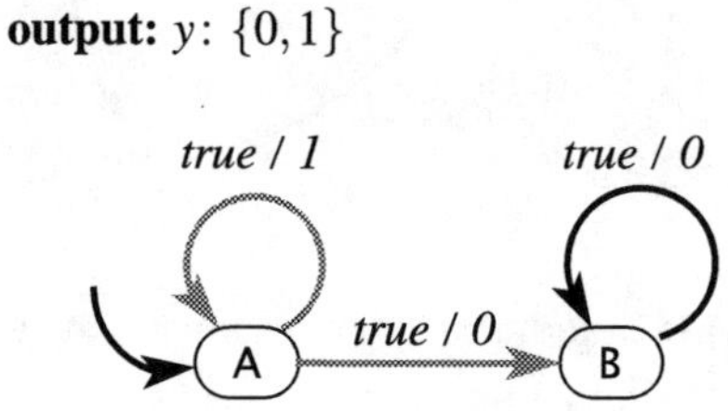

Determine whether the following statement is true or false, and give a supporting argument:

> The output will eventually be a constant 0, or it will eventually be a constant 1. That is, for some $n \in \mathbb{N}$, after the n-th reaction, either the output will be 0 in every subsequent reaction, or it will be 1 in every subsequent reaction.

Note that Chapter 13 gives mechanisms for making such statements precise and for reasoning about them.

4. How many reachable states does the following state machine have?

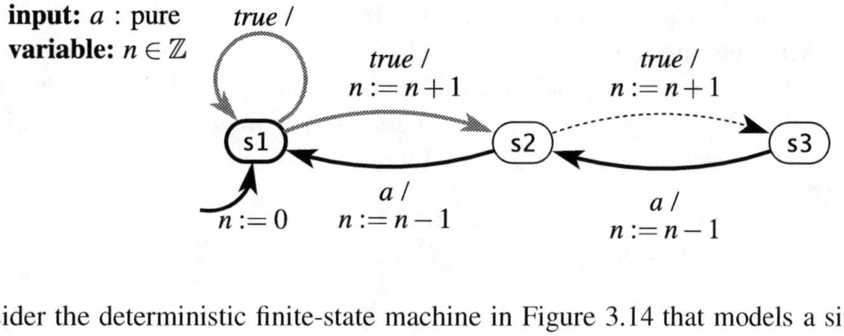

5. Consider the deterministic finite-state machine in Figure 3.14 that models a simple traffic light.

(a) Formally write down the description of this FSM as a 5-tuple:

$$(States, Inputs, Outputs, update, initialState).$$

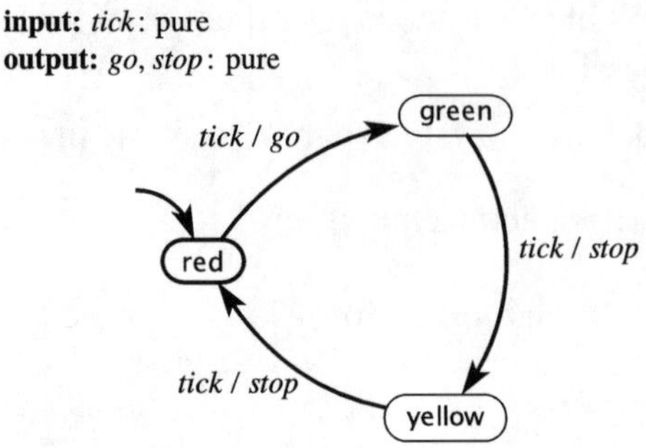

Figure 3.14: Deterministic finite-state machine for Exercise 5

p.67

(b) Give an execution trace of this FSM of length 4 assuming the input *tick* is *present* on each reaction.

(c) Now consider merging the red and yellow states into a single stop state. Transitions that pointed into or out of those states are now directed into or out of the new stop state. Other transitions and the inputs and outputs stay the same. The new stop state is the new initial state. Is the resulting state machine deterministic? Why or why not? If it is deterministic, give a prefix of the trace of length 4. If it is nondeterministic, draw the computation tree up to depth 4.

p.50

6. This problem considers variants of the FSM in Figure 3.11, which models arrivals of pedestrians at a crosswalk. We assume that the traffic light at the crosswalk is controlled by the FSM in Figure 3.10. In all cases, assume a time triggered model, where both the pedestrian model and the traffic light model react once per second. Assume further that in each reaction, each machine sees as inputs the output produced by the other machine *in the same reaction* (this form of composition, which is called synchronous composition, is studied further in Chapter 6).

   (a) Suppose that instead of Figure 3.11, we use the following FSM to model the arrival of pedestrians:

inputs: *sigR, sigG, sigY* : pure
outputs: *pedestrian* : pure

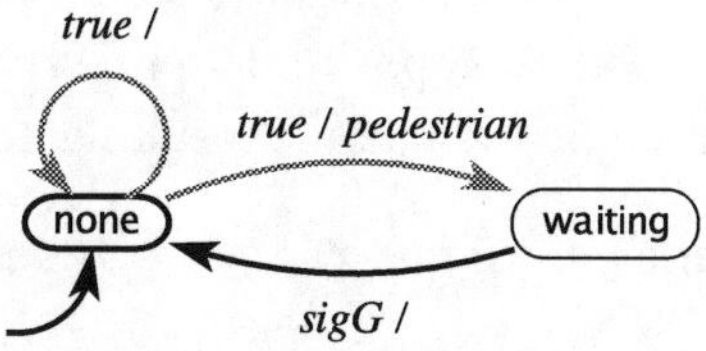

Find a trace whereby a pedestrian arrives (the above machine transitions to waiting) but the pedestrian is never allowed to cross. That is, at no time after the pedestrian arrives is the traffic light in state red.

(b) Suppose that instead of Figure 3.11, we use the following FSM to model the arrival of pedestrians:

inputs: *sigR, sigG, sigY* : pure
outputs: *pedestrian* : pure

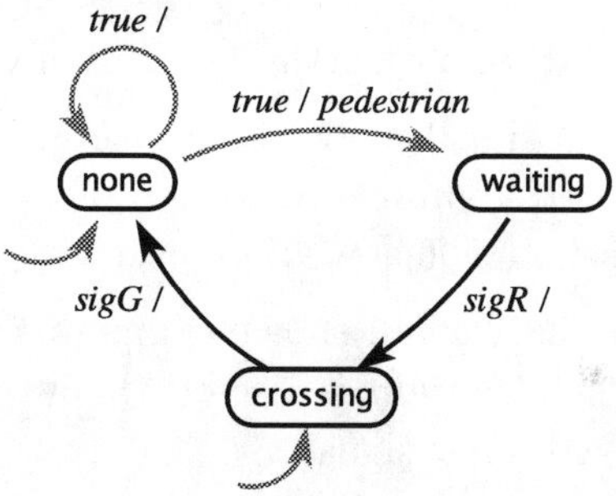

Here, the initial state is nondeterministically chosen to be one of none or crossing. Find a trace whereby a pedestrian arrives (the above machine transitions from none to waiting) but the pedestrian is never allowed to cross. That is, at no time after the pedestrian arrives is the traffic light in state red.

7. Consider the state machine in Figure 3.15. State whether each of the following is a behavior for this machine. In each of the following, the ellipsis "$\cdots$" means that the last symbol is repeated forever. Also, for readability, *absent* is denoted by the shorthand a and *present* by the shorthand p. p.66

   (a) $x = (p, p, p, p, p, \cdots), \quad y = (0, 1, 1, 0, 0, \cdots)$
   (b) $x = (p, p, p, p, p, \cdots), \quad y = (0, 1, 1, 0, a, \cdots)$
   (c) $x = (a, p, a, p, a, \cdots), \quad y = (a, 1, a, 0, a, \cdots)$
   (d) $x = (p, p, p, p, p, \cdots), \quad y = (0, 0, a, a, a, \cdots)$
   (e) $x = (p, p, p, p, p, \cdots), \quad y = (0, a, 0, a, a, \cdots)$

8. (NOTE: This exercise is rather advanced.) This exercise studies properties of discrete signals as formally defined in the sidebar on page 44. Specifically, we will show that discreteness is not a compositional property. That is, when combining two discrete behaviors in a single system, the resulting combination is not necessarily discrete.

p.43

(a) Consider a pure signal $x\colon \mathbb{R} \to \{present, absent\}$ given by

$$x(t) = \begin{cases} present & \text{if } t \text{ is a non-negative integer} \\ absent & \text{otherwise} \end{cases}$$

for all $t \in \mathbb{R}$. Show that this signal is discrete.

p.43

(b) Consider a pure signal $y\colon \mathbb{R} \to \{present, absent\}$ given by

$$y(t) = \begin{cases} present & \text{if } t = 1 - 1/n \text{ for any positive integer } n \\ absent & \text{otherwise} \end{cases}$$

for all $t \in \mathbb{R}$. Show that this signal is discrete.

(c) Consider a signal w that is the merge of x and y in the previous two parts. That is, $w(t) = present$ if either $x(t) = present$ or $y(t) = present$, and is *absent* otherwise. Show that w is not discrete.

(d) Consider the example shown in Figure 3.1. Assume that each of the two signals *arrival* and *departure* is discrete. Show that this does not imply that the output *count* is a discrete signal.

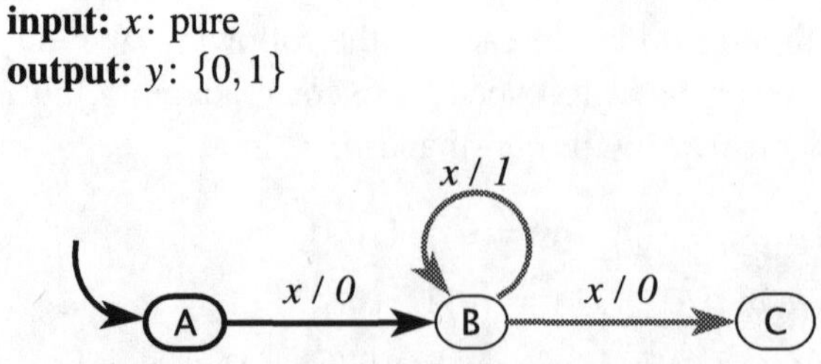

Figure 3.15: State machine for Exercise 7.

4

Hybrid Systems

Contents

Chapters 2 and 3 describe two very different modeling strategies, one focused on continuous dynamics and one on discrete dynamics. For continuous dynamics, we use differential equations and their corresponding actor models. For discrete dynamics, we use state machines. p.26

Cyber-physical systems integrate physical dynamics and computational systems, so they commonly combine both discrete and continuous dynamics. In this chapter, we show that the modeling techniques of Chapters 2 and 3 can be combined, yielding what are known as **hybrid systems**. Hybrid system models are often much simpler and more understandable than brute-force models that constrain themselves to only one of the two styles in Chapters 2 and 3. They are a powerful tool for understanding real-world systems.

4.1 Modal Models

In this section, we show that state machines can be generalized to admit continuous inputs and outputs and to combine discrete and continuous dynamics.

4.1.1 Actor Model for State Machines

In Section 3.3.1 we explain that state machines have inputs defined by the set *Inputs* that p.43 may be pure signals or may carry a value. In either case, the state machine has a number of input ports, which in the case of pure signals are either present or absent, and in the case of p.44 valued signals have a value at each reaction of the state machine.

p.48 We also explain in Section 3.3.1 that actions on transitions set the values of outputs. The outputs can also be represented by ports, and again the ports can carry pure signals or valued signals. In the case of pure signals, a transition that is taken specifies whether the output is present or absent, and in the case of valued signals, it assigns a value or asserts that the signal is absent. Outputs are presumed to be absent between transitions.

Given this input/output view of state machines, it is natural to think of a state machine as an actor, as illustrated in Figure 4.1. In that figure, we assume some number n of input ports named $i_1 \cdots i_n$. At each reaction, these ports have a value that is either *present* or *absent* (if the port carries a pure signal) or a member of some set of values (if the port carries a valued signal). The outputs are similar. The guards on the transitions define subsets of possible values on input ports, and the actions assign values to output ports. Given such an actor p.21 model, it is straightforward to generalize FSMs to admit continuous-time signals as inputs.

4.1.2 Continuous Inputs

p.44 We have so far assumed that state machines operate in a sequence of discrete reactions. We have assumed that inputs and outputs are absent between reactions. We will now generalize p.21 this to allow inputs and outputs to be continuous-time signals.

In order to get state machine models to coexist with time-based models, we need to interpret state transitions to occur on the same timeline used for the time-based portion of the system. The notion of discrete reactions described in Section 3.1 suffices for this purpose, but we will no longer require inputs and outputs to be absent between reactions. Instead, we will define a transition to occur when a guard on an outgoing transition from the current state becomes enabled. As before, during the time between reactions, a state machine is un-

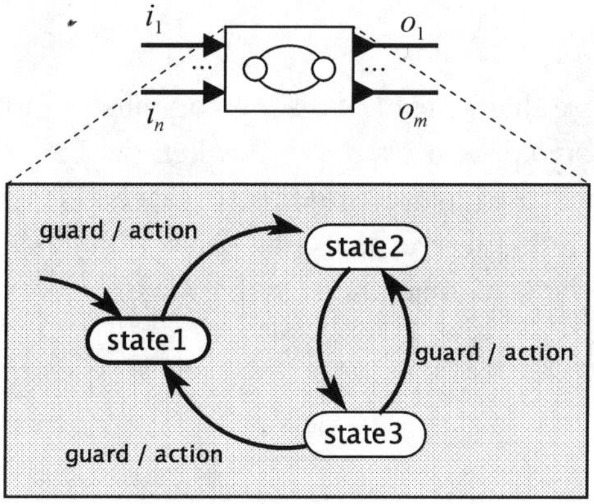

Figure 4.1: An FSM represented as an actor.

derstood to not transition between modes. But the inputs and outputs are no longer required to be absent during that time.

Example 4.1: Consider a thermostat modeled as a state machine with states $\Sigma = \{\text{heating}, \text{cooling}\}$, shown in Figure 4.2. This is a variant of the model of Example 3.5 where instead of a discrete input that provides a temperature at each reaction, the input is a continuous-time signal $\tau \colon \mathbb{R} \to \mathbb{R}$ where $\tau(t)$ represents the temperature at time t. The initial state is cooling, and the transition out of this state is enabled at the earliest time t after the start time when $\tau(t) \leq 18$. In this example, we assume the outputs are pure signals *heatOn* and *heatOff*.

In the above example, the outputs are present only at the times the transitions are taken. We can also generalize FSMs to support continuous-time outputs, but to do this, we need the notion of state refinements.

4.1.3 State Refinements

A hybrid system associates with each state of an FSM a dynamic behavior. Our first (very simple) example uses this capability merely to produce continuous-time outputs.

Example 4.2: Suppose that instead of discrete outputs as in Example 4.1 we wish to produce a control signal whose value is 1 when the heat is on and 0 when the heat is off. Such a control signal could directly drive a heater. The thermostat in Figure 4.3 does this. In that figure, each state has a refinement that gives the value of the output h while the state machine is in that state.

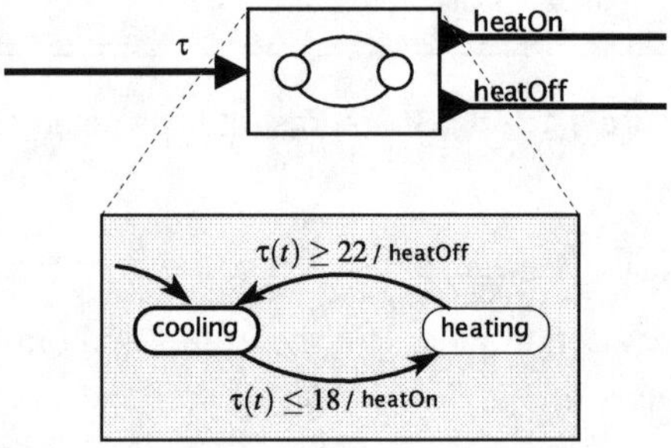

Figure 4.2: A thermostat modeled as an FSM with a continuous-time input signal.

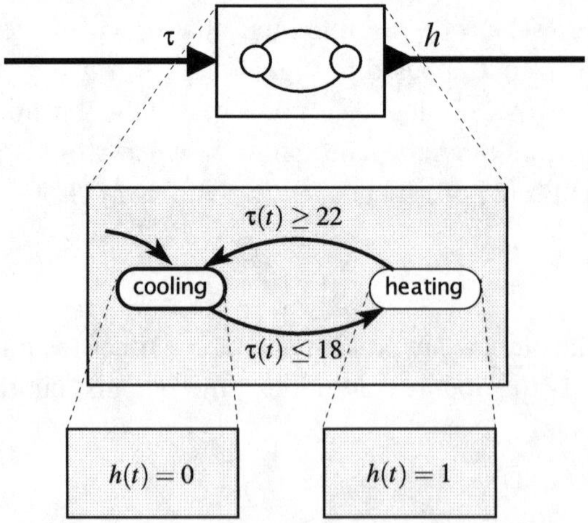

Figure 4.3: A thermostat with continuous-time output.

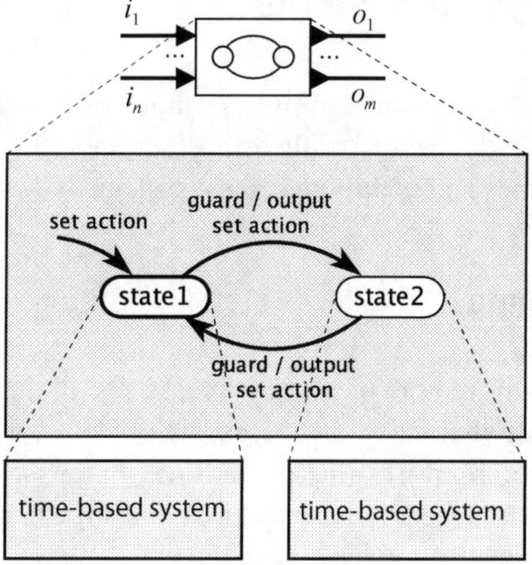

Figure 4.4: Notation for hybrid systems.

In a hybrid system, the current state of the state machine has a **state refinement** that gives the dynamic behavior of the output as a function of the input. In the above simple example, the output is constant in each state, which is rather trivial dynamics. Hybrid systems can get much more elaborate.

The general structure of a hybrid system model is shown in Figure 4.4. In that figure, there is a two-state finite-state machine. Each state is associated with a state refinement labeled in the figure as a "time-based system." The state refinement defines dynamic behavior of the outputs and (possibly) additional continuous state variables. In addition, each transition can optionally specify set actions, which set the values of such additional state variables when p.59
a transition is taken. The example of Figure 4.3 is rather trivial, in that it has no continuous state variables, no output actions, and no set actions.

A hybrid system is sometimes called a **modal model** because it has a finite number of **mode**s, one for each state of the FSM, and when it is in a mode, it has dynamics specified by the state refinement. The states of the FSM may be referred to as modes rather than states, which as we will see, helps prevent confusion with states of the refinements.

The next simplest such dynamics, besides the rather trivial constant outputs of Example 4.2 is found in timed automata, which we discuss next.

4.2 Classes of Hybrid Systems

Hybrid systems can be quite elaborate. In this section, we first describe a relatively simple form known as timed automata. We then illustrate more elaborate forms that model nontrivial physical dynamics and nontrivial control systems.

4.2.1 Timed Automata

Most cyber-physical systems require measuring the passage of time and performing actions at specific times. A device that measures the passage of time, a **clock**, has a particularly
p.11 simple dynamics: its state progresses linearly in time. In this section, we describe **timed automata**, a formalism introduced by Alur and Dill (1994), which enable the construction of more complicated systems from such simple clocks.

Timed automata are the simplest non-trivial hybrid systems. They are modal models where the time-based refinements have very simple dynamics; all they do is measure the passage of time. A clock is modeled by a first-order differential equation,

$$\forall\, t \in T_m, \quad \dot{s}(t) = a,$$

where $s\colon \mathbb{R} \to \mathbb{R}$ is a continuous-time signal, $s(t)$ is the value of the clock at time t, and $T_m \subset \mathbb{R}$ is the subset of time during which the hybrid system is in mode m. The rate of the clock, a, is a constant while the system is in this mode.[1]

p.52 **Example 4.3:** Recall the thermostat of Example 4.1, which uses hysteresis to prevent chattering. An alternative implementation that would also prevent chattering would use a single temperature threshold, but instead would require that the heater remain on or off for at least a minimum amount of time, regardless of the temperature. This design would not have the hysteresis property, but may be useful nonetheless. This can be modeled as a timed automaton as shown in Figure 4.5. In that figure, each state refinement has a clock, which is a continuous-time signal s with dynamics given by

$$\dot{s}(t) = 1\ .$$

[1] The variant of timed automata we describe in this chapter differs from the original model of Alur and Dill (1994) in that the rates of clocks in different modes can be different. This variant is sometimes described in the literature as *multi-rate* timed automata.

The value $s(t)$ increases linearly with t. Note that in that figure, the state refinement is shown directly with the name of the state in the state bubble. This shorthand is convenient when the refinement is relatively simple.

Notice that the initial state cooling has a set action on the dangling transition indi- p.59 cating the initial state, written as

$$s(t) := T_c \, .$$

As we did with extended state machines, we use the notation ":=" to emphasize p.56 that this is an assignment, not a predicate. This action ensures that when the ther- p.51 mostat starts, it can immediately transition to the heating mode if the temperature $\tau(t)$ is less than or equal to 20 degrees. The other two transitions each have set actions that reset the clock s to zero. The portion of the guard that specifies $s(t) \geq T_h$ ensures that the heater will always be on for at least time T_h. The portion of the guard that specifies $s(t) \geq T_c$ specifies that once the heater goes off, it will remain off for at least time T_c.

A possible execution of this timed automaton is shown in Figure 4.6. In that figure, we assume that the temperature is initially above the setpoint of 20 degrees, so the p.50 FSM remains in the cooling state until the temperature drops to 20 degrees. At that time t_1, it can take the transition immediately because $s(t_1) > T_c$. The transition resets s to zero and turns on the heater. The heater will remain on until time $t_1 + T_h$, assuming that the temperature only rises when the heater is on. At time $t_1 + T_h$, it will transition back to the cooling state and turn the heater off. (We assume here that a transition is taken as soon as it is enabled. Other transition semantics are possible.) It will cool until at least time T_c elapses and until the temperature drops again to 20 degrees, at which point it will turn the heater back on.

In the previous example the state of the system at any time t is not only the mode, heating or cooling, but also the current value $s(t)$ of the clock. We call s a **continuous state** variable, whereas heating and cooling are **discrete states**. Thus, note that the term "state" for such a hybrid system can become confusing. The FSM has states, but so do the refinement systems (unless they are memoryless). When there is any possibility of confusion we explicitly refer to the states of the machine as modes.

Transitions between modes have actions associated with them. Sometimes, it is useful to have transitions from one mode back to itself, just so that the action can be realized. This

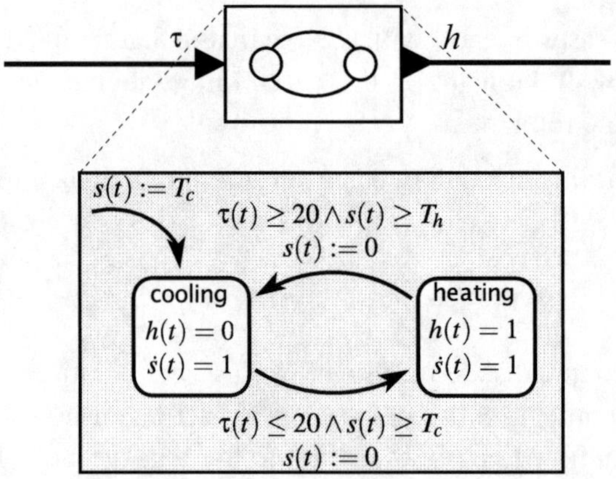

Figure 4.5: A timed automaton modeling a thermostat with a single temperature threshold, 20, and minimum times T_c and T_h in each mode.

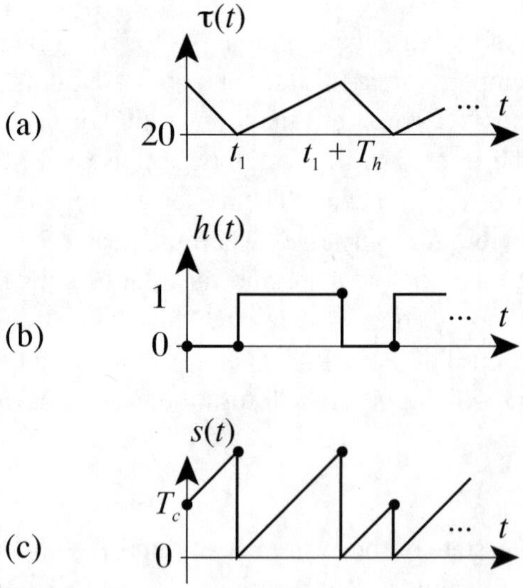

Figure 4.6: (a) A temperature input to the hybrid system of Figure 4.5, (b) the output h, and (c) the refinement state s.

is illustrated in the next example, which also shows a timed automaton that produces a pure output.

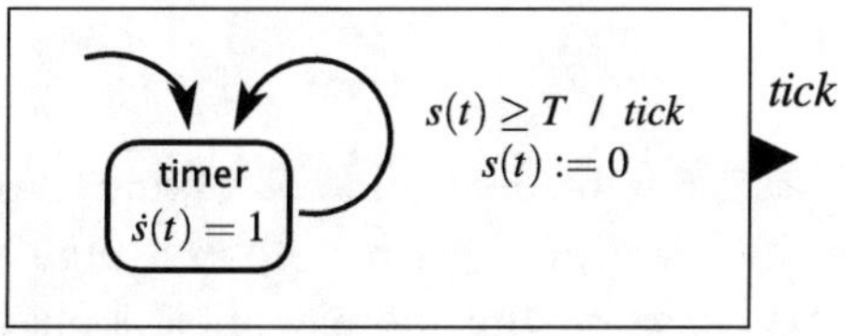

Figure 4.7: A timed automaton that generates a pure output event every T time units.

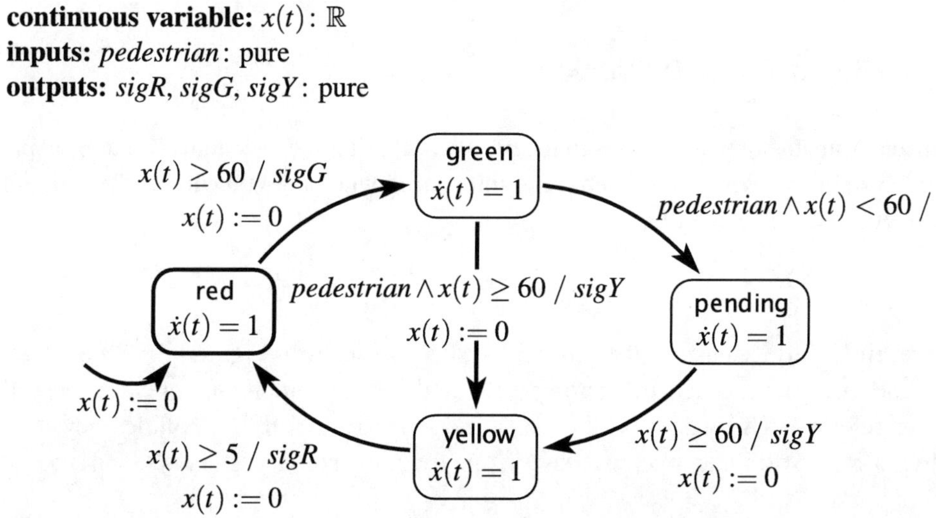

Figure 4.8: A timed automaton variant of the traffic light controller of Figure 3.10.

Example 4.4: The timed automaton in Figure 4.7 produces a pure output that will be present every T time units, starting at the time when the system begins executing. Notice that the guard on the transition, $s(t) \geq T$, is followed by an output action, *tick*, and a set action, $s(t) := 0$.

Figure 4.7 shows another notational shorthand that works well for simple diagrams. The automaton is shown directly inside the icon for its actor model.

p.50 **Example 4.5:** The traffic light controller of Figure 3.10 is a time triggered machine that assumes it reacts once each second. Figure 4.8 shows a timed automaton with the same behavior. It is more explicit about the passage of time in that its temporal dynamics do not depend on unstated assumptions about when the machine will react.

4.2.2 Higher-Order Dynamics

In timed automata, all that happens in the time-based refinement systems is that time passes. Hybrid systems, however, are much more interesting when the behavior of the refinements is more complex. Specifically,

Example 4.6: Consider the physical system depicted in Figure 4.9. Two sticky round masses are attached to springs. The springs are compressed or extended and then released. The masses oscillate on a frictionless table. If they collide, they stick together and oscillate together. After some time, the stickiness decays, and masses pull apart again.

A plot of the displacement of the two masses as a function of time is shown in Figure 4.9. Both springs begin compressed, so the masses begin moving towards one another. They almost immediately collide, and then oscillate together for a brief period until they pull apart. In this plot, they collide two more times, and almost collide a third time.

The physics of this problem is quite simple if we assume idealized springs. Let $y_1(t)$ denote the right edge of the left mass at time t, and $y_2(t)$ denote the left edge of the right mass at time t, as shown in Figure 4.9. Let p_1 and p_2 denote the neutral positions of the two masses, i.e., when the springs are neither extended nor compressed, so the force is zero. For an ideal spring, the force at time t on the mass is proportional to $p_1 - y_1(t)$ (for the left mass) and $p_2 - y_2(t)$ (for the right mass). The force is positive to the right and negative to the left.

Let the spring constants be k_1 and k_2, respectively. Then the force on the left spring is $k_1(p_1 - y_1(t))$, and the force on the right spring is $k_2(p_2 - y_2(t))$. Let the masses

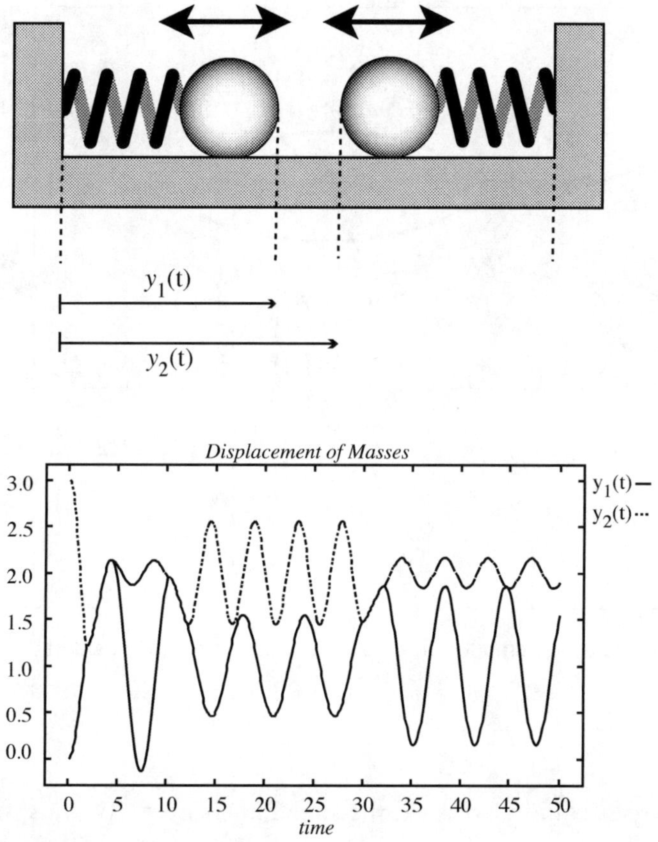

Figure 4.9: Sticky masses system considered in Example 4.6.

be m_1 and m_2 respectively. Now we can use Newton's second law, which relates force, mass, and acceleration,

$$f = ma.$$

The acceleration is the second derivative of the position with respect to time, which we write $\ddot{y_1}(t)$ and $\ddot{y_2}(t)$. Thus, as long as the masses are separate, their dynamics are given by

$$\ddot{y}_1(t) = k_1(p_1 - y_1(t))/m_1 \quad (4.1)$$

$$\ddot{y}_2(t) = k_2(p_2 - y_2(t))/m_2. \quad (4.2)$$

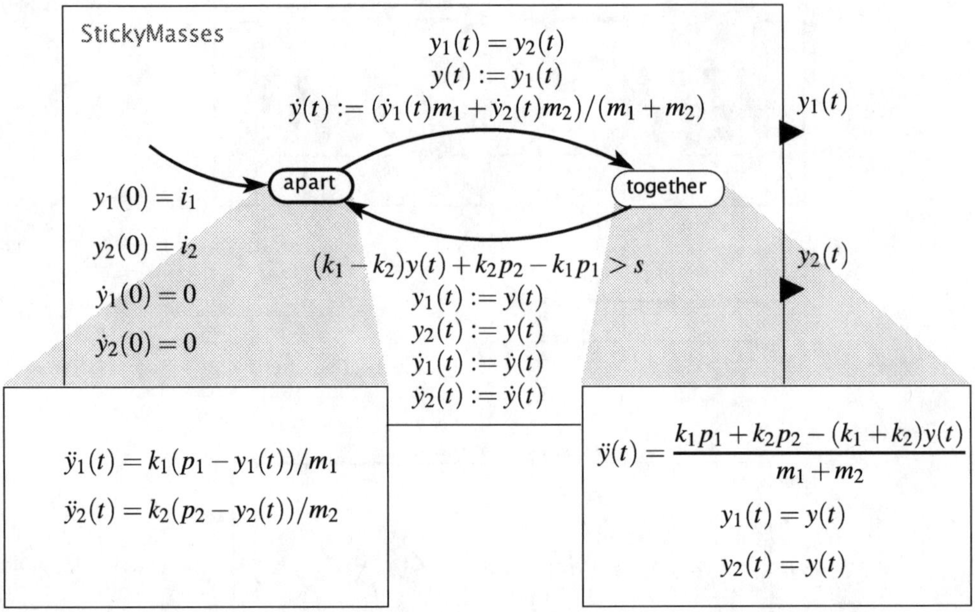

Figure 4.10: Hybrid system model for the sticky masses system considered in Example 4.6.

When the masses collide, however, the situation changes. With the masses stuck together, they behave as a single object with mass $m_1 + m_2$. This single object is pulled in opposite directions by two springs. While the masses are stuck together, $y_1(t) = y_2(t)$. Let

$$y(t) = y_1(t) = y_2(t).$$

The dynamics are then given by

$$\ddot{y}(t) = \frac{k_1 p_1 + k_2 p_2 - (k_1 + k_2) y(t)}{m_1 + m_2}. \tag{4.3}$$

It is easy to see now how to construct a hybrid systems model for this physical system. The model is shown in Figure 4.10. It has two modes, apart and together. The refinement of the apart mode is given by (4.1) and (4.2), while the refinement of the together mode is given by (4.3).

We still have work to do, however, to label the transitions. The initial transition is shown in Figure 4.10 entering the apart mode. Thus, we are assuming the masses begin apart. Moreover, this transition is labeled with a set action that sets the initial positions of the two masses to i_1 and i_2 and the initial velocities to zero. p.59

The transition from apart to together has the guard

$$y_1(t) = y_2(t) .$$

This transition has a set action which assigns values to two continuous state variables $y(t)$ and $\dot{y}(t)$, which will represent the motion of the two masses stuck together. The value it assigns to $\dot{y}(t)$ conserves momentum. The momentum of the left mass is $\dot{y}_1(t)m_1$, the momentum of the right mass is $\dot{y}_2(t)m_2$, and the momentum of the combined masses is $\dot{y}(t)(m_1 + m_2)$. To make these equal, it sets p.81

$$\dot{y}(t) = \frac{\dot{y}_1(t)m_1 + \dot{y}_2(t)m_2}{m_1 + m_2}.$$

The refinement of the together mode gives the dynamics of y and simply sets $y_1(t) = y_2(t) = y(t)$, since the masses are moving together. The transition from apart to together sets $y(t)$ equal to $y_1(t)$ (it could equally well have chosen $y_2(t)$, since these are equal).

The transition from together to apart has the more complicated guard

$$(k_1 - k_2)y(t) + k_2 p_2 - k_1 p_1 > s,$$

where s represents the stickiness of the two masses. This guard is satisfied when the right-pulling force on the right mass exceeds the right-pulling force on the left mass by more than the stickiness. The right-pulling force on the right mass is simply

$$f_2(t) = k_2(p_2 - y(t))$$

and the right-pulling force on the left mass is

$$f_1(t) = k_1(p_1 - y(t)).$$

Thus,

$$f_2(t) - f_1(t) = (k_1 - k_2)y(t) + k_2 p_2 - k_1 p_1.$$

When this exceeds the stickiness s, then the masses pull apart.

An interesting elaboration on this example, considered in problem 11, modifies the together mode so that the stickiness is initialized to a starting value, but then decays according to the differential equation

$$\dot{s}(t) = -as(t)$$

where $s(t)$ is the stickiness at time t, and a is some positive constant. In fact, it is the dynamics of such an elaboration that is plotted in Figure 4.9.

As in Example 4.4, it is sometimes useful to have hybrid system models with only one state. The actions on one or more state transitions define the discrete event behavior that combines with the time-based behavior.

Example 4.7: Consider a bouncing ball. At time $t = 0$, the ball is dropped from a height $y(0) = h_0$, where h_0 is the initial height in meters. It falls freely. At some later time t_1 it hits the ground with a velocity $\dot{y}(t_1) < 0$ m/s (meters per second). A *bump* event is produced when the ball hits the ground. The collision is **inelastic** (meaning that kinetic energy is lost), and the ball bounces back up with velocity $-a\dot{y}(t_1)$, where a is constant with $0 < a < 1$. The ball will then rise to a certain height and fall back to the ground repeatedly.

The behavior of the bouncing ball can be described by the hybrid system of Figure 4.11. There is only one mode, called free. When it is not in contact with the ground, we know that the ball follows the second-order differential equation,

$$\ddot{y}(t) = -g, \tag{4.4}$$

p.81 where $g = 9.81$ m/sec^2 is the acceleration imposed by gravity. The continuous state variables of the free mode are

$$s(t) = \begin{bmatrix} y(t) \\ \dot{y}(t) \end{bmatrix}$$

with the initial conditions $y(0) = h_0$ and $\dot{y}(0) = 0$. It is then a simple matter to rewrite (4.4) as a first-order differential equation,

$$\dot{s}(t) = f(s(t)) \tag{4.5}$$

for a suitably chosen function f.

At the time $t = t_1$ when the ball first hits the ground, the guard

$$y(t) = 0$$

is satisfied, and the self-loop transition is taken. The output *bump* is produced, and p.59 the set action $\dot{y}(t) := -a\dot{y}(t)$ changes $\dot{y}(t_1)$ to have value $-a\dot{y}(t_1)$. Then (4.4) is followed again until the guard becomes true again.

By integrating (4.4) we get, for all $t \in (0, t_1)$,

$$\begin{aligned} \dot{y}(t) &= -gt, \\ y(t) &= y(0) + \int_0^t \dot{y}(\tau)d\tau = h_0 - \frac{1}{2}gt^2. \end{aligned}$$

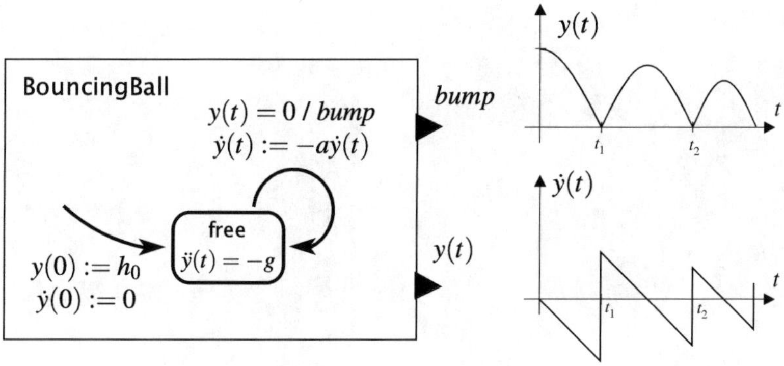

Figure 4.11: The motion of a bouncing ball may be described as a hybrid system with only one mode. The system outputs a *bump* each time the ball hits the ground, and also outputs the position of the ball. The position and velocity are plotted versus time at the right.

So $t_1 > 0$ is determined by $y(t_1) = 0$. It is the solution to the equation

$$h_0 - \frac{1}{2}gt^2 = 0.$$

Thus,

$$t_1 = \sqrt{2h_0/g}.$$

Figure 4.11 plots the continuous state versus time.

The bouncing ball example above has an interesting difficulty that is explored in Exercise 10. Specifically, the time between bounces gets smaller as time increases. In fact, it gets smaller fast enough that an infinite number of bounces occur in a finite amount of time. A system with an infinite number of discrete events in a finite amount of time is called a **Zeno** system, after Zeno of Elea, a pre-Socratic Greek philosopher famous for his paradoxes. In the physical world, of course, the ball will eventually stop bouncing; the Zeno behavior is an artifact of the model. Another example of a Zeno hybrid system is considered in Exercise 13.

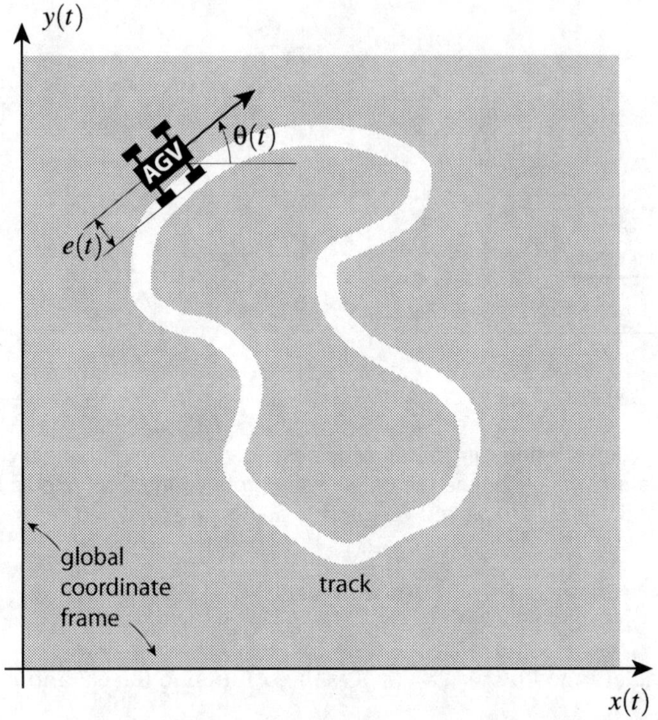

Figure 4.12: Illustration of the automated guided vehicle of Example 4.8. The vehicle is following a curved painted track, and has deviated from the track by a distance $e(t)$. The coordinates of the vehicle at time t with respect to the global coordinate frame are $(x(t), y(t), \theta(t))$.

4.2.3 Supervisory Control

A control system involves four components: a system called the **plant**, the physical process that is to be controlled; the environment in which the plant operates; the sensors that measure some variables of the plant and the environment; and the controller that determines the mode transition structure and selects the time-based inputs to the plant. The controller has two levels: the **supervisory control** that determines the mode transition structure, and the **low-level control** that determines the time-based inputs to the plant. Intuitively, the supervisory controller determines which of several strategies should be followed, and the low-level controller implements the selected strategy. Hybrid systems are ideal for modeling such two-level controllers. We show how through a detailed example.

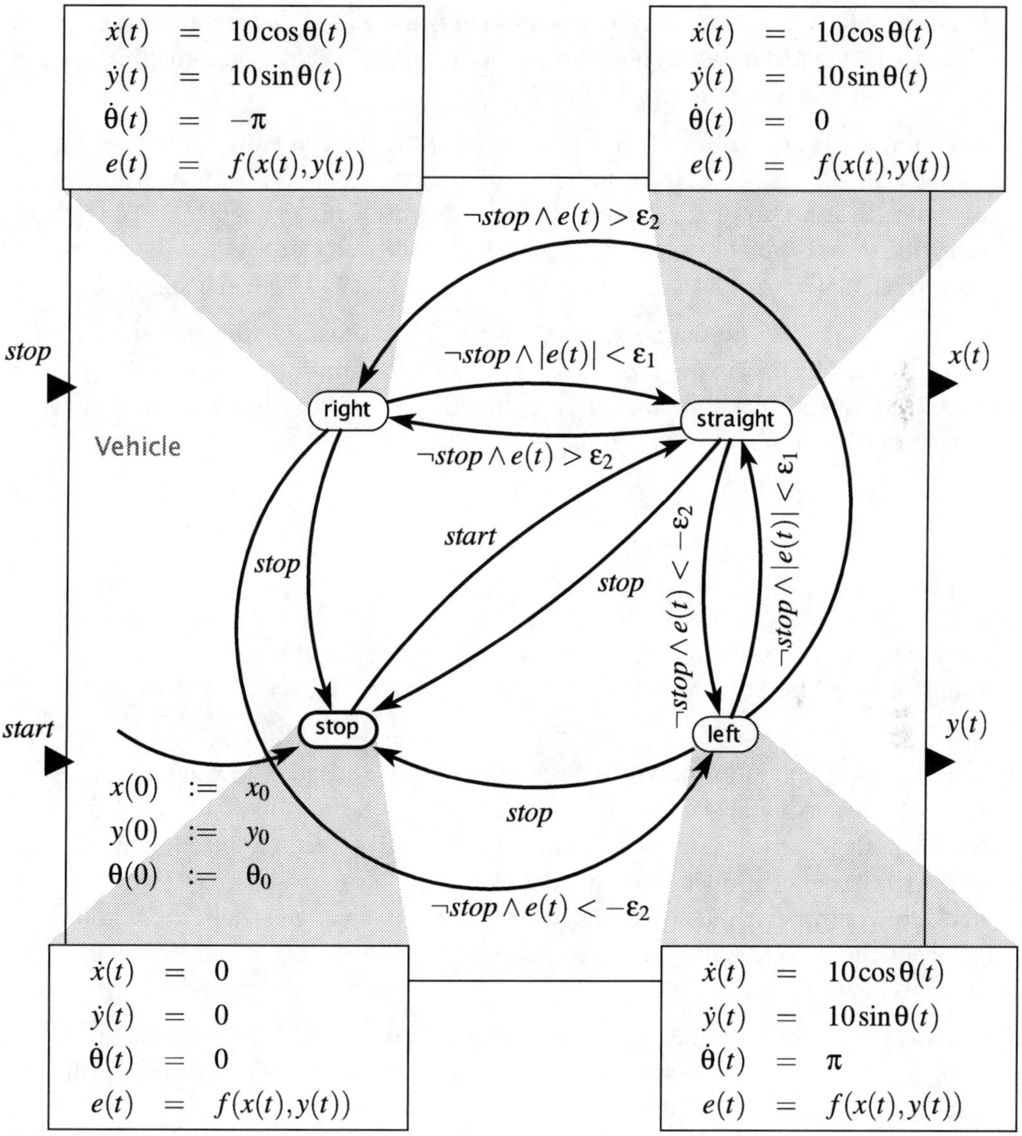

Figure 4.13: The automatic guided vehicle of Example 4.8 has four modes: stop, straight, left, right.

Example 4.8: Consider an **automated guided vehicle** (**AGV**) that moves along a closed track painted on a warehouse or factory floor. We will design a controller so that the vehicle closely follows the track.

The vehicle has two degrees of freedom. At any time t, it can move forward along its body axis with speed $u(t)$ with the restriction that $0 \leq u(t) \leq 10$ mph (miles per hour). It can also rotate about its center of gravity with an angular speed $\omega(t)$ restricted to $-\pi \leq \omega(t) \leq \pi$ radians/second. We ignore the inertia of the vehicle, so we assume that we can instantaneously change the velocity or angular speed.

Let $(x(t), y(t)) \in \mathbb{R}^2$ be the position relative to some fixed coordinate frame and $\theta(t) \in (-\pi, \pi]$ be the angle (in radians) of the vehicle at time t, as shown in Figure 4.12. In terms of this coordinate frame, the motion of the vehicle is given by a system of three differential equations,

$$\begin{aligned} \dot{x}(t) &= u(t)\cos\theta(t), \\ \dot{y}(t) &= u(t)\sin\theta(t), \\ \dot{\theta}(t) &= \omega(t). \end{aligned} \tag{4.6}$$

Equations (4.6) describe the plant. The environment is the closed painted track. It could be described by an equation. We will describe it indirectly below by means of a sensor.

The two-level controller design is based on a simple idea. The vehicle always moves at its maximum speed of 10 mph. If the vehicle strays too far to the left of the track, the controller steers it towards the right; if it strays too far to the right of the track, the controller steers it towards the left. If the vehicle is close to the track, the controller maintains the vehicle in a straight direction. Thus the controller guides the vehicle in four modes, left, right, straight, and stop. In stop mode, the vehicle comes to a halt.

The following differential equations govern the AGV's motion in the refinements of the four modes. They describe the low-level controller, i.e., the selection of the time-based plant inputs in each mode.

straight

$$\begin{aligned} \dot{x}(t) &= 10\cos\theta(t) \\ \dot{y}(t) &= 10\sin\theta(t) \\ \dot{\theta}(t) &= 0 \end{aligned}$$

left

$$\begin{aligned}\dot{x}(t) &= 10\cos\theta(t)\\ \dot{y}(t) &= 10\sin\theta(t)\\ \dot{\theta}(t) &= \pi\end{aligned}$$

right

$$\begin{aligned}\dot{x}(t) &= 10\cos\theta(t)\\ \dot{y}(t) &= 10\sin\theta(t)\\ \dot{\theta}(t) &= -\pi\end{aligned}$$

stop

$$\begin{aligned}\dot{x}(t) &= 0\\ \dot{y}(t) &= 0\\ \dot{\theta}(t) &= 0\end{aligned}$$

In the stop mode, the vehicle is stopped, so $x(t)$, $y(t)$, and $\theta(t)$ are constant. In the left mode, $\theta(t)$ increases at the rate of π radians/second, so from Figure 4.12 we see that the vehicle moves to the left. In the right mode, it moves to the right. In the straight mode, $\theta(t)$ is constant, and the vehicle moves straight ahead with a constant heading. The refinements of the four modes are shown in the boxes of Figure 4.13.

We design the supervisory control governing transitions between modes in such a way that the vehicle closely follows the track, using a sensor that determines how far the vehicle is to the left or right of the track. We can build such a sensor using photodiodes. Let's suppose the track is painted with a light-reflecting color, whereas the floor is relatively dark. Underneath the AGV we place an array of photodiodes as shown in Figure 4.14. The array is perpendicular to the AGV body axis. As the AGV passes over the track, the diode directly above the track generates more current than the other diodes. By comparing the magnitudes of the currents through the different diodes, the sensor estimates the displacement $e(t)$ of the center of the array (hence, the center of the AGV) from the track. We adopt the convention that $e(t) < 0$ means that the AGV is to the right of the track and $e(t) > 0$ means it is to the left. We model the sensor output as a function f of the AGV's position,

$$\forall t, \quad e(t) = f(x(t), y(t)).$$

The function f of course depends on the environment—the track. We now specify the supervisory controller precisely. We select two thresholds, $0 < \varepsilon_1 < \varepsilon_2$, as shown in Figure 4.14. If the magnitude of the displacement is small, $|e(t)| < \varepsilon_1$, we consider that the AGV is close enough to the track, and the AGV can move straight ahead, in straight mode. If $e(t) > \varepsilon_2$ ($e(t)$ is large and positive), the AGV has strayed too far to the left and must be steered to the right, by switching to right mode. If $e(t) < -\varepsilon_2$ ($e(t)$ is large and negative), the AGV has strayed too far to the right and must be steered to the left, by switching to left mode. This control logic is captured in the mode transitions of Figure 4.13. The inputs are pure signals *stop* and *start*. These model an operator that can stop or start the AGV. There is no continuous-time input. The outputs represent the position of the vehicle, $x(t)$ and $y(t)$. The initial mode is stop, and the initial values of its refinement are (x_0, y_0, θ_0).

We analyze how the AGV will move. Figure 4.15 sketches one possible trajectory. Initially the vehicle is within distance ε_1 of the track, so it moves straight. At some later time, the vehicle goes too far to the left, so the guard

$$\neg stop \land e(t) > \varepsilon_2$$

is satisfied, and there is a mode switch to right. After some time, the vehicle will again be close enough to the track, so the guard

$$\neg stop \land |e(t)| < \varepsilon_1$$

is satisfied, and there is a mode switch to straight. Some time later, the vehicle is too far to the right, so the guard

$$\neg stop \land e(t) < -\varepsilon_2$$

is satisfied, and there is a mode switch to left. And so on.

The example illustrates the four components of a control system. The plant is described by the differential equations (4.6) that govern the evolution of the continuous state at time t, $(x(t), y(t), \theta(t))$, in terms of the plant inputs u and ω. The second component is the environment—the closed track. The third component is the sensor, whose output at time t, $e(t) = f(x(t), y(t))$, gives the position of the AGV relative to the track. The fourth component is the two-level controller. The supervisory controller comprises the four modes and the guards that determine when to switch between modes. The low-level controller specifies how the time-based inputs to the plant, u and ω, are selected in each mode.

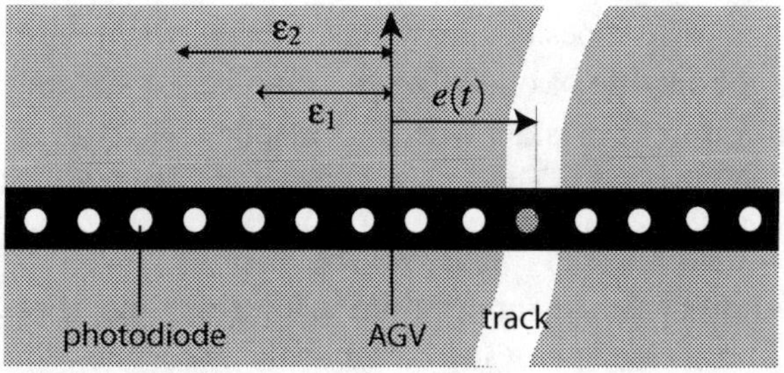

Figure 4.14: An array of photodiodes under the AGV is used to estimate the displacement e of the AGV relative to the track. The photodiode directly above the track generates more current.

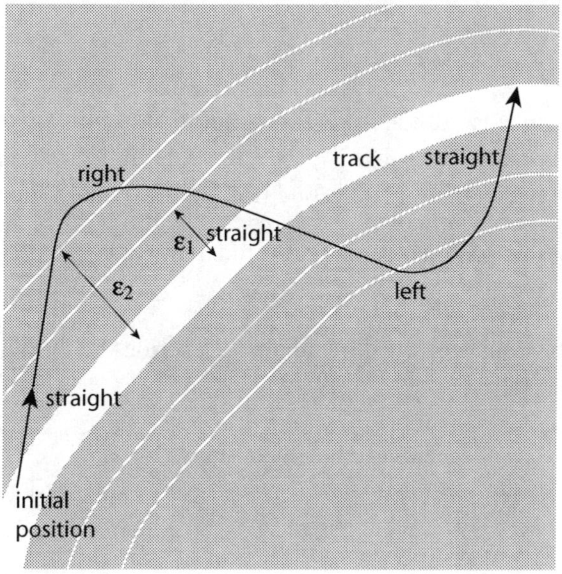

Figure 4.15: A trajectory of the AGV, annotated with modes.

4.3 Summary

Hybrid systems provide a bridge between time-based models and state-machine models. The combination of the two families of models provides a rich framework for describing real-world systems. There are two key ideas. First, discrete events (state changes in a state machine) get embedded in a time base. Second, a hierarchical description is particularly useful, where the system undergoes discrete transitions between different modes of operation. Associated with each mode of operation is a time-based system called the refinement of the mode. Mode transitions are taken when guards that specify the combination of inputs and continuous states are satisfied. The action associated with a transition, in turn, sets the continuous state in the destination mode.

The behavior of a hybrid system is understood using the tools of state machine analysis for mode transitions and the tools of time-based analysis for the refinement systems. The design of hybrid systems similarly proceeds on two levels: state machines are designed to achieve the appropriate logic of mode transitions, and continuous refinement systems are designed to secure the desired time-based behavior in each mode.

Exercises

1. Construct (on paper is sufficient) a timed automaton similar to that of Figure 4.7 which produces *tick* at times $1, 2, 3, 5, 6, 7, 8, 10, 11, \cdots$. That is, ticks are produced with intervals between them of 1 second (three times) and 2 seconds (once).

2. The objective of this problem is to understand a timed automaton, and then to modify it as specified.

   (a) For the timed automaton shown below, describe the output y. Avoid imprecise or sloppy notation.

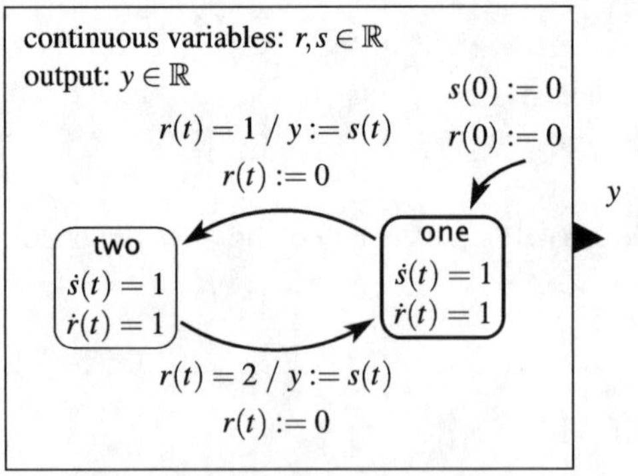

   (b) Assume there is a new pure input *reset*, and that when this input is present, the hybrid system starts over, behaving as if it were starting at time 0 again. Modify the hybrid system from part (a) to do this.

3. In Exercise 6 of Chapter 2, we considered a DC motor that is controlled by an input voltage. Controlling a motor by varying an input voltage, in reality, is often not practical. It requires analog circuits that are capable of handling considerable power. Instead, it is common to use a fixed voltage, but to turn it on and off periodically to vary the amount of power delivered to the motor. This technique is called pulse width modulation (PWM). p.194 p.193

   Construct a timed automaton that provides the voltage input to the motor model from Exercise 6. Your hybrid system should assume that the PWM circuit delivers a 25 kHz square wave with a duty cycle between zero and 100%, inclusive. The input to your hybrid system should be the duty cycle, and the output should be the voltage.

4. Consider the following timed automaton:

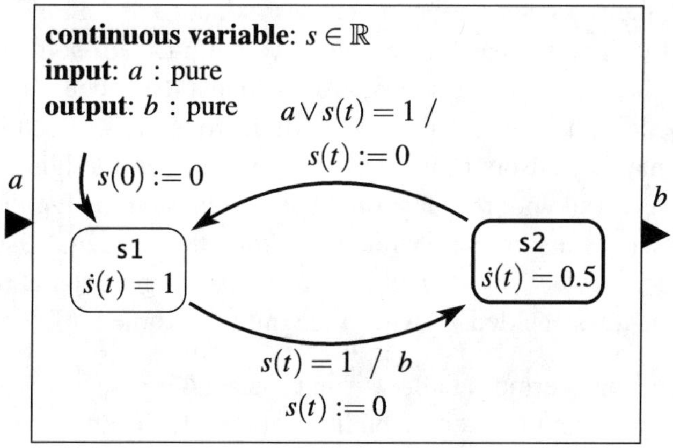

   Assume that the input signals a and b are discrete continuous-time signals, meaning that each can be given as a function of form $a\colon \mathbb{R} \to \{present, absent\}$, where at almost all times $t \in \mathbb{R}$, $a(t) = absent$. Assume that the state machine can take at most one transition at each distinct time t, and that machine begins executing at time $t = 0$.

   (a) Sketch the output b if the input a is present only at times

   $$t = 0.75, 1.5, 2.25, 3, 3.75, 4.5, \cdots$$

   Include at least times from $t = 0$ to $t = 5$.

   (b) Sketch the output b if the input a is present only at times $t = 0, 1, 2, 3, \cdots$.

   (c) Assuming that the input a can be any discrete signal at all, find a lower bound on the amount of time between events b. What input signal a (if any) achieves this lower bound?

5. You have an analog source that produces a pure tone. You can switch the source on or off by the input event *on* or *off*. Construct a timed automaton that provides the *on* and *off* signals as outputs, to be connected to the inputs of the tone generator. Your system should behave as follows. Upon receiving an input event *ring*, it should produce an 80 ms-long sound consisting of three 20 ms-long bursts of the pure tone separated by two 10 ms intervals of silence. What does your system do if it receives two *ring* events that are 50 ms apart?

6. Automobiles today have the features listed below. Implement each feature as a timed automaton.

   (a) The dome light is turned on as soon as any door is opened. It stays on for 30 seconds after all doors are shut. What sensors are needed?

   (b) Once the engine is started, a beeper is sounded and a red light warning is indicated if there are passengers that have not buckled their seat belt. The beeper stops sounding after 30 seconds, or as soon the seat belts are buckled, whichever is sooner. The warning light is on all the time the seat belt is unbuckled. **Hint:** Assume the sensors provide a *warn* event when the ignition is turned on and there is a seat with passenger not buckled in, or if the ignition is already on and a passenger sits in a seat without buckling the seatbelt. Assume further that the sensors provide a *noWarn* event when a passenger departs from a seat, or when the buckle is buckled, or when the ignition is turned off.

7. A programmable thermostat allows you to select 4 times, $0 \leq T_1 \leq \cdots \leq T_4 < 24$ (for p.50 a 24-hour cycle) and the corresponding setpoint temperatures $a_1, \cdots, a_4$. Construct a timed automaton that sends the event a_i to the heating systems controller. The controller maintains the temperature close to the value a_i until it receives the next event. How many timers and modes do you need?

8. Consider the following timed automaton:

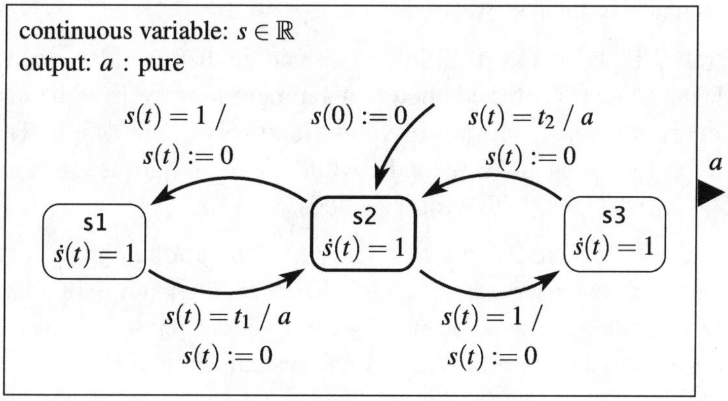

Assume t_1 and t_2 are positive real numbers. What is the minimum amount of time between events a? That is, what is the smallest possible time between two times when the signal a is present?

9. Figure 4.16 depicts the intersection of two one-way streets, called Main and Secondary. A light on each street controls its traffic. Each light goes through a cycle

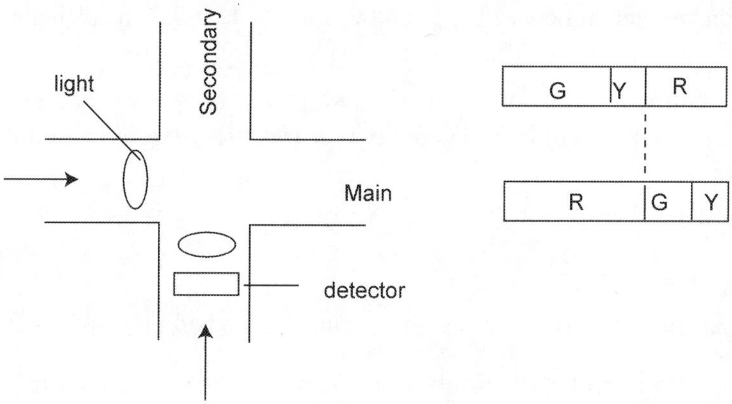

Figure 4.16: Traffic lights control the intersection of a main street and a secondary street. A detector senses when a vehicle crosses it. The red phase of one light must coincide with the green and yellow phases of the other light.

consisting of a red (R), green (G), and yellow (Y) phases. It is a safety requirement that when one light is in its green or yellow phase, the other is in its red phase. The yellow phase is always 5 seconds long.

The traffic lights operate as follows. A sensor in the secondary road detects a vehicle. While no vehicle is detected, there is a 4 minute-long cycle with the main light having 3 minutes of green, 5 seconds of yellow, and 55 seconds of red. The secondary light is red for 3 minutes and 5 seconds (while the main light is green and yellow), green for 50 seconds, then yellow for 5 seconds.

If a vehicle is detected on the secondary road, the traffic light quickly gives a right of way to the secondary road. When this happens, the main light aborts its green phase and immediately switches to its 5 second yellow phase. If the vehicle is detected while the main light is yellow or red, the system continues as if there were no vehicle.

Design a hybrid system that controls the lights. Let this hybrid system have six pure outputs, one for each light, named *mG*, *mY*, and *mR*, to designate the main light being green, yellow, or red, respectively, and *sG*, *sY*, and *sR*, to designate the secondary light being green, yellow, or red, respectively. These signals should be generated to turn on a light. You can implicitly assume that when one light is turned on, whichever has been on is turned off.

10. For the bouncing ball of Example 4.7, let t_n be the time when the ball hits the ground for the n-th time, and let $v_n = \dot{y}(t_n)$ be the velocity at that time.

    (a) Find a relation between v_{n+1} and v_n for $n > 1$, and then calculate v_n in terms of v_1.

    (b) Obtain t_n in terms of v_1 and a. Use this to show that the bouncing ball is a p.89 Zeno system. **Hint:** The **geometric series identity** might be useful, where for $|b| < 1$,

$$\sum_{m=0}^{\infty} b^m = \frac{1}{1-b}.$$

    (c) Calculate the maximum height reached by the ball after successive bumps.

11. Elaborate the hybrid system model of Figure 4.10 so that in the *together* mode, the stickiness decays according to the differential equation

$$\dot{s}(t) = -as(t)$$

where $s(t)$ is the stickiness at time t, and a is some positive constant. On the transition into this mode, the stickiness should be initialized to some starting stickiness b.

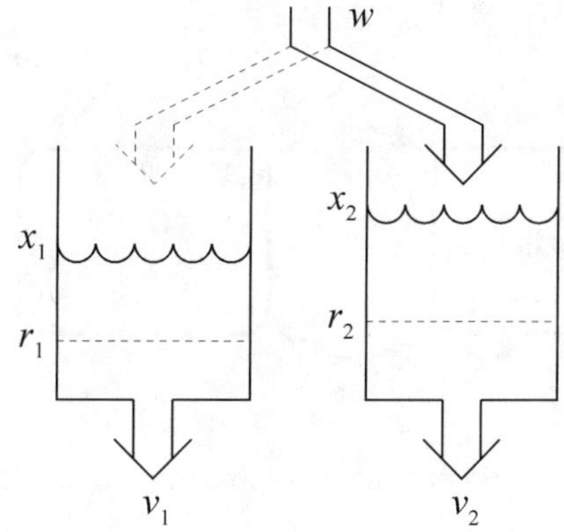

Figure 4.17: Water tank system.

12. Show that the trajectory of the AGV of Figure 4.13 while it is in *left* or *right* mode is a circle. What is the radius of this circle, and how long does it take to complete a circle?

13. Consider Figure 4.17 depicting a system comprising two tanks containing water. Each tank is leaking at a constant rate. Water is added at a constant rate to the system through a hose, which at any point in time is filling either one tank or the other. It is assumed that the hose can switch between the tanks instantaneously. For $i \in \{1,2\}$, let x_i denote the volume of water in Tank i and $v_i > 0$ denote the constant flow of water out of Tank i. Let w denote the constant flow of water into the system. The objective is to keep the water volumes above r_1 and r_2, respectively, assuming that the water volumes are above r_1 and r_2 initially. This is to be achieved by a controller that switches the inflow to Tank 1 whenever $x_1(t) \leq r_1(t)$ and to Tank 2 whenever $x_2(t) \leq r_2(t)$.

    The hybrid automaton representing this two-tank system is given in Figure 4.18.

    Answer the following questions:

    (a) Construct a model of this hybrid automaton in Ptolemy II, LabVIEW, or Simulink. Use the following parameter values: $r_1 = r_2 = 0$, $v_1 = v_2 = 0.5$, and $w = 0.75$. Set the initial state to be $(q_1, (0,1))$. (That is, initial value $x_1(0)$ is 0 and $x_2(0)$ is 1.)

    Verify that this hybrid automaton is Zeno. What is the reason for this Zeno behavior? Simulate your model and plot how x_1 and x_2 vary as a function of time t, simulating long enough to illustrate the Zeno behavior. p.89

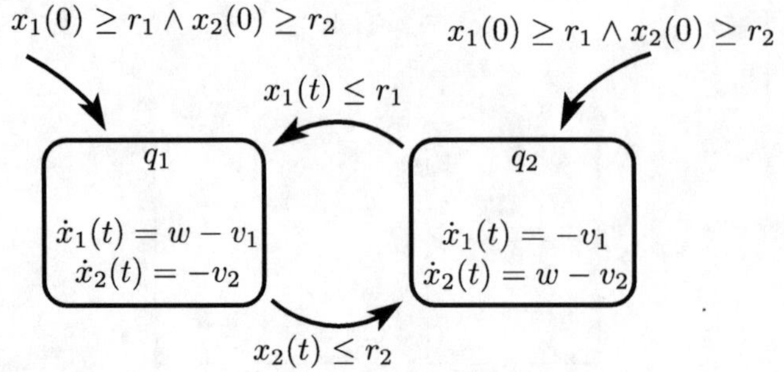

Figure 4.18: Hybrid automaton representing water tank system.

(b) A Zeno system may be **regularized** by ensuring that the time between transitions is never less than some positive number ε. This can be emulated by inserting extra modes in which the hybrid automaton dwells for time ε. Use regularization to make your model from part (a) non-Zeno. Again, plot x_1 and x_2 for the same length of time as in the first part. State the value of ε that you used.

Include printouts of your plots with your answer.

5

Composition of State Machines

Contents

State machines provide a convenient way to model behaviors of systems. One disadvantage that they have is that for most interesting systems, the number of states is very large, often even infinite. Automated tools can handle large state spaces, but humans have more difficulty with any direct representation of a large state space.

A time-honored principle in engineering is that complicated systems should be described as compositions of simpler systems. This chapter gives a number of ways to do this with state machines. The reader should be aware, however, that there are many subtly different ways to compose state machines. Compositions that look similar on the surface may mean different things to different people. The rules of notation of a model are called its **syntax**, and the meaning of the notation is called its **semantics**.

Example 5.1: In the standard syntax of arithmetic, a plus sign + has a number or expression before it, and a number or expression after it. Hence, $1+2$, a sequence of three symbols, is a valid arithmetic expression, but $1+$ is not. The semantics of the expression $1+2$ is the addition of two numbers. This expression means "the number three, obtained by adding 1 and 2." The expression $2+1$ is syntactically different, but semantically identical (because addition is commutative).

The models in this book predominantly use a visual syntax, where the elements are boxes, circles, arrows, etc., rather than characters in a character set, and where the positioning of the elements is not constrained to be a sequence. Such syntaxes are less standardized than, for example, the syntax of arithmetic. We will see that the same syntax can have many different semantics, which can cause no end of confusion.

Example 5.2: A now popular notation for concurrent composition of state machines called Statecharts was introduced by Harel (1987). Although they are all based on the same original paper, many variants of Statecharts have evolved (von der Beeck, 1994). These variants often assign different semantics to the same syntax.

p.56 In this chapter, we assume an actor model for extended state machines using the syntax summarized in Figure 5.1. The semantics of a single such state machine is described in Chapter 3. This chapter will discuss the semantics that can be assigned to compositions of multiple such machines.

The first composition technique we consider is concurrent composition. Two or more state machines react either simultaneously or independently. If the reactions are simultaneous, we call it **synchronous composition**. If they are independent, then we call it **asynchronous composition**. But even within these classes of composition, many subtle variations in the semantics are possible. These variations mostly revolve around whether and how the state machines communicate and share variables.

The second composition technique we will consider is hierarchy. Hierarchical state machines can also enable complicated systems to be described as compositions of simpler systems. Again, we will see that subtle differences in semantics are possible.

5.1 Concurrent Composition

To study concurrent composition of state machines, we will proceed through a sequence of patterns of composition. These patterns can be combined to build arbitrarily complicated systems. We begin with the simplest case, side-by-side composition, where the state machines being composed do not communicate. We then consider allowing communication through shared variables, showing that this creates significant subtleties that can complicate modeling. We then consider communication through ports, first looking at serial composition, then expanding to arbitrary interconnections. We consider both synchronous and asynchronous composition for each type of composition.

5.1.1 Side-by-Side Synchronous Composition

The first pattern of composition that we consider is **side-by-side composition**, illustrated for two actors in Figure 5.2. In this pattern, we assume that the inputs and outputs of the two actors are disjoint, i.e., that the state machines do not communicate. In the figure, actor

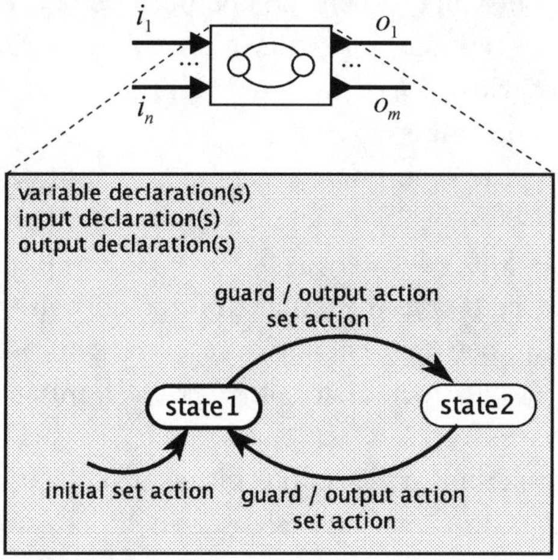

Figure 5.1: Summary of notation for state machines used in this chapter.

About Synchrony

The term **synchronous** means (1) occurring or existing at the same time or (2) moving or operating at the same rate. In engineering and computer science, the term has a number of meanings that are mostly consistent with these definitions, but oddly inconsistent with one another. p.286 In referring to concurrent software constructed using threads p.154 or processes, synchronous communication refers to a rendezvous style of communication, where the sender of a message must wait for the receiver to be ready to receive, and the receiver must wait for the sender. Conceptually, the two threads see the communication occurring at the same time, consistent with definition (1). In Java, the keyword `synchronized` defines blocks of code that are not permitted to execute simultaneously. Oddly, two code blocks that are synchronized *cannot* "occur" (execute) at the same time, which is inconsistent with both definitions.

In the world of software, there is yet a third meaning of the word synchronous, and it is this third meaning that we use in this chapter. This third meaning underlies the p.141 synchronous languages (see box on page 141). Two key ideas govern these languages. First, the outputs of components in a program are (conceptually) simultaneous with their inputs (this is called the **synchrony hypothesis**). Second, components in a program exe- p.116 cute (conceptually) simultaneously and instantaneously. Real executions do not literally occur simultaneously nor instantaneously, and outputs are not really simultaneous with the inputs, but a correct execution must behave as if they were. This use of the word synchronous is consistent with *both* definitions above; executions of components occur at the same time and operate at the same rate.

In circuit design, the word synchronous refers to a style of design where a clock that is distributed throughout a circuit drives latches that record their inputs on edges of the clock. The time between clock edges needs to be sufficient for circuits between latches to settle. Conceptually, this model is similar to the model in synchronous languages. Assuming that the circuits between latches have zero delay is equivalent to the synchrony hypothesis, and global clock distribution gives simultaneous and instantaneous execution.

In power systems engineering, synchronous means that electrical waveforms have the same frequency and phase. In signal processing, synchronous means that signals have the same sample rate, or that their sample rates are fixed multiples of one another. The p.145 term synchronous dataflow, described in Section 6.3.2, is based on this latter meaning of the word synchronous. This usage is consistent with definition (2).

A has input i_1 and output o_1, and actor B has input i_2 and output o_2. The composition of the two actors is itself an actor C with inputs i_1 and i_2 and outputs o_1 and o_2.[1]

In the simplest scenario, if the two actors are extended state machines with variables, then those variables are also disjoint. We will later consider what happens when the two state machines share variables. Under **synchronous composition**, a reaction of C is a simultaneous reaction of A and B.

Example 5.3: Consider FSMs A and B in Figure 5.3. A has a single pure output a, and B has a single pure output b. The side-by-side composition C has two pure outputs, a and b. If the composition is synchronous, then on the first reaction, a will be *absent* and b will be *present*. On the second reaction, it will be the reverse. On subsequent reactions, a and b will continue to alternate being present.

Synchronous side-by-side composition is simple for several reasons. First, recall from Section 3.3.2 that the environment determines when a state machine reacts. In synchronous side-by-side composition, the environment need not be aware that C is a composition of two state machines. Such compositions are **modular** in the sense that the composition itself becomes a component that can be further composed as if it were itself an atomic component.

[1] The composition actor C may rename these input and output ports, but here we assume it uses the same names as the component actors. p.25

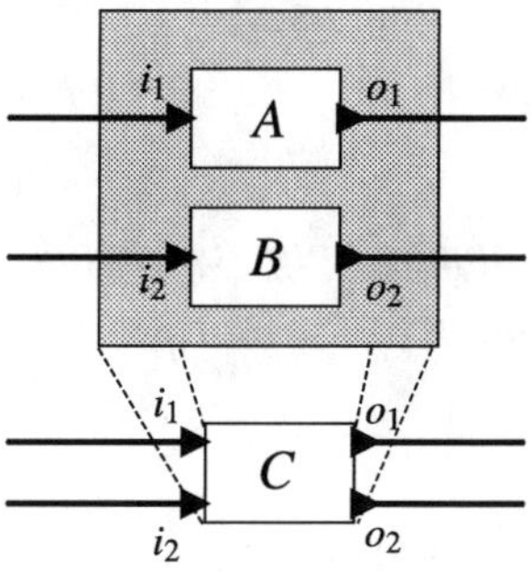

Figure 5.2: Side-by-side composition of two actors.

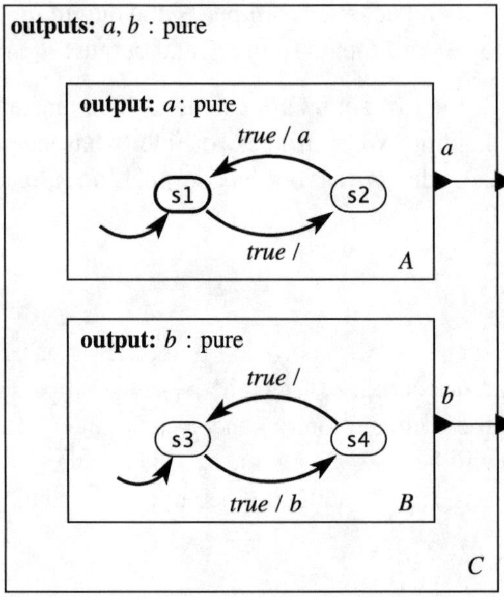

Figure 5.3: Example of side-by-side composition of two actors.

p.56 Moreover, if the two state machines A and B are deterministic, then the synchronous side-by-side composition is also deterministic. We say that a property is **compositional** if a property held by the components is also a property of the composition. For synchronous side-by-side composition, determinism is a compositional property.

In addition, a synchronous side-by-side composition of finite state machines is itself an FSM. A rigorous way to give the semantics of the composition is to define a single state machine for the composition. Suppose that as in Section 3.3.3, state machines A and B are given by the five tuples,

$$\begin{aligned} A &= (States_A, Inputs_A, Outputs_A, update_A, initialState_A) \\ B &= (States_B, Inputs_B, Outputs_B, update_B, initialState_B)\ . \end{aligned}$$

Then the synchronous side-by-side composition C is given by

$$States_C = States_A \times States_B \tag{5.1}$$
$$Inputs_C = Inputs_A \times Inputs_B \tag{5.2}$$
$$Outputs_C = Outputs_A \times Outputs_B \tag{5.3}$$
$$initialState_C = (initialState_A, initialState_B) \tag{5.4}$$

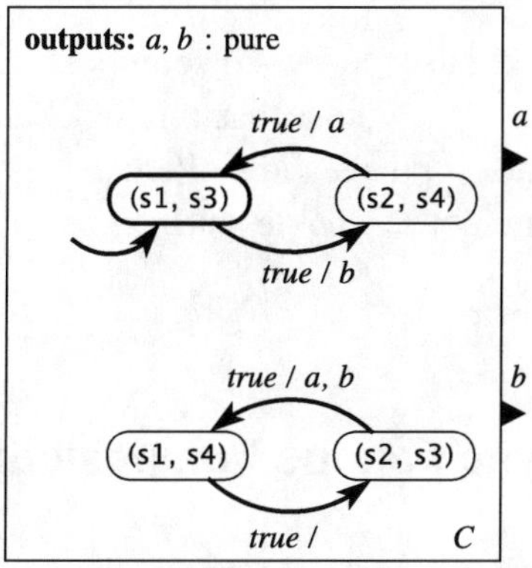

Figure 5.4: Single state machine giving the semantics of synchronous side-by-side composition of the state machines in Figure 5.3.

and the update function is defined by

$$update_C((s_A, s_B), (i_A, i_B)) = ((s'_A, s'_B), (o_A, o_B)),$$

where

$$(s'_A, o_A) = update_A(s_A, i_A),$$

and

$$(s'_B, o_B) = update_B(s_B, i_B),$$

for all $s_A \in States_A$, $s_B \in States_B$, $i_A \in Inputs_A$, and $i_B \in Inputs_B$.

Recall that $Inputs_A$ and $Inputs_B$ are sets of valuations. Each valuation in the set is an assignment of values to ports. What we mean by p.45

$$Inputs_C = Inputs_A \times Inputs_B$$

is that a valuation of the inputs of C must include *both* valuations for the inputs of A and the inputs of B.

As usual, the single FSM C can be given pictorially rather than symbolically, as illustrated in the next example.

Example 5.4: The synchronous side-by-side composition C in Figure 5.3 is given as a single FSM in Figure 5.4. Notice that this machine behaves exactly as described in Example 5.3. The outputs a and b alternate being present. Notice further p.61 that (s1, s4) and (s2, s3) are not reachable states.

5.1.2 Side-by-Side Asynchronous Composition

In an **asynchronous composition** of state machines, the component machines react independently. This statement is rather vague, and in fact, it has several different interpretations. p.104 Each interpretation gives a semantics to the composition. The key to each semantics is how p.44 to define a reaction of the composition C in Figure 5.2. Two possibilities are:

- **Semantics 1.** A reaction of C is a reaction of one of A or B, where the choice is p.63 nondeterministic.
- **Semantics 2.** A reaction of C is a reaction of A, B, or both A and B, where the choice is nondeterministic. A variant of this possibility might allow *neither* to react.

Semantics 1 is referred to as an **interleaving semantics**, meaning that A or B never react simultaneously. Their reactions are interleaved in some order.

A significant subtlety is that under these semantics machines A and B may completely miss input events. That is, an input to C destined for machine A may be present in a reaction where the nondeterministic choice results in B reacting rather than A. If this is not desirable, p.107 then some control over scheduling (see sidebar on page 112) or synchronous composition becomes a better choice.

Example 5.5: For the example in Figure 5.3, semantics 1 results in the composition state machine shown in Figure 5.5. This machine is nondeterministic. From state (s1, s3), when C reacts, it can move to (s2, s3) and emit no output, or it can move to (s1, s4) and emit b. Note that if we had chosen semantics 2, then it would also be able to move to (s2, s4).

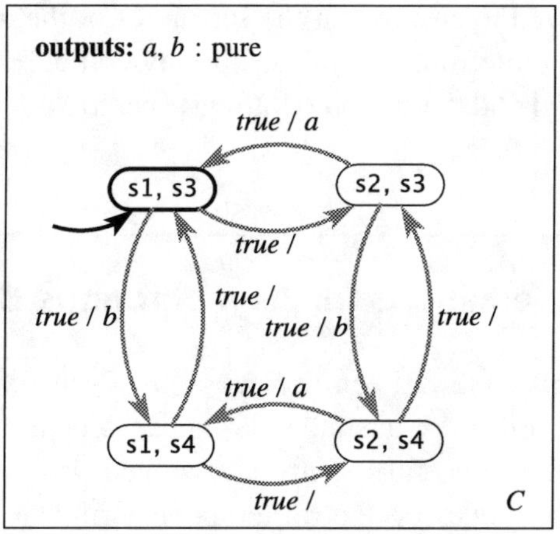

Figure 5.5: State machine giving the semantics of asynchronous side-by-side composition of the state machines in Figure 5.3.

For asynchronous composition under semantics 1, the symbolic definition of C has the same definitions of $States_C$, $Inputs_C$, $Outputs_C$, and $initialState_C$ as for synchronous composition, given in (5.1) through (5.4). But the update function differs, becoming

$$update_C((s_A, s_B), (i_A, i_B)) = ((s'_A, s'_B), (o'_A, o'_B)),$$

where either

$$(s'_A, o'_A) = update_A(s_A, i_A) \text{ and } s'_B = s_B \text{ and } o'_B = absent$$

or

$$(s'_B, o'_B) = update_B(s_B, i_B) \text{ and } s'_A = s_A \text{ and } o'_A = absent$$

for all $s_A \in States_A$, $s_B \in States_B$, $i_A \in Inputs_A$, and $i_B \in Inputs_B$. What we mean by $o'_B = absent$ is that all outputs of B are absent. Semantics 2 can be similarly defined (see Exercise 2).

5.1.3 Shared Variables

An extended state machine has local variables that can be read and written as part of taking p.56
a transition. Sometimes it is useful when composing state machines to allow these variables to be shared among a group of machines. In particular, such shared variables can be useful for modeling interrupts, studied in Chapter 10, and threads, studied in Chapter 11. However, p.266

considerable care is required to ensure that the semantics of the model conforms with that of the program containing interrupts or threads. Many complications arise, including the p.295 memory consistency model and the notion of atomic operations.

Scheduling Semantics for Asynchronous Composition

In the case of semantics 1 and 2 given in Section 5.1.2, the choice of which component machine reacts is nondeterministic. The model does not express any particular constraints. It is often more useful to introduce some scheduling policies, where the environment is able to influence or control the nondeterministic choice. This leads to two additional possible semantics for asynchronous composition:

- **Semantics 3.** A reaction of C is a reaction of one of A or B, where the environment chooses which of A or B reacts.
- **Semantics 4.** A reaction of C is a reaction of A, B, or both A and B, where the choice is made by the environment.

p.110 Like semantics 1, semantics 3 is an interleaving semantics.

p.108 In one sense, semantics 1 and 2 are more compositional than semantics 3 and 4. To implement semantics 3 and 4, a composition has to provide some mechanism for the environment to choose which component machine should react (for scheduling the component machines). This means that the hierarchy suggested in Figure 5.2 does not quite work. Actor C has to expose more of its internal structure than just the ports and the ability to react.

In another sense, semantics 1 and 2 are less compositional than semantics 3 and 4 because determinism is not preserved by composition. A composition of deterministic state machines is not a deterministic state machine.

p.361 Notice further that semantics 1 is an abstraction of semantics 3 in the sense that every behavior under semantics 3 is also a behavior under semantics 1. This notion of abstraction is studied in detail in Chapter 14.

The subtle differences between these choices make asynchronous composition rather treacherous. Considerable care is required to ensure that it is clear which semantics is used.

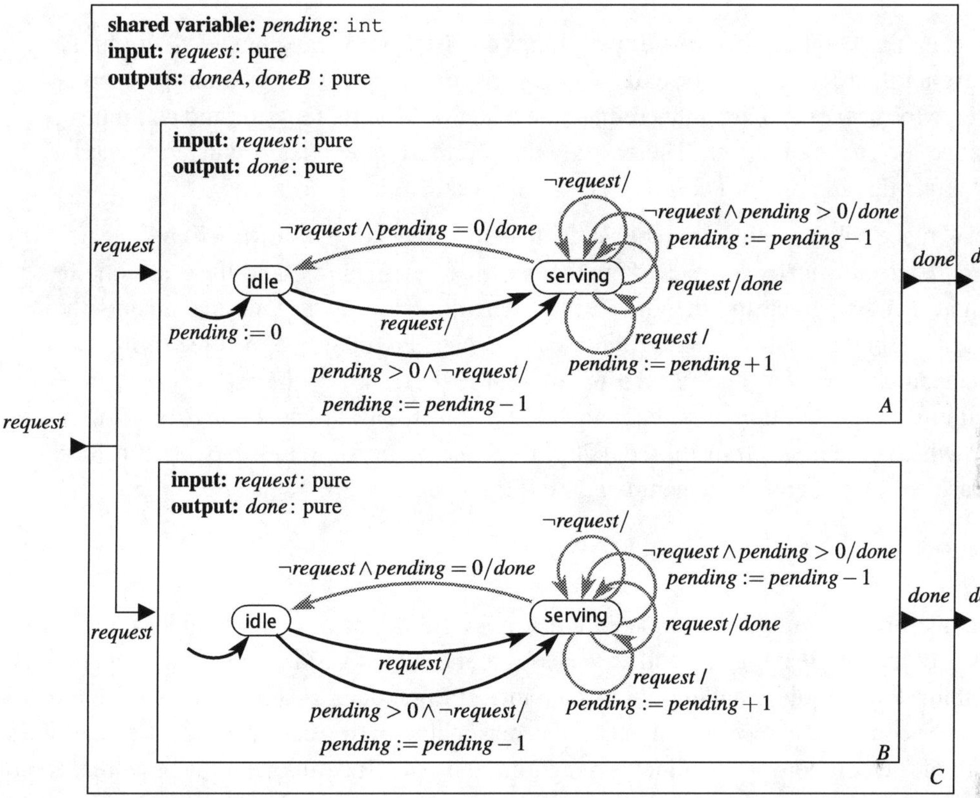

Figure 5.6: Model of two servers with a shared task queue, assuming asynchronous composition under semantics 1.

Example 5.6: Consider two servers that can receive requests from a network. Each request requires an unknown amount of time to service, so the servers share a queue of requests. If one server is busy, the other server can respond to a request, even if the request arrives at the network interface of the first server.

This scenario fits a pattern similar to that in Figure 5.2, where A and B are the servers. We can model the servers as state machines as shown in Figure 5.6. In this model, a shared variable *pending* counts the number of pending job requests. When a request arrives at the composite machine C, one of the two servers is nondeterministically chosen to react, assuming asynchronous composition under semantics 1. If that server is idle, then it proceeds to serve the request. If the server is serving another request, then one of two things can happen: it can coincidentally finish serving the request it is currently serving, issuing the output *done*, and proceed to serve the new one, or it can increment the count of pending requests and continue to serve the current request. The choice between these is nondeterministic, to model the fact that the time it takes to service a request is unknown.

If C reacts when there is no request, then again either server A or B will be selected nondeterministically to react. If the server that reacts is idle and there are one or more pending requests, then the server transitions to serving and decrements the variable *pending*. If the server that reacts is not idle, then one of three things can happen. It may continue serving the current request, in which case it simply tran-
p.48 sitions on the self transition back to serving. Or it may finish serving the request, in which case it will transition to idle if there are no pending requests, or transition back to serving and decrement *pending* if there are pending requests.

The model in the previous example exhibits many subtleties of concurrent systems. First,
p.260 because of the interleaving semantics, accesses to the shared variable are atomic operations, something that is quite challenging to guarantee in practice, as discussed in Chapters 10 and 11. Second, the choice of semantics 1 is reasonable in this case because the input goes to both of the component machines, so regardless of which component machine reacts, no input event will be missed. However, this semantics would not work if the two machines had independent inputs, because then requests could be missed. Semantics 2 can help prevent that, but what strategy should be used by the environment to determine which machine reacts? What if the two independent inputs both have requests present at the same reaction of C? If we choose semantics 4 in the sidebar on page 112 to allow both machines to react simultaneously, then what is the meaning when both machines update the shared variable? The updates are no longer atomic, as they are with an interleaving semantics.

Note further that choosing asynchronous composition under semantics 1 allows behaviors that do not make good use of idle machines. In particular, suppose that machine A is serving, machine B is idle, and a *request* arrives. If the nondeterministic choice results in machine A reacting, then it will simply increment *pending*. Not until the nondeterministic choice results in B reacting will the idle machine be put to use. In fact, semantics 1 allows behaviors that never use one of the machines.

p.107 Shared variables may be used in synchronous compositions as well, but sophisticated subtleties again emerge. In particular, what should happen if in the same reaction one machine reads a shared variable to evaluate a guard and another machine writes to the shared variable? Do we require the write before the read? What if the transition doing the write to the shared variable also reads the same variable in its guard expression? One possibility is to choose a **synchronous interleaving semantics**, where the component machines react in arbitrary order, chosen nondeterministically. This strategy has the disadvantage that a composition of two deterministic machines may be nondeterministic. An alternative version of the synchronous interleaving semantics has the component machines react in a fixed order determined by the environment or by some additional mechanism such as priority. p.313

The difficulties of shared variables, particularly with asynchronous composition, reflect the inherent complexity of concurrency models with shared variables. Clean solutions require a more sophisticated semantics, to be discussed in Chapter 6. In that chapter, we will explain the synchronous-reactive model of computation, which gives a synchronous composition p.132 semantics that is reasonably compositional.

So far, we have considered composition of machines that do not directly communicate. We next consider what happens when the outputs of one machine are the inputs of another.

5.1.4 Cascade Composition

Consider two state machines A and B that are composed as shown in Figure 5.7. The output of machine A feeds the input of B. This style of composition is called cascade composition p.26 or **serial composition**.

In the figure, output port o_1 from A feeds events to input port i_2 of B. Assume the data type p.45 of o_1 is V_1 (meaning that o_1 can take values from V_1 or be *absent*), and the data type of i_2 is V_2. Then a requirement for this composition to be valid is that

$$V_1 \subseteq V_2 \, .$$

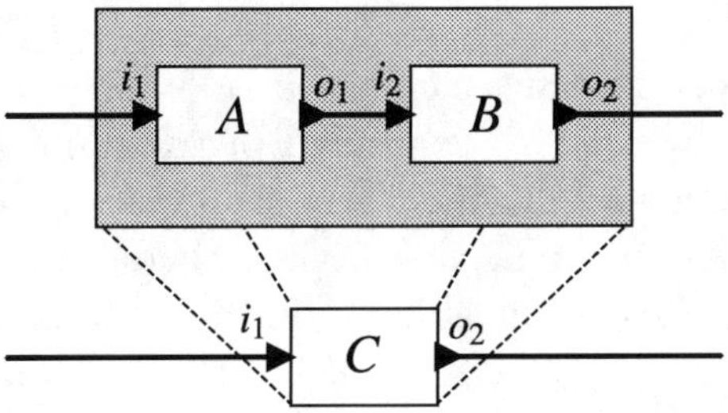

Figure 5.7: Cascade composition of two actors.

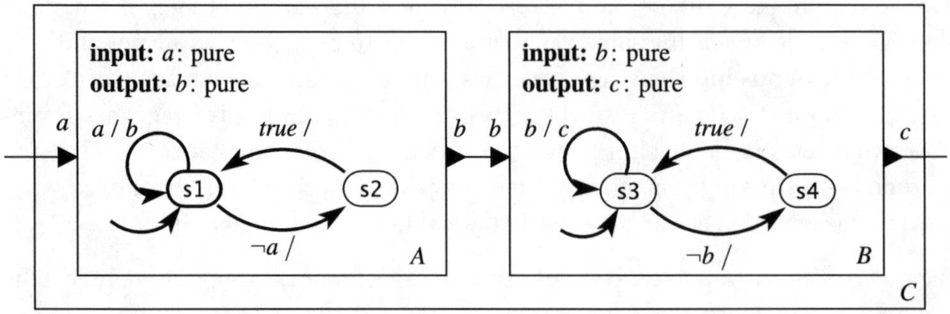

Figure 5.8: Example of a cascade composition of two FSMs.

This asserts that any output produced by A on port o_1 is an acceptable input to B on port i_2. The composition **type checks**.

For cascade composition, if we wish the composition to be asynchronous, then we need to introduce some machinery for buffering the data that is sent from A to B. We defer discus-
p.142 sion of such asynchronous composition to Chapter 6, where dataflow and process network models of computation will provide such asynchronous composition. In this chapter, we will only consider synchronous composition for cascade systems.

In synchronous composition of the cascade structure of Figure 5.7, a reaction of C consists of a reaction of both A and B, where A reacts first, produces its output (if any), and then B reacts. Logically, we view this as occurring in zero time, so the two reactions are in a sense **simultaneous and instantaneous**. But they are causally related in that the outputs of A can affect the behavior of B.

Example 5.7: Consider the cascade composition of the two FSMs in Figure 5.8. Assuming synchronous semantics, the meaning of a reaction of C is given in Figure 5.9. That figure makes it clear that the reactions of the two machines are simultaneous and instantaneous. When moving from the initial state (s1, s3) to (s2, s4) (which occurs when the input a is absent), the composition machine C does not
p.61 pass through (s2, s3)! In fact, (s2, s3) is not a reachable state! In this way, a *single* reaction of C encompasses a reaction of both A *and* B.

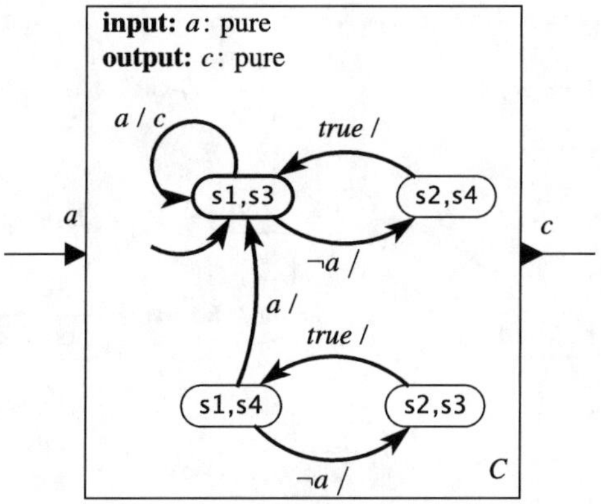

Figure 5.9: Semantics of the cascade composition of Figure 5.8, assuming synchronous composition.

To construct the composition machine as in Figure 5.9, first form the state space as the cross product of the state spaces of the component machines, and then determine which transitions are taken under what conditions. It is important to remember that the transitions are simultaneous, even when one logically causes the other.

Example 5.8: Recall the traffic light model of Figure 3.10. Suppose that we wish to compose this with a model of a pedestrian crossing light, like that shown in Fig-

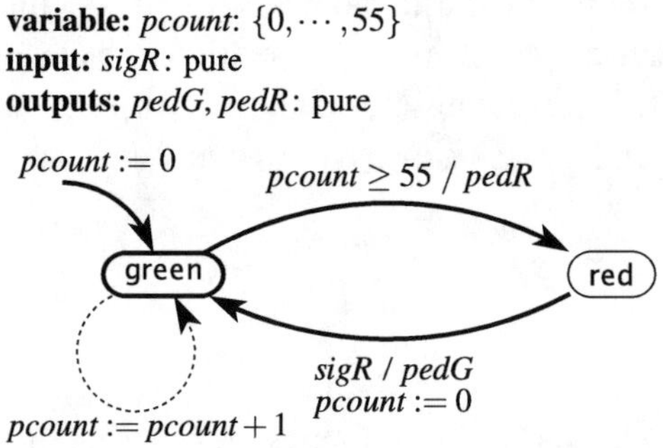

Figure 5.10: A model of a pedestrian crossing light, to be composed in a synchronous cascade composition with the traffic light model of Figure 3.10.

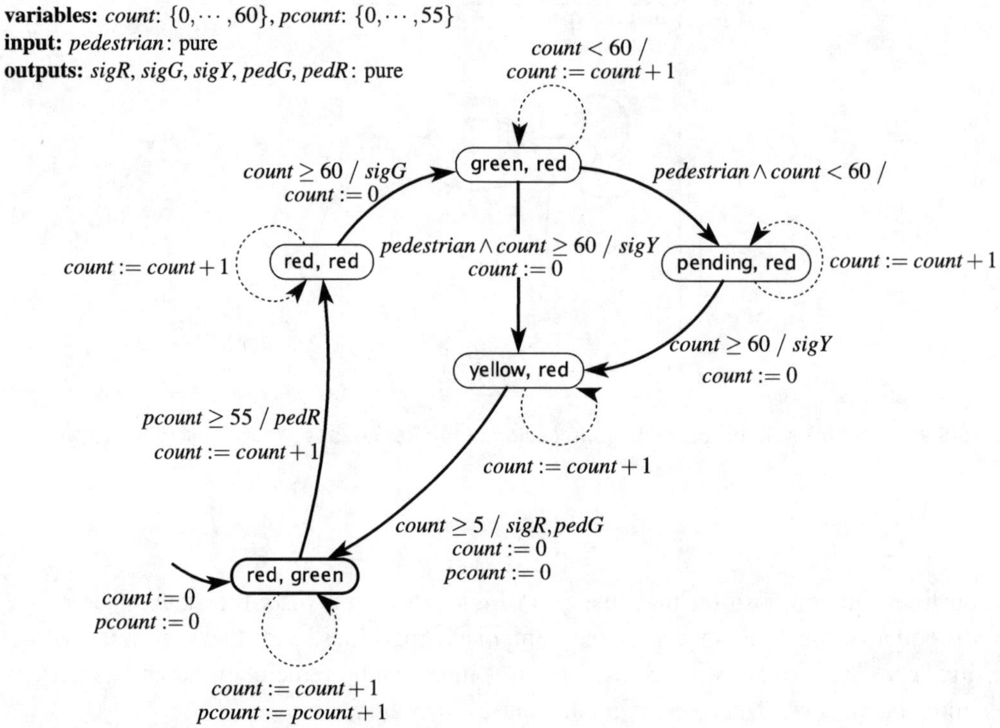

Figure 5.11: Semantics of a synchronous cascade composition of the traffic light model of Figure 3.10 with the pedestrian light model of Figure 5.10.

ure 5.10. The output *sigR* of the traffic light can provide the input *sigR* of the pedestrian light. Under synchronous cascade composition, the meaning of the composite is given in Figure 5.11. Note that unsafe states, such as (green, green), which is the state when both cars and pedestrians have a green light, are not reachable states, and hence are not shown. p.61

In its simplest form, cascade composition implies an ordering of the reactions of the components. Since this ordering is well defined, we do not have as much difficulty with shared variables as we did with side-by-side composition. However, we will see that in more general compositions, the ordering is not so simple.

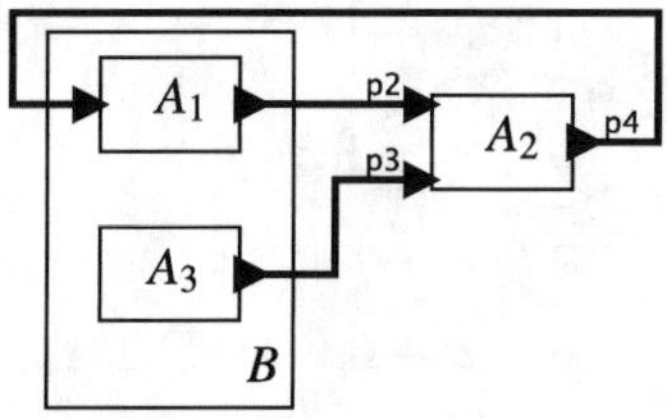

Figure 5.12: Arbitrary interconnections of state machines are combinations of side-by-side and cascade compositions, possibly creating cycles, as in this example.

5.1.5 General Composition

Side-by-side and cascade composition provide the basic building blocks for building more complex compositions of machines. Consider for example the composition in Figure 5.12. A_1 and A_3 are a side-by-side composition that together define a machine B. B and A_2 are a cascade composition, with B feeding events to A_2. However, B and A_2 are also a cascade composition in the opposite order, with A_2 feeding events to B. Cycles like this are called feedback, and they introduce a conundrum; which machine should react first, B or A_2? This p.31
conundrum will be resolved in the next chapter when we explain the synchronous-reactive p.132
model of computation.

5.2 Hierarchical State Machines

In this section, we consider **hierarchical FSMs**, which date back to Statecharts (Harel, 1987). There are many variants of Statecharts, often with subtle semantic differences between them (von der Beeck, 1994). Here, we will focus on some of the simpler aspects only, and we will pick a particular semantic variant.

The key idea in hierarchical state machines is state refinement. In Figure 5.13, state B has a p.79
refinement that is another FSM with two states, C and D. What it means for the machine to be in state B is that it is in one of states C or D.

The meaning of the hierarchy in Figure 5.13 can be understood by comparing it to the equivalent flattened FSM in Figure 5.14. The machine starts in state A. When guard g_2 evaluates to true, the machine transitions to state B, which means a transition to state C, the initial state of the refinement. Upon taking this transition to C, the machine performs action a_2, which may produce an output event or set a variable (if this is an extended state machine). p.56

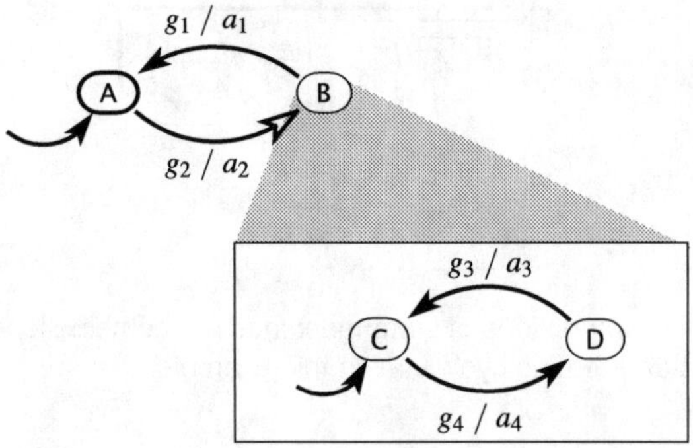

Figure 5.13: In a hierarchical FSM, a state may have a refinement that is another state machine.

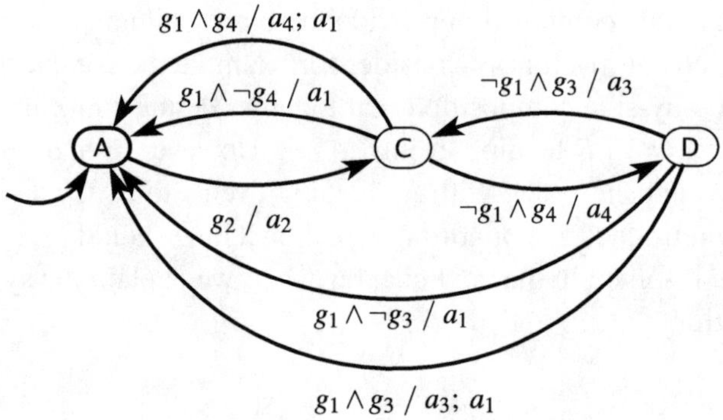

Figure 5.14: Semantics of the hierarchical FSM in Figure 5.13.

There are then two ways to exit C. Either guard g_1 evaluates to true, in which case the machine exits B and returns to A, or guard g_4 evaluates to true and the machine transitions to D. A subtle question is what happens if both guards g_1 and g_4 evaluate to true. Different variants of Statecharts may make different choices at this point. It seems reasonable that the machine should end up in state A, but which of the actions should be performed, a_4, a_1, or both? Such subtle questions help account for the proliferation of different variants of Statecharts.

We choose a particular semantics that has attractive modularity properties (Lee and Tripakis, 2010). In this semantics, a reaction of a hierarchical FSM is defined in a depth-first fashion. The deepest refinement of the current state reacts first, then its container state machine, then its container, etc. In Figure 5.13, this means that if the machine is in state

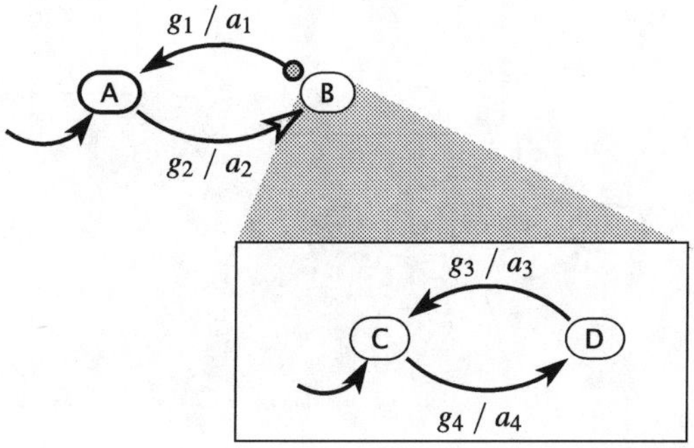

Figure 5.15: Variant of Figure 5.13 that uses a preemptive transition.

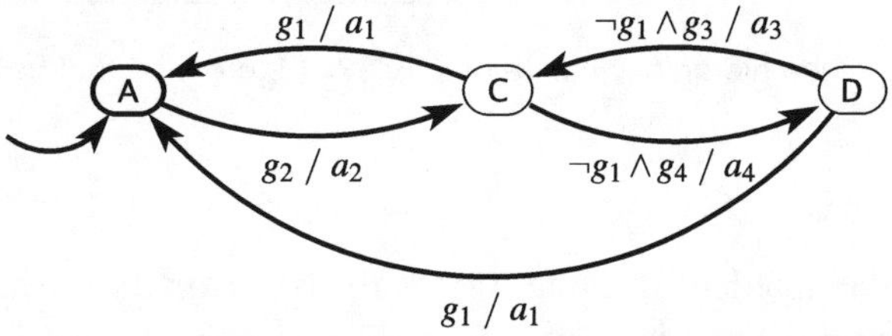

Figure 5.16: Semantics of Figure 5.15 with a preemptive transition.

B (which means that it is in either C or D), then the refinement machine reacts first. If it is C, and guard g_4 is true, the transition is taken to D and action a_4 is performed. But then, as part of the same reaction, the top-level FSM reacts. If guard g_1 is also true, then the machine transitions to state A. It is important that logically these two transitions are simultaneous and instantaneous, so the machine does not actually go to state D. Nonetheless, action a_4 is performed, and so is action a_1. This combination corresponds to the topmost transition of Figure 5.14. p.116

Another subtlety is that if two (non-absent) actions are performed in the same reaction, they may conflict. For example, two actions may write different values to the same output port. Or they may set the same variable to different values. Our choice is that the actions are performed in sequence, as suggested by the semicolon in the action a_4; a_1. As in an imperative language like C, the semicolon denotes a sequence. If the two actions conflict, the later one dominates. p.212

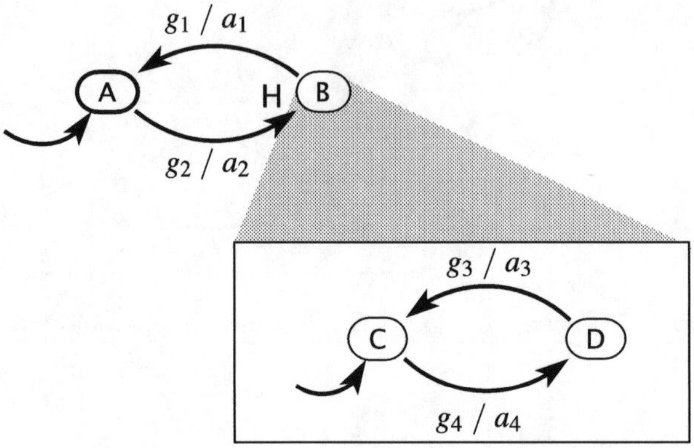

Figure 5.17: Variant of the hierarchical state machine of Figure 5.13 that has a history transition.

Such subtleties can be avoided by using a **preemptive transition**, shown in Figure 5.15, which has the semantics shown in Figure 5.16. The guards of a preemptive transition are evaluated *before* the refinement reacts, and if any guard evaluates to true, the refinement does not react. As a consequence, if the machine is in state B and g_1 is true, then neither action a_3 nor a_4 is performed. A preemptive transition is shown with a (red) circle at the originating end of the transition.

Notice in Figures 5.13 and 5.14 that whenever the machine enters B, it always enters C, never D, even if it was previously in D when leaving B. The transition from A to B is called a **reset transition** because the destination refinement is reset to its initial state, regardless of where it had previously been. A reset transition is indicated in our notation with a hollow arrowhead at the destination end of a transition.

In Figure 5.17, the transition from A to B is a **history transition**, an alternative to a reset transition. In our notation, a solid arrowhead denotes a history transition. It may also be marked with an "H" for emphasis. When a history transition is taken, the destination refinement resumes in whatever state it was last in (or its initial state on the first entry).

The semantics of the history transition is shown in Figure 5.18. The initial state is labeled (A, C) to indicate that the machine is in state A, and if and when it next enters B it will go to C. The first time it goes to B, it will be in the state labeled (B, C) to indicate that it is in state B and, more specifically, C. If it then transitions to (B, D), and then back to A, it will

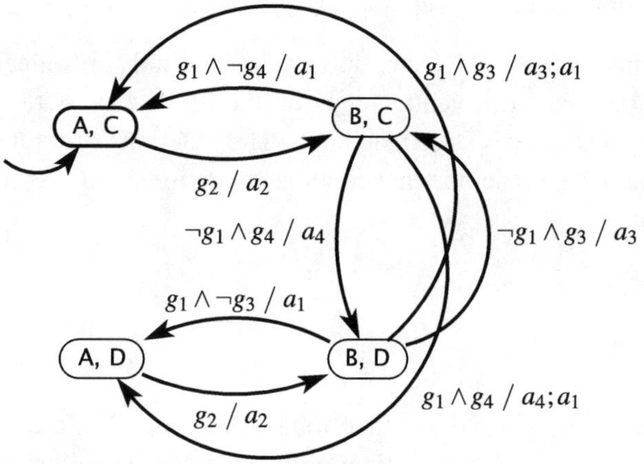

Figure 5.18: Semantics of the hierarchical state machine of Figure 5.17 that has a history transition.

end up in the state labeled (A, D), which means it is in state A, but if and when it next enters B it will go to D. That is, it remembers its history, specifically where it was when it left B.

As with concurrent composition, hierarchical state machines admit many possible meanings. The differences can be subtle. Considerable care is required to ensure that models are clear and that their semantics match what is being modeled.

5.3 Summary

Any well-engineered system is a composition of simpler components. In this chapter, we have considered two forms of composition of state machines, concurrent composition and hierarchical composition.

For concurrent composition, we introduced both synchronous and asynchronous composition, but did not complete the story. We have deferred dealing with feedback to the next chapter, because for synchronous composition, significant subtleties arise. For asynchronous composition, communication via ports requires additional mechanisms that are not (yet) part of our model of state machines. Even without communication via ports, significant subtleties arise because there are several possible semantics for asynchronous composition, and each has strengths and weaknesses. One choice of semantics may be suitable for one application and not for another. These subtleties motivate the topic of the next chapter, which provides more structure to concurrent composition and resolves most of these questions (in a variety of ways).

For hierarchical composition, we focus on a style originally introduced by Harel (1987) known as Statecharts. We specifically focus on the ability for states in an FSM to have refinements that are themselves state machines. The reactions of the refinement FSMs are composed with those of the machine that contains the refinements. As usual, there are many possible semantics.

Exercises

1. Consider the extended state machine model of Figure 3.8, the garage counter. Suppose that the garage has two distinct entrance and exit points. Construct a side-by-side concurrent composition of two counters that share a variable c that keeps track of the number of cars in the garage. Specify whether you are using synchronous or asynchronous composition, and define exactly the semantics of your composition by giving a single machine modeling the composition. If you choose synchronous semantics, explain what happens if the two machines simultaneously modify the shared variable. If you choose asynchronous composition, explain precisely which variant of asynchronous semantics you have chosen and why. Is your composition machine deterministic?

2. For semantics 2 in Section 5.1.2, give the five tuple for a single machine representing the composition C,

$$(States_C, Inputs_C, Outputs_C, update_C, initialState_C)$$

for the side-by-side asynchronous composition of two state machines A and B. Your answer should be in terms of the five-tuple definitions for A and B,

$$(States_A, Inputs_A, Outputs_A, update_A, initialState_A)$$

and

$$(States_B, Inputs_B, Outputs_B, update_B, initialState_B)$$

3. Consider the following synchronous composition of two state machines A and B:

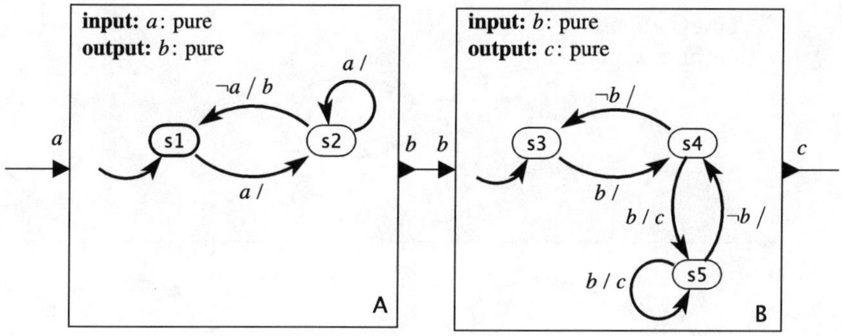

Construct a single state machine C representing the composition. Which states of the composition are unreachable?

4. Consider the following synchronous composition of two state machines A and B:

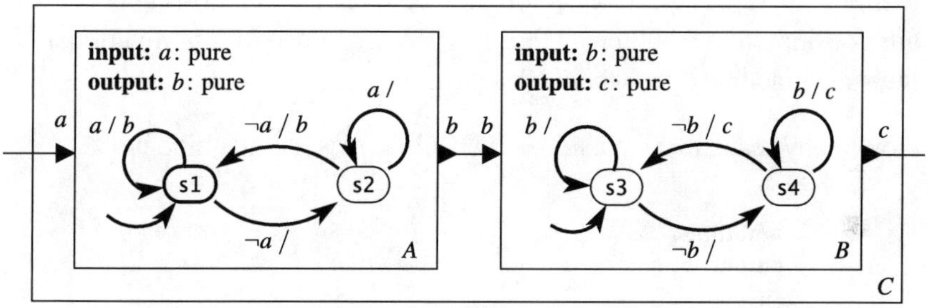

Construct a single state machine C representing the composition. Which states of the composition are unreachable?

5. Consider the following hierarchical state machine:

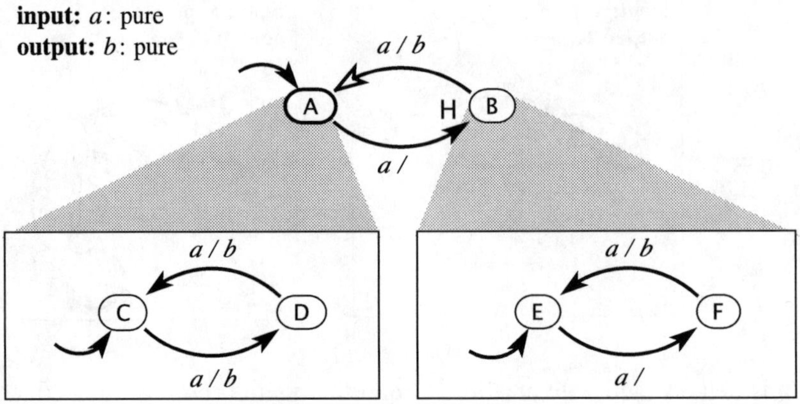

Construct an equivalent flat FSM giving the semantics of the hierarchy. Describe in words the input/output behavior of this machine. Is there a simpler machine that exhibits the same behavior? (Note that equivalence relations between state machines are considered in Chapter 14, but here, you can use intuition and just consider what the state machine does when it reacts.)

6. How many reachable states does the following state machine have?

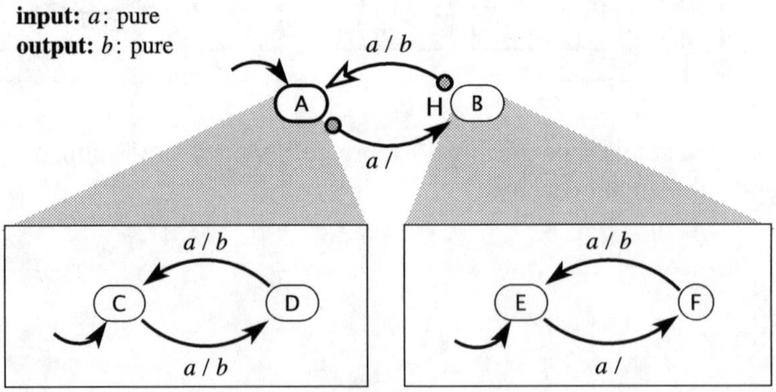

7. Suppose that the machine of Exercise 8 of Chapter 4 is composed in a synchronous side-by-side composition with the following machine:

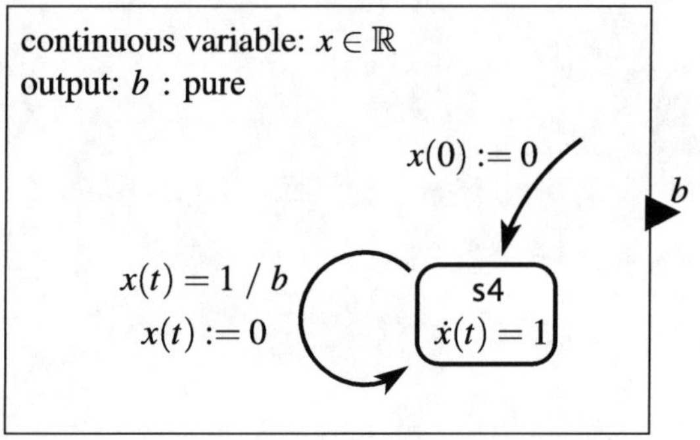

Find a tight lower bound on the time between events a and b. That is, find a lower bound on the time gap during which there are no events in signals a or b. Give an argument that your lower bound is tight.

6

Concurrent Models of Computation

Contents

In sound engineering practice, systems are built by composing components. In order for the composition to be well understood, we need first for the individual components to be well understood, and then for the meaning of the interaction between components to be well understood. The previous chapter dealt with composition of finite state machines. p.47 With such composition, the components are well defined (they are FSMs), but there are many possible interpretations for the interaction between components. The meaning of a p.104 composition is referred to as its semantics.

This chapter focuses on the semantics of **concurrent** composition. The word "concurrent" literally means "running together." A system is said to be concurrent if different parts of the system (components) conceptually operate at the same time. There is no particular order to their operations. The semantics of such concurrent operation can be quite subtle, however.

p.26 The components we consider in this chapter are actors, which react to stimuli at input ports and produce stimuli on output ports. In this chapter, we will be only minimally concerned with how the actors themselves are defined. They may be FSMs, hardware, or programs p.212 specified in an imperative programming language. We will need to impose some constraints on what these actors can do, but we need not constrain how they are specified.

The semantics of a concurrent composition of actors is governed by three sets of rules that we collectively call a **model of computation** (**MoC**). The first set of rules specifies what constitutes a component. The second set specifies the concurrency mechanisms. The third specifies the communication mechanisms.

In this chapter, a component will be an actor with ports and a set of **execution actions**. An execution action defines how the actor reacts to inputs to produce outputs and change state. The ports will be interconnected to provide for communication between actors, and the execution actions will be invoked by the environment of the actor to cause the actor to perform its function. For example, for FSMs, one action is provided that causes a reaction. The focus of this chapter is on introducing a few of the possible concurrency and communication mechanisms that can govern the interactions between such actors.

We begin by laying out the common structure of models that applies to all MoCs studied in this chapter. We then proceed to describe a suite of MoCs.

6.1 Structure of Models

In this chapter, we assume that models consist of fixed interconnections of actors like that shown in Figure 6.1(a). The interconnections between actors specify communication paths.

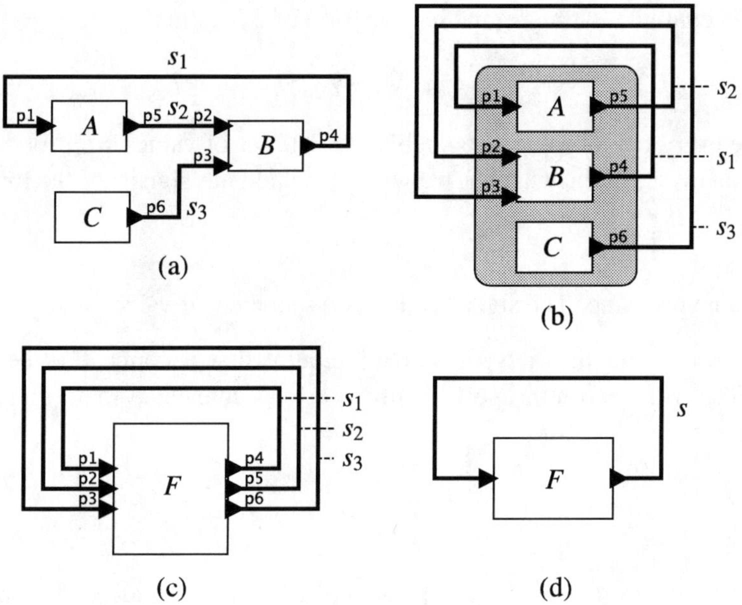

Figure 6.1: Any interconnection of actors can be modeled as a single (side-by-side composite) actor with feedback.

The communication itself takes the form of a **signal**, which consists of one or more **communication events**. For the discrete signals of Section 3.1, for example, a signal s has the form of a function p.44

$$s\colon \mathbb{R} \to V_s \cup \{absent\},$$

where V_s is a set of values called the type of the signal s. A communication event in this case is a non-absent value of s. p.45

Example 6.1: Consider a pure signal s that is a discrete signal given by p.43

$$s(t) = \begin{cases} present & \text{if } t \text{ is a multiple of } P \\ absent & \text{otherwise} \end{cases}$$

for all $t \in \mathbb{R}$ and some $P \in \mathbb{R}$. Such a signal is called a **clock signal** with period P. Communication events occur every P time units.

In Chapter 2, a continuous-time signal has the form of a function

$$s\colon \mathbb{R} \to V_s,$$

in which case every one of the (uncountably) infinite set of values $s(t)$, for all $t \in \mathbb{R}$, is a communication event. In this chapter, we will also encounter signals of the form

$$s\colon \mathbb{N} \to V_s,$$

where there is no time line. The signal is simply a sequence of values.

A communication event has a type, and we require that a connection between actors p.116 type check. That is, if an output port y with type V_y is connected to an input port x with type V_x, then

$$V_y \subseteq V_x.$$

As suggested in Figure 6.1(b-d), any actor network can be reduced to a rather simple form. If we rearrange the actors as shown in Figure 6.1(b), then the actors form a p.105 side-by-side composition indicated by the box with rounded corners. This box is itself an actor F as shown in Figure 6.1(c) whose input is a three-tuple (s_1, s_2, s_3) of signals and whose output is *the same* three-tuple of signals. If we let $s = (s_1, s_2, s_3)$, then the actor can be depicted as in Figure 6.1(d), which hides all the complexity of the model.

p.31 Notice that Figure 6.1(d) is a feedback system. By following the procedure that we used to build it, every interconnection of actors can be structured as a similar feedback system (see Exercise 1).

6.2 Synchronous-Reactive Models

In Chapter 5 we studied synchronous composition of state machines, but we avoided the nuances of feedback compositions. For a model described as the feedback system of Figure 6.1(d), the conundrum discussed in Section 5.1.5 takes a particularly simple form. If F in Figure 6.1(d) is realized by a state machine, then in order for it to react, we need to know its inputs at the time of the reaction. But its inputs are the same as its outputs, so in order for F to react, we need to know its outputs. But we cannot know its outputs until after it reacts.

As shown in Section 6.1 above and Exercise 1, all actor networks can be viewed as feedback systems, so we really do have to resolve the conundrum. We do that now by giving a model of computation known as the **synchronous-reactive** (**SR**) MoC.

Actor Networks as a System of Equations

In a model, if the actors are determinate, then each actor is a function that maps input signals to output signals. For example, in Figure 6.1(a), actor A may be a function relating signals s_1 and s_2 as follows, p.56

$$s_2 = A(s_1).$$

Similarly, actor B relates three signals by

$$s_1 = B(s_2, s_3).$$

Actor C is a bit more subtle, since it has no input ports. How can it be a function? What is the domain of the function? If the actor is determinate, then its output signal s_3 is a constant signal. The function C needs to be a constant function, one that yields the same output for every input. A simple way to ensure this is to define C so that its domain is a singleton set (a set with only one element). Let $\{\emptyset\}$ be the singleton set, so C can only be applied to $\emptyset$. The function C is then given by p.472 p.472

$$C(\emptyset) = s_3.$$

Hence, Figure 6.1(a) gives a system of equations

$$\begin{aligned} s_1 &= B(s_2, s_3) \\ s_2 &= A(s_1) \\ s_3 &= C(\emptyset). \end{aligned}$$

The semantics of such a model, therefore, is a solution to such a system of equations. This can be represented compactly using the function F in Figure 6.1(d), which is

$$F(s_1, s_2, s_3) = (B(s_2, s_3), A(s_1), C(\emptyset)).$$

All actors in Figure 6.1(a) have output ports; if we had an actor with no output port, then we could similarly define it as a function whose codomain is $\{\emptyset\}$. The output of such function is $\emptyset$ for all inputs. p.472

Fixed-Point Semantics

p.56 In a model, if the actors are determinate, then each actor is a function that maps input signals to output signals. The semantics of such a model is a system of equations (see sidebar on page 133) and the reduced form of Figure 6.1(d) becomes

$$s = F(s), \tag{6.1}$$

where $s = (s_1, s_2, s_3)$. Of course, this equation only *looks* simple. Its complexity lies in the definition of the function F and the structure of the domain and range of F.

Given any function $F : X \to X$ for any set X, if there is an $x \in X$ such that $F(x) = x$, then x is called a **fixed point**. Equation (6.1) therefore asserts that the semantics of a determinate actor network is a fixed point. Whether a fixed point exists, whether the fixed point is unique, and how to find the fixed point, all become interesting questions that are central to the model of computation.

p.132 In the SR model of computation, the execution of all actors is simultaneous and instantaneous and occurs at ticks of the global clock. If the actor is determinate, then each such execution implements a function called a **firing function**. For example, in the n-th tick of the global clock, actor A in Figure 6.1 implements a function of the form

$$a_n \colon V_1 \cup \{absent\} \to V_2 \cup \{absent\}$$

where V_i is the type of signal s_i. Hence, if $s_i(n)$ is the value of s_i at the n-th tick, then

$$s_2(n) = a_n(s_1(n)).$$

Given such a firing function f_n for each actor F we can, just as in Figure 6.1(d) define the execution at a single tick by a fixed point,

$$s(n) = f_n(s(n)),$$

where $s(n) = (s_1(n), s_2(n), s_3(n))$ and f_n is a function where

$$f_n(s_1(n), s_2(n), s_3(n)) = (b_n(s_2(n), s_3(n)), a_n(s_1(n)), c_n(\emptyset)).$$

Thus, for SR, the semantics at each tick of the global clock is a fixed point of the function f_n, just as its execution over all ticks is a fixed point of the function F.

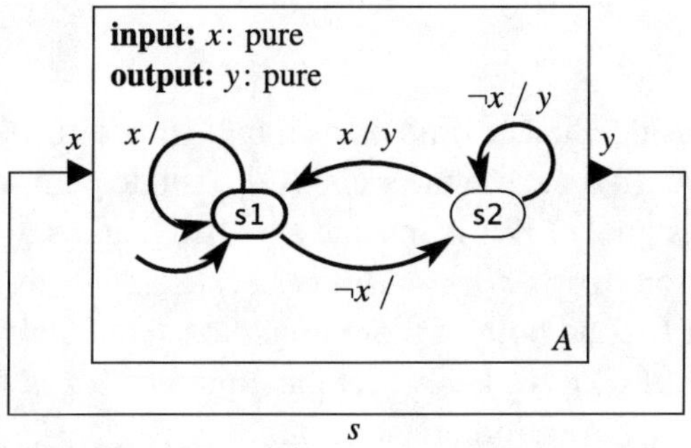

Figure 6.2: A simple well-formed feedback model.

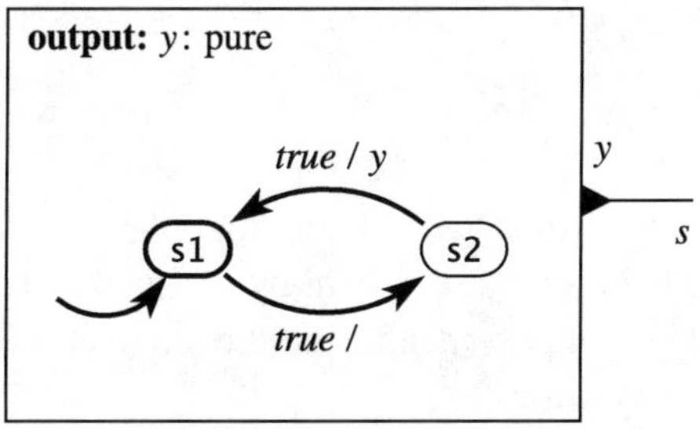

Figure 6.3: The semantics of the model in Figure 6.2.

An SR model is a discrete system where signals are absent at all times except (possibly) at **ticks** of a **global clock**. Conceptually, execution of a model is a sequence of global reactions that occur at discrete times, and at each such reaction, the reaction of all actors is simultaneous and instantaneous.

p.42
p.116

6.2.1 Feedback Models

We focus first on feedback models of the form of Figure 6.1(d), where F in the figure is realized as a state machine. At the n-th tick of the global clock, we have to find the value of the signal s so that it is both a valid input and a valid output of the state machine, given its current state. Let $s(n)$ denote the value of the signal s at the n-th reaction. The goal is to determine, at each tick of the global clock, the value of $s(n)$.

Example 6.2: Consider first a simpler example shown in Figure 6.2. (This is simpler than Figure 6.1(d) because the signal s is a single pure signal rather than an aggregation of three signals.) If A is in state `s1` when that reaction occurs, then the only possible value for $s(n)$ is $s(n) = absent$ because a reaction must take one of the transitions out of `s1`, and both of these transitions emit absent. Moreover, once we know that $s(n) = absent$, we know that the input port x has value *absent*, so we can determine that A will transition to state `s2`.

If A is in state `s2` when the reaction occurs, then the only possible value for $s(n)$ is $s(n) = present$, and the machine will transition to state `s1`. Therefore, s alternates between *absent* and *present*. The semantics of machine A in the feedback model is therefore given by Figure 6.3.

In the previous example, it is important to note that the input x and output y have the *same value* in every reaction. This is what is meant by the feedback connection. Any connection from an output port to an input port means that the value at the input port is the same as the value at the output port at all times.

p.56 Given a deterministic state machine in a feedback model like that of Figure 6.2, in each state i we can define a function a_i that maps input values to output values,

$$a_i\colon \{present, absent\} \to \{present, absent\},$$

where the function depends on the state the machine is in. This function is defined by the
p.54 update function.

Example 6.3: For the example in Figure 6.2, if the machine is in state `s1`, then $a_{\texttt{s1}}(x) = absent$ for all $x \in \{present, absent\}$.

The function a_i is called the firing function for state i (see box on page 134). Given a firing function, to find the value $s(n)$ at the n-th reaction, we simply need to find a value $s(n)$ such that

$$s(n) = a_i(s(n)).$$

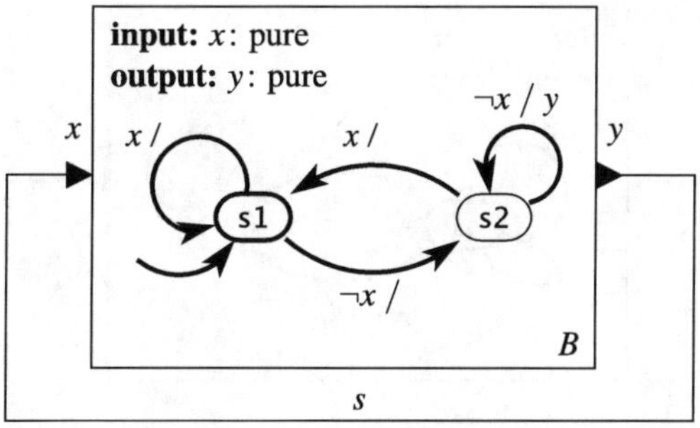

Figure 6.4: An ill-formed feedback model that has no fixed point in state s2.

Such a value $s(n)$ is called a **fixed point** of the function a_i. It is easy to see how to generalize this so that the signal s can have any type. Signal s can even be an aggregation of signals, as in Figure 6.1(d) (see box on page 134).

6.2.2 Well-Formed and Ill-Formed Models

There are two potential problems that may occur when seeking a fixed point. First, there may be no fixed point. Second, there may be more than one fixed point. If either case occurs in a reachable state, we call the system **ill formed**. Otherwise, it is **well formed**. p.61

Example 6.4: Consider machine B shown in Figure 6.4. In state s1, we get the unique fixed point $s(n) = absent$. In state s2, however, there is no fixed point. If we attempt to choose $s(n) = present$, then the machine will transition to s1 and its output will be *absent*. But the output has to be the same as the input, and the input is *present*, so we get a contradiction. A similar contradiction occurs if we attempt to choose $s(n) = absent$.

Since state s2 is reachable, this feedback model is ill formed.

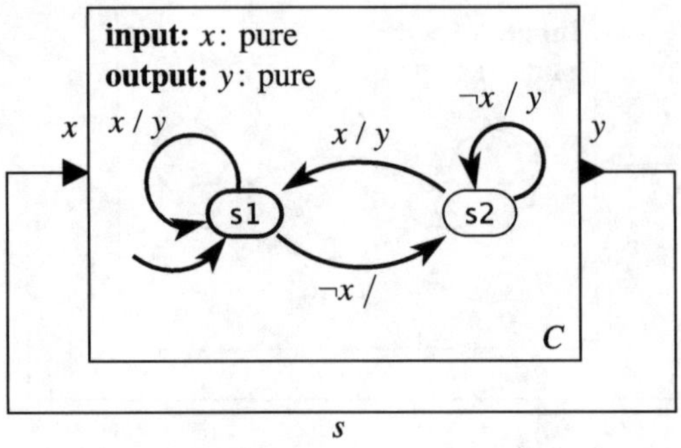

Figure 6.5: An ill-formed feedback model that has more than one fixed point in state s1.

Example 6.5: Consider machine C shown in Figure 6.5. In state s1, both $s(n) = absent$ and $s(n) = present$ are fixed points. Either choice is valid. Since state s1 is reachable, this feedback model is ill formed.

If in a reachable state there is more than one fixed point, we declare the machine to be ill formed. An alternative semantics would not reject such a model, but rather would declare it to be nondeterministic. This would be a valid semantics, but it would have the disadvantage that a composition of deterministic state machines is not assured of being deterministic. In fact, C in Figure 6.5 is deterministic, and under this alternative semantics, the feedback composition in the figure would not be deterministic. Determinism would not p.108 be a compositional property. Hence, we prefer to reject such models.

6.2.3 Constructing a Fixed Point

If the type V_s of the signal s or the signals it is an aggregate of is finite, then one way to find a fixed point is by **exhaustive search**, which means to try all values. If exactly one fixed point is found, then the model is well formed. However, exhaustive search is expensive (and impossible if the types are not finite). We can develop instead a systematic procedure that for most, but not all, well-formed models will find a fixed point. The procedure is as follows. For each reachable state i,

1. Start with $s(n)$ *unknown*.

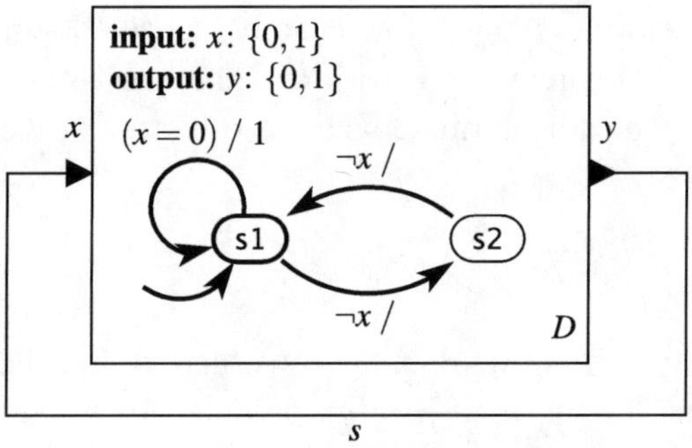

Figure 6.6: A well-formed feedback model that is not constructive.

2. Determine as much as you can about $f_i(s(n))$, where f_i is the firing function in state i. Note that in this step, you should use only the firing function, which is given by the state machine; you should not use knowledge of how the state machine is connected on the outside.

3. Repeat step 2 until all values in $s(n)$ become known (whether they are present and what their values are), or until no more progress can be made.

4. If unknown values remain, then reject the model.

This procedure may reject models that have a unique fixed point, as illustrated by the following example.

Example 6.6: Consider machine D shown in Figure 6.6. In state `s1`, if the input is unknown, we cannot immediately tell what the output will be. We have to try all the possible values for the input to determine that in fact $s(n) = absent$ for all n.

A state machine for which the procedure works in all reachable states is said to be **constructive** (Berry, 1999). The example in Figure 6.6 is not constructive. For non-constructive machines, we are forced to do exhaustive search or to invent some more elaborate solution technique. Since exhaustive search is often too expensive for practical use, many SR languages and modeling tools (see box on page 141) reject non-constructive models.

Step 2 of the above procedure is key. How exactly can we determine the outputs if the inputs are not all known? This requires what is called a **must-may analysis** of the model. Examining the machine, we can determine what *must* be true of the outputs and what *may* be true of the outputs.

Example 6.7: The model in Figure 6.2 is constructive. In state s1, we can immediately determine that the machine *may not* produce an output. Therefore, we can immediately conclude that the output is *absent*, even though the input is unknown. Of course, once we have determined that the output is absent, we now know that the input is absent, and hence the procedure concludes.

In state s2, we can immediately determine that the machine *must* produce an output, so we can immediately conclude that the output is *present*.

The above procedure can be generalized to an arbitrary model structure. Consider for example Figure 6.1(a). There is no real need to convert it to the form of Figure 6.1(d). Instead, we can just begin by labeling all signals unknown, and then in arbitrary order, examine each actor to determine whatever can be determined about the outputs, given its initial state. We repeat this until no further progress can be made, at which point either all signals become known, or we reject the model as either ill-formed or non-constructive. Once we know all signals, then all actors can make state transitions, and we repeat the procedure in the new state for the next reaction.

The constructive procedure above can be adapted to support nondeterministic machines (see Exercise 4). But now, things become even more subtle, and there are variants to the semantics. One way to handle nondeterminism is that when executing the constructive procedure, when encountering a nondeterministic choice, make an arbitrary choice. If the result leads to a failure of the procedure to find a fixed point, then we could either reject the model (not all choices lead to a well-formed or constructive model) or reject the choice and try again.

In the SR model of computation, actors react simultaneously and instantaneously, at least conceptually. Achieving this with realistic computation requires tight coordination of the computation. We consider next a family of models of computation that require less coordination.

Synchronous-Reactive Languages

The synchronous-reactive MoC has a history dating at least back to the mid 1980s when a suite of programming languages were developed. The term "reactive" comes from a distinction in computational systems between **transformational systems**, which accept input data, perform computation, and produce output data, and **reactive systems**, which engage in an ongoing dialog with their environment (Harel and Pnueli, 1985). Manna and Pnueli (1992) state

> "The role of a reactive program ... is not to produce a final result but to maintain some ongoing interaction with its environment."

The distinctions between transformational and reactive systems led to the development of a number of innovative programming languages. The **synchronous languages** (Benveniste and Berry, 1991) take a particular approach to the design of reactive systems, in which pieces of the program react simultaneously and instantaneously at each tick of a global clock. First among these languages are Lustre (Halbwachs et al., 1991), Esterel (Berry and Gonthier, 1992), and Signal (Le Guernic et al., 1991). Statecharts (Harel, 1987) and its implementation in Statemate (Harel et al., 1990) also have a strongly synchronous flavor.

SCADE (Berry, 2003) (Safety Critical Application Development Environment), a commercial product of Esterel Technologies, builds on Lustre, borrows concepts from Esterel, and provides a graphical syntax, where state machines are drawn and actor models are composed in a similar manner to the figures in this text. One of the main attractions of synchronous languages is their strong formal properties that yield quite effectively to formal analysis and verification techniques. For this reason, SCADE models are used in the design of safety-critical flight control software systems for commercial aircraft made by Airbus.

The principles of synchronous languages can also be used in the style of a **coordination language** rather than a programming language, as done in Ptolemy II (Edwards and Lee, 2003) and ForSyDe (Sander and Jantsch, 2004). This allows for "primitives" in a system to be complex components rather than built-in language primitives. This approach allows heterogeneous combinations of MoCs, since the complex components may themselves be given as compositions of further subcomponents under some other MoC.

6.3 Dataflow Models of Computation

In this section, we consider MoCs that are much more asynchronous than SR. Reactions may occur simultaneously, or they may not. Whether they do or do not is not an essential part of the semantics. The decision as to when a reaction occurs can be much more decentralized, and can in fact reside with each individual actor. When reactions are dependent on one another, the dependence is due to the flow of data, rather than to the synchrony of events. If a reaction of actor A requires data produced by a reaction of actor B, then the reaction of A must occur after the reaction of B. A MoC where such data dependencies are the key constraints on reactions is called a **dataflow** model of computation. There are several variants of dataflow MoCs, a few of which we consider here.

6.3.1 Dataflow Principles

In dataflow models, the signals providing communication between the actors are *sequences* p.472 of message, where each message is called a **token**. That is, a signal s is a partial function of the form

$$s\colon \mathbb{N} \rightharpoonup V_s,$$

p.45 where V_s is the type of the signal, and where the signal is defined on an **initial segment** $\{0, 1, \cdots, n\} \subset \mathbb{N}$, or (for infinite executions) on the entire set $\mathbb{N}$. Each element $s(n)$ of this sequence is a token. A (determinate) actor will be described as a function that maps input sequences to output sequences. We will actually use two functions, an **actor function**, which maps *entire* input sequences to *entire* output sequences, and a **firing function**, which maps a finite portion of the input sequences to output sequences, as illustrated in the following example.

Example 6.8: Consider an actor that has one input and one output port as shown below

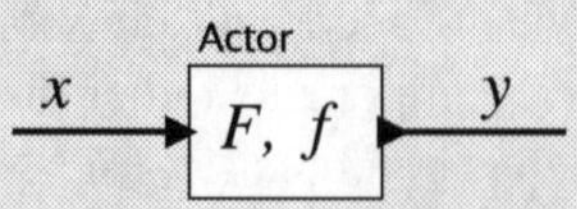

Suppose that the input type is $V_x = \mathbb{R}$. Suppose that this is a Scale actor parameterized by a parameter $a \in \mathbb{R}$, similar to the one in Example 2.3, which multiplies

inputs by a. Then

$$F(x_1, x_2, x_3, \cdots) = (ax_1, ax_2, ax_3, \cdots).$$

Suppose that when the actor fires, it performs one multiplication in the firing. Then the firing function f operates only on the first element of the input sequence, so

$$f(x_1, x_2, x_3, \cdots) = f(x_1) = (ax_1).$$

The output is a sequence of length one.

As illustrated in the previous example, the actor function F combines the effects of multiple invocations of the firing function f. Moreover, the firing function can be invoked with only partial information about the input sequence to the actor. In the above example, the firing function can be invoked if one or more tokens are available on the input. The rule requiring one token is called a **firing rule** for the Scale actor. A firing rule specifies the number of tokens required on each input port in order to fire the actor.

The Scale actor in the above example is particularly simple because the firing rule and the firing function never vary. Not all actors are so simple.

Example 6.9: Consider now a different actor Delay with parameter $d \in \mathbb{R}$. The actor function is

$$D(x_1, x_2, x_3, \cdots) = (d, x_1, x_2, x_3, \cdots).$$

This actor prepends a sequence with a token with value d. This actor has two firing functions, d_1 and d_2, and two firing rules. The first firing rule requires no input tokens at all and produces an output sequence of length one, so

$$d_1(s) = (d),$$

where s is a sequence of any length, including length zero (the empty sequence). This firing rule is initially the one used, and it is used exactly once. The second firing rule requires one input token and is used for all subsequent firings. It triggers the firing function

$$d_2(x_1, \cdots) = (x_1).$$

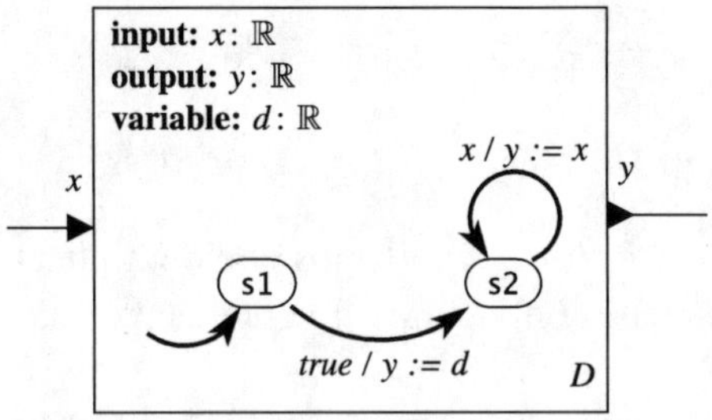

Figure 6.7: An FSM model for the Delay actor in Example 6.9.

The actor consumes one input token and produces on its output the same token. The actor can be modeled by a state machine, as shown in Figure 6.7. In that figure, the firing rules are implicit in the guards. The tokens required to fire are exactly those required to evaluate the guards. The firing function d_1 is associated with state s1, and d_2 with s2.

When dataflow actors are composed, with an output of one going to an input of another, the communication mechanism is quite different from that of the previous MoCs considered in this chapter. Since the firing of the actors is asynchronous, a token sent from one actor to another must be buffered; it needs to be saved until the destination actor is ready to consume it. When the destination actor fires, it **consumes** one or more input tokens. After being consumed, a token may be discarded (meaning that the memory in which it is buffered can be reused for other purposes).

Dataflow models pose a few interesting problems. One question is how to ensure that the memory devoted to buffering of tokens is bounded. A dataflow model may be able to execute forever (or for a very long time); this is called an **unbounded execution**. For an unbounded execution, we may have to take measures to ensure that buffering of unconsumed tokens does not overflow the available memory.

p.26 **Example 6.10:** Consider the following cascade composition of dataflow actors:

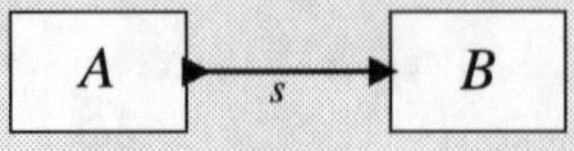

Since *A* has no input ports, its firing rule is simple. It can fire at any time. Suppose that on each firing, *A* produces one token. What is to keep *A* from firing at a faster rate than *B*? Such faster firing could result in an unbounded build up of unconsumed tokens on the buffer between *A* and *B*. This will eventually exhaust available memory.

In general, for dataflow models that are capable of unbounded execution, we will need scheduling policies that deliver **bounded buffers**.

A second problem that may arise is **deadlock**. Deadlock occurs when there are cycles, as in Figure 6.1, and a directed loop has insufficient tokens to satisfy any of the firing rules of the actors in the loop. The Delay actor of Example 6.9 can help prevent deadlock because it is able to produce an initial output token without having any input tokens available. Dataflow models with feedback will generally require Delay actors (or something similar) in every cycle. p.31

For general dataflow models, it can be difficult to tell whether the model will deadlock, and whether there exists an unbounded execution with bounded buffers. In fact, these two questions are undecidable, meaning that there is no algorithm that can answer the question in bounded time for all dataflow models (Buck, 1993). Fortunately, there are useful constraints that we can impose on the design of actors that make these questions decidable. We examine those constraints next. p.489

6.3.2 Synchronous Dataflow

Synchronous dataflow (**SDF**) is a constrained form of dataflow where for each actor, every firing consumes a fixed number of input tokens on each input port and produces a fixed number of output tokens on each output port (Lee and Messerschmitt, 1987).[1]

Consider a single connection between two actors, *A* and *B*, as shown in Figure 6.8. The notation here means that when *A* fires, it produces *M* tokens on its output port, and when *B*

[1] Despite the term, synchronous dataflow is not synchronous in the sense of SR. There is no global clock in SDF models, and firings of actors are asynchronous. For this reason, some authors use the term **static dataflow** rather than synchronous dataflow. This does not avoid all confusion, however, because Dennis (1974) had previously coined the term "static dataflow" to refer to dataflow graphs where buffers could hold at most one token. Since there is no way to avoid a collision of terminology, we stick with the original "synchronous dataflow" terminology used in the literature. The term SDF arose from a signal processing concept, where two signals with sample rates that are related by a rational multiple are deemed to be synchronous. p.208

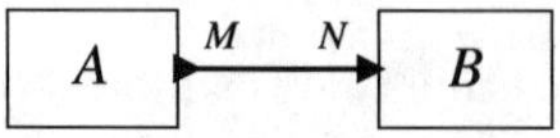

Figure 6.8: SDF actor A produces M tokens when it fires, and actor B consumes N tokens when it fires.

fires, it consumes N tokens on its input port. M and N are positive integers. Suppose that A fires q_A times and B fires q_B times. All tokens that A produces are consumed by B if and only if the following **balance equation** is satisfied,

$$q_A M = q_B N. \tag{6.2}$$

Given values q_A and q_B satisfying (6.2), we can find a schedule that delivers unbounded execution with bounded buffers. An example of such a schedule fires A repeatedly, q_A times, followed by B repeatedly, q_B times. It can repeat this sequence forever without exhausting available memory.

Example 6.11: Suppose that in Figure 6.8, $M = 2$ and $N = 3$. Then $q_A = 3$ and $q_B = 2$ satisfy (6.2). Hence, the following schedule can be repeated forever,

$$A, A, A, B, B.$$

An alternative schedule is also available,

$$A, A, B, A, B.$$

In fact, this latter schedule has an advantage over the former one in that it requires less memory. B fires as soon as there are enough tokens, rather than waiting for A to complete its entire cycle.

Another solution to (6.2) is $q_A = 6$ and $q_B = 4$. This solution includes more firings in the schedule than are strictly needed to keep the system in balance.

The equation is also satisfied by $q_A = 0$ and $q_B = 0$, but if the number of firings of actors is zero, then no useful work is done. Clearly, this is not a solution we want. Negative solutions are also not desirable.

Generally we will be interested in finding the least positive integer solution to the balance equations.

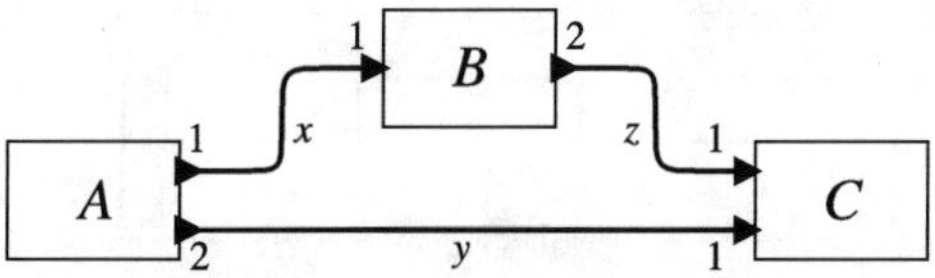

Figure 6.9: A consistent SDF model.

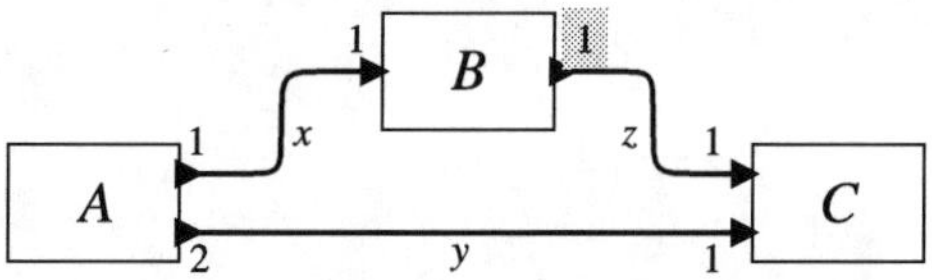

Figure 6.10: An inconsistent SDF model.

In a more complicated SDF model, every connection between actors results in a balance equation. Hence, the model defines a system of equations.

Example 6.12: Figure 6.9 shows a network with three SDF actors. The connections x, y, and z, result in the following system of balance equations,

$$\begin{aligned} q_A &= q_B \\ 2q_B &= q_C \\ 2q_A &= q_C. \end{aligned}$$

The least positive integer solution to these equations is $q_A = q_B = 1$, and $q_C = 2$, so the following schedule can be repeated forever to get an unbounded execution with bounded buffers,

$$A, B, C, C.$$

The balance equations do not always have a non-trivial solution, as illustrated in the following example.

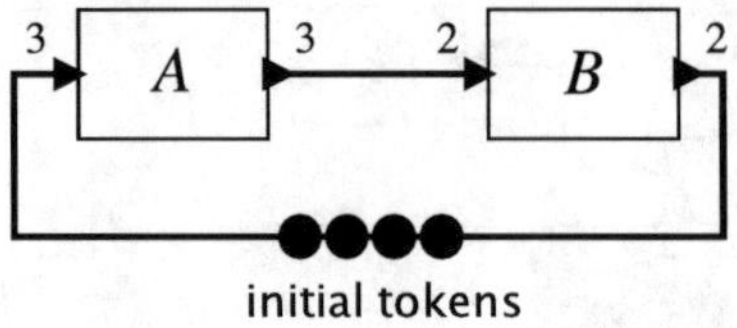

Figure 6.11: An SDF model with initial tokens on a feedback loop.

Example 6.13: Figure 6.10 shows a network with three SDF actors where the only solution to the balance equations is the trivial one, $q_A = q_B = q_C = 0$. A consequence is that there is no unbounded execution with bounded buffers for this model. It cannot be kept in balance.

An SDF model that has a non-zero solution to the balance equations is said to be **consistent**. If the only solution is zero, then it is **inconsistent**. An inconsistent model has no unbounded execution with bounded buffers.

Lee and Messerschmitt (1987) showed that if the balance equations have a non-zero solution, then they also have a solution where q_i is a positive integer for all actors i. Moreover, for connected models (where there is a communication path between any two actors), they gave a procedure for finding the least positive integer solution. Such a procedure forms the foundation for a scheduler for SDF models.

Consistency is sufficient to ensure bounded buffers, but it is not sufficient to ensure that an unbounded execution exists. In particular, when there is feedback, as in Figure 6.1, then p.294 deadlock may occur. Deadlock bounds an execution.

To allow for feedback, the SDF model treats Delay actors specially. Recall from Example 6.9, that the Delay actor is able to produce output tokens before it receives any input tokens, and then it subsequently behaves like a trivial SDF actor that copies inputs to outputs. But such a trivial actor is not really needed, and the cost of copying inputs to outputs is unnecessary. The Delay actor can be implemented very efficiently as a connection with initial tokens (those tokens that the actor is able to produce before receiving inputs). No actor is actually needed at run time. The scheduler must take the initial tokens into account.

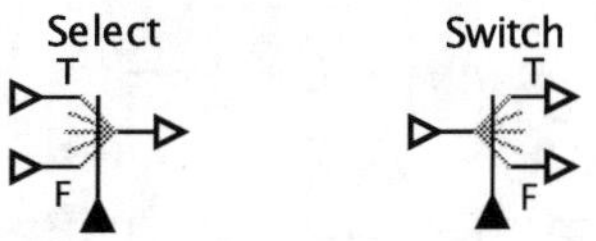

Figure 6.12: Dynamic dataflow actors.

Example 6.14: Figure 6.11 shows an SDF model with initial tokens on a feedback loop. These replace a Delay actor that is able to initially produce four tokens. The balance equations are

$$\begin{aligned} 3q_A &= 2q_B \\ 2q_B &= 3q_A. \end{aligned}$$

The least positive integer solution is $q_A = 2$, and $q_B = 3$, so the model is consistent. With four initial tokens on the feedback connection, as shown, the following schedule can be repeated forever,

$$A, B, A, B, B.$$

Were there any fewer than four initial tokens, however, the model would deadlock. If there were only three tokens, for example, then A could fire, followed by B, but in the resulting state of the buffers, neither could fire again.

In addition to the procedure for solving the balance equations, Lee and Messerschmitt (1987) gave a procedure that will either provide a schedule for an unbounded execution or will prove that no such schedule exists. Hence, both bounded buffers and deadlock are decidable for SDF models. p.489

6.3.3 Dynamic Dataflow

Although the ability to guarantee bounded buffers and rule out deadlock is valuable, it comes at a price. SDF is not very expressive. It cannot directly express, for example, conditional firing, where an actor fires only if, for example, a token has a particular value.

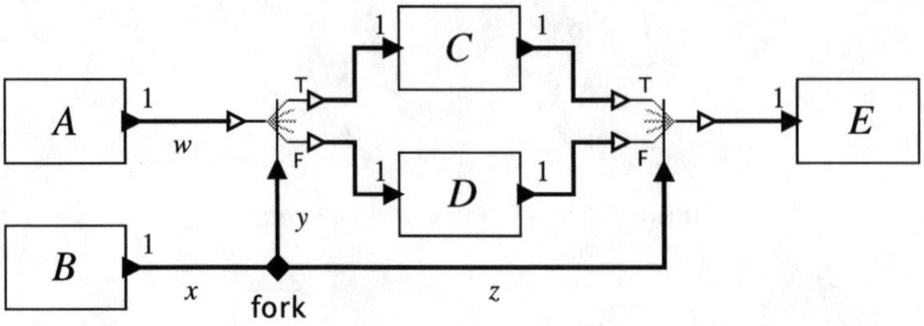

Figure 6.13: A DDF model that accomplishes conditional firing.

Such conditional firing is supported by a more general dataflow MoC known as **dynamic dataflow** (**DDF**). Unlike SDF actors, DDF actors can have multiple firing rules, and they are not constrained to produce the same number of output tokens on each firing. The Delay actor of Example 6.9 is directly supported by the DDF MoC, without any need for special treatment of initial tokens. So are two basic actors known as Switch and Select, shown in Figure 6.12.

The Select actor on the left has three firing rules. Initially, it requires one token on the bottom input port. The type of that port is Boolean, so the value of the token must be *true* or *false*. If a token with value *true* is received on that input port, then the actor produces no output, but instead activates the next firing rule, which requires one token for the top left input port, labeled T. When the actor next fires, it consumes the token on the T port and sends it to the output port. If a token with value *false* is received on the bottom input port, then the actor activates a firing rule that requires a token on the bottom left input port labeled F. When it consumes that token, it again sends it to the output port. Thus, it fires twice to produce one output.

The Switch actor performs a complementary function. It has only one firing rule, which requires a single token on both input ports. The token on the left input port will be sent to either the T or the F output port, depending on the Boolean value of the token received on the bottom input port. Hence, Switch and Select accomplish conditional routing of tokens, as illustrated in the following example.

Example 6.15: Figure 6.13 uses Switch and Select to accomplish conditional firing. Actor B produces a stream of Boolean-valued tokens x. This stream is replicated by the fork to provide the control inputs y and z to the Switch and Select

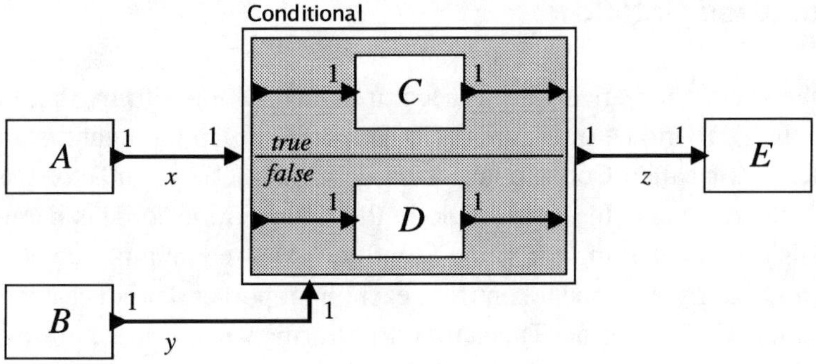

Figure 6.14: Structured dataflow approach to conditional firing.

actors. Based on the value of the control tokens on these streams, the tokens produced by actor A are sent to either C or D, and the resulting outputs are collected and sent to E. This model is the DDF equivalent of the familiar `if-then-else` programming construct in imperative languages. p.212

Addition of Switch and Select to the actor library means that we can no longer always find a bounded buffer schedule, nor can we provide assurances that the model will not deadlock. Buck (1993) showed that bounded buffers and deadlock are undecidable for DDF models. p.489 Thus, in exchange for the increased expressiveness and flexibility, we have paid a price. The models are not as readily analyzed.

Switch and Select are dataflow analogs of the **goto** statement in imperative languages. p.212 They provide low-level control over execution by conditionally routing tokens. Like goto statements, using them can result in models that are very difficult to understand. Dijkstra (1968) indicted the goto statement, discouraging its use, advocating instead the use of **structured programming**. Structured programming replaces goto statements with nested `for` loops, `if-then-else`, `do-while`, and recursion. Fortunately, structured programming is also available for dataflow models, as we discuss next.

6.3.4 Structured Dataflow

Figure 6.14 shows an alternative way to accomplish conditional firing that has many advantages over the DDF model in Figure 6.13. The grey box in the figure is an example of a **higher-order actor** called Conditional. A higher-order actor is an actor that has one or more models as parameters. In the example in the figure, Conditional is parameterized by two sub-models, one containing the actor C and the other containing the actor D. When Conditional fires, it consumes one token from each input port and produces one token on its output port, so it is an SDF actor. The action it performs when it fires, however, is dependent on the value of the token that arrives at the lower input port. If that value is true, then actor C fires. Otherwise, actor D fires.

This style of conditional firing is called **structured dataflow**, because, much like structured programming, control constructs are nested hierarchically. Arbitrary data-dependent token routing is avoided (which is analogous to avoiding arbitrary branches using goto instructions). Moreover, when using such Conditional actors, the overall model is still an SDF model. In the example in Figure 6.14, every actor consumes and produces exactly one token on every port. Hence, the model is analyzable for deadlock and bounded buffers.

This style of structured dataflow was introduced in LabVIEW, a design tool developed by National Instruments (Kodosky et al., 1991). In addition to a conditional similar to that in Figure 6.14, LabVIEW provides structured dataflow constructs for iterations (analogous to `for` and `do-while` loops in an imperative language), for `case` statements (which have an arbitrary number of conditionally executed submodels), and for sequences (which cycle through a finite set of submodels). It is also possible to support recursion using structured dataflow (Lee and Parks, 1995), but without careful constraints, boundedness again becomes undecidable.

6.3.5 Process Networks

A model of computation that is closely related to dataflow models is **Kahn process networks** (or simply, **process networks** or **PN**), named after Gilles Kahn, who introduced them (Kahn, 1974). The relationship between dataflow and PN is studied in detail by Lee and Parks (1995) and Lee and Matsikoudis (2009), but the short story is quite simple. In p.298 PN, each actor executes concurrently in its own process. That is, instead of being defined by its firing rules and firing functions, a PN actor is defined by a (typically non-terminating) program that reads data tokens from input ports and writes data tokens to output ports. All actors execute simultaneously (conceptually, whether they actually execute simultaneously or are interleaved is irrelevant).

In the original paper, Kahn (1974) gave very elegant mathematical conditions on the actors that would ensure that a network of such actors was determinate in the sense that the sequence of tokens on every connection between actors is unique, and specifically independent of how the processes are scheduled. Thus, Kahn showed that concurrent execution was possible without nondeterminacy.

Three years later, Kahn and MacQueen (1977) gave a simple, easily implemented mechanism for programs that ensures that the mathematical conditions are met to ensure determinacy. A key part of the mechanism is to perform **blocking reads** on input ports whenever a process is to read input data. Specifically, blocking reads mean that if the process chooses to access data through an input port, it issues a read request and blocks until the data becomes available. It cannot test the input port for the availability of data and then perform a conditional branch based on whether data are available, because such a branch would introduce schedule-dependent behavior.

Blocking reads are closely related to firing rules. Firing rules specify the tokens required to continue computing (with a new firing function). Similarly, a blocking read specifies a single token required to continue computing (by continuing execution of the process).

When a process writes to an output port, it performs a **nonblocking write**, meaning that the write succeeds immediately and returns. The process does not block to wait for the receiving process to be ready to receive data. This is exactly how writes to output ports work in dataflow MoCs as well. Thus, the only material difference between dataflow and PN is that with PN, the actor is not broken down into firing functions. It is designed as a continuously executing program.

Kahn and MacQueen (1977) called the processes in a PN network **coroutines** for an interesting reason. A **routine** or **subroutine** is a program fragment that is "called" by another program. The subroutine executes to completion before the calling fragment can continue executing. The interactions between processes in a PN model are more symmetric, in that there is no caller and callee. When a process performs a blocking read, it is in a sense invoking a routine in the upstream process that provides the data. Similarly, when it performs a write, it is in a sense invoking a routine in the downstream process to process the data. But the relationship between the producer and consumer of the data is much more symmetric than with subroutines.

Just like dataflow, the PN MoC poses challenging questions about boundedness of buffers and about deadlock. PN is expressive enough that these questions are undecidable. An p.489
elegant solution to the boundedness question is given by Parks (1995) and elaborated by Geilen and Basten (2003).

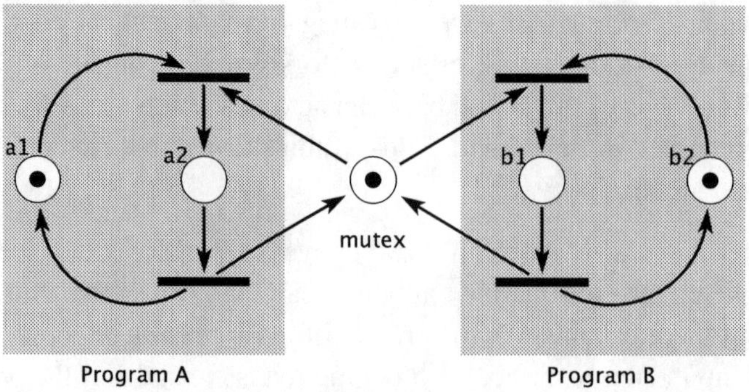

Figure 6.15: A Petri net model of two concurrent programs with a mutual exclusion protocol.

An interesting variant of process networks performs **blocking writes** rather than nonblocking writes. That is, when a process writes to an output port, it blocks until the receiving process is ready to receive the data. Such an interaction between processes is called a **rendezvous**. Rendezvous forms the basis for well known process formalisms such as **communicating sequential processes** (**CSP**) (Hoare, 1978) and the **calculus of communicating systems** (**CCS**) (Milner, 1980). It also forms the foundation for the **Occam** programming language (Galletly, 1996), which enjoyed some success for a period of time in the 1980s and 1990s for programming parallel computers.

In both the SR and dataflow models of computation considered so far, time plays a minor role. In dataflow, time plays no role. In SR, computation occurs simultaneously and instantaneously at each of a sequence of ticks of a global clock. Although the term "clock" implies that time plays a role, it actually does not. In the SR MoC, all that matters is the sequence. The physical time at which the ticks occur is irrelevant to the MoC. It is just a *sequence* of ticks. Many modeling tasks, however, require a more explicit notion of time. We examine next MoCs that have such a notion.

6.4 Timed Models of Computation

p.1 For cyber-physical systems, the time at which things occur in software can matter, because the software interacts with physical processes. In this section, we consider a few concurrent MoCs that explicitly refer to time. We describe three timed MoCs, each of which have many variants. Our treatment here is necessarily brief. A complete study of these MoCs would require a much bigger volume.

Petri Nets

Petri nets, named after Carl Adam Petri, are a popular modeling formalism related to dataflow (Murata, 1989). They have two types of elements, **places** and **transitions**, depicted as white circles and rectangles, respectively: p.142

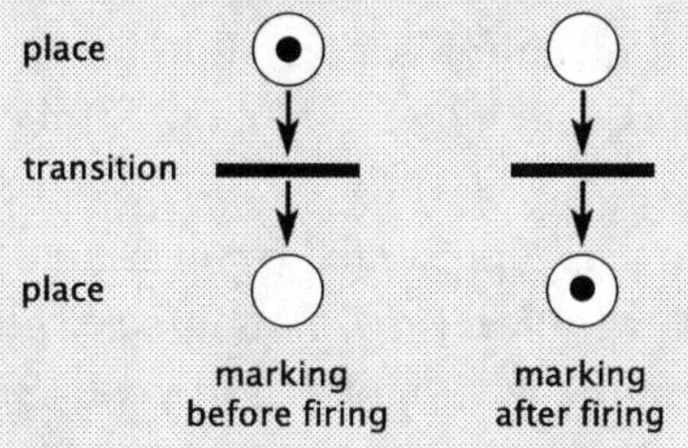

A place can contain any number of tokens, depicted as black circles. A transition is **enabled** if all places connected to it as inputs contain at least one token. Once a transition is enabled, it can **fire**, consuming one token from each input place and putting one token on each output place. The state of a network, called its **marking**, is the number of tokens on each place in the network. The figure above shows a simple network with its marking before and after the firing of the transition. If a place provides input to more than one transition, then the network is nondeterministic. A token on that place may trigger a firing of either destination transition.

An example of a Petri net model is shown in Figure 6.15, which models two concurrent programs with a mutual exclusion protocol. Each of the two programs has a critical section, meaning that only one of the programs can be in its critical section at any time. In the model, program A is in its critical section if there is a token on place a2, and program B is in its critical section if there is a token on place b1. The job of the mutual exclusion protocol is to ensure that these two places cannot simultaneously have a token. p.291

If the initial marking of the model is as shown in the figure, then both top transitions are enabled, but only one can fire (there is only one token in the place labeled mutex). Which one fires is chosen nondeterministically. Suppose program A fires. After this firing, there will be a token in place a2, so the corresponding bottom transition becomes enabled. Once that transition fires, the model returns to its initial marking. It is easy to see that the mutual exclusion protocol is correct in this model.

Unlike dataflow buffers, places do not preserve an ordering of tokens. Petri nets with a finite number of markings are equivalent to FSMs. p.47

Models of Time

How to model physical time is surprisingly subtle. How should we define simultaneity across a distributed system? A thoughtful discussion of this question is considered by Galison (2003). What does it mean for one event to cause another? Can an event that causes another be simultaneous with it? Several thoughtful essays on this topic are given in Price and Corry (2007).

In Chapter 2, we assume time is represented by a variable $t \in \mathbb{R}$ or $t \in \mathbb{R}_+$. This model is sometimes referred to as **Newtonian time**. It assumes a globally shared absolute time, where any reference anywhere to the variable t will yield the same value. This notion of time is often useful for modeling even if it does not perfectly reflect physical realities, but it has its deficiencies. Consider for example Newton's cradle, a toy with five steel balls suspended by strings. If you lift one ball and release it, it strikes the second ball, which does not move. Instead, the fifth ball reacts by rising. Consider the momentum of the middle ball as a function of time. The middle ball does not move, so its momentum must be everywhere zero. But the momentum of the first ball is somehow transfered to the fifth ball, passing through the middle ball. So the momentum cannot be always zero. Let $m\colon \mathbb{R} \to \mathbb{R}$ represent the momentum of this ball and τ be the time of the collision. Then

$$m(t) = \begin{cases} M & \text{if } t = \tau \\ 0 & \text{otherwise} \end{cases}$$

for all $t \in \mathbb{R}$. In a cyber-physical system, we may, however, want to represent this function in software, in which case a sequence of samples will be needed. But how can such sample unambiguously represent the rather unusual structure of this signal?

One option is to use **superdense time** (Manna and Pnueli, 1993; Maler et al., 1992; Lee and Zheng, 2005; Cataldo et al., 2006), where instead of $\mathbb{R}$, time is represented by a set $\mathbb{R} \times \mathbb{N}$. A time value is a tuple (t,n), where t represents Newtonian time and n represents a sequence index within an instant. In this representation, the momentum of the middle ball can be unambiguously represented by a sequence where $m(\tau,0) = 0$, $m(\tau,1) = M$, and $m(\tau,2) = 0$. Such a representation also handles events that are
p.116 simultaneous and instantaneous but also causally related.

Another alternative is **partially ordered time**, where two time values may or may not be ordered relative to each other. When there is a chain of causal relationships between them, then they must be ordered, and otherwise not.

6.4.1 Time-Triggered Models

Kopetz and Grunsteidl (1994) introduced mechanisms for periodically triggering distributed computations according to a distributed clock that measures the passage of time. The result is a system architecture called a **time-triggered architecture** (**TTA**). A key contribution was to show how a TTA could tolerate certain kinds of faults, such that failures in part of the system could not disrupt the behaviors in other parts of the system (see also Kopetz (1997) and Kopetz and Bauer (2003)). Henzinger et al. (2003) lifted the key idea of TTA to the programming language level, providing a well-defined semantics for modeling distributed time-triggered systems. Since then, these techniques have come into practical use in the design of safety-critical avionics and automotive systems, becoming a key part of standards such as FlexRay, a networking standard developed by a consortium of automotive companies.

A time-triggered MoC is similar to SR p.132 in that there is a global clock that coordinates the computation. But computations take time instead of being simultaneous and instantaneous p.116. Specifically, time-triggered MoCs associate with a computation a **logical execution time**. The inputs to the computation are provided at ticks of the global clock, but the outputs are not visible to other computations until the *next* tick of the global clock. Between ticks, there is no interaction between the computations, so concurrency difficulties such as race conditions p.291 do not exist. Since the computations are not (logically) instantaneous, there are no difficulties with feedback, and all models are constructive p.139.

The Simulink modeling system, sold by The MathWorks, supports a time-triggered MoC, and in conjunction with another product called Real-Time Workshop, can translate such models in embedded C code. In LabVIEW, from National Instruments, timed loops accomplish a similar capability within a dataflow p.142 MoC.

In the simplest form, a time-triggered model specifies periodic computation with a fixed time interval (the **period**) between ticks of the clock. Giotto (Henzinger et al., 2003) supports modal models p.79, where the periods differ in different modes. Some authors have further extended the concept of logical execution time to non-periodic systems (Liu and Lee, 2003; Ghosal et al., 2004).

Time triggered models are conceptually simple, but computations are tied closely to a periodic clock. The model becomes awkward when actions are not periodic. DE systems, considered next, encompass a richer set of timing behaviors.

6.4.2 Discrete Event Systems

Discrete-event systems (**DE** systems) have been used for decades as a way to build simulations for an enormous variety of applications, including for example digital networks, military systems, and economic systems. A pioneering formalism for DE models is due to Zeigler (1976), who called the formalism **DEVS**, abbreviating discrete event system specification. p.57 DEVS is an extension of Moore machines that associates a non-zero lifespan with each state, thus endowing the Moore machines with an explicit notion of the passage of time (vs. a sequence of reactions).

The key idea in a DE MoC is that events are endowed with a **time stamp**, a value in some model of time (see box on page 156). Normally, two distinct time stamps must be comparable. That is, they are either equal, or one is earlier than the other. A DE model is a network of actors where each actor reacts to input events in time-stamp order and produces output events in time-stamp order.

p.131 **Example 6.16:** The clock signal with period P of Example 6.1 consists of events with time stamps nP for all $n \in \mathbb{Z}$.

To execute a DE model, we can use an **event queue**, which is a list of events sorted by time stamp. The list begins empty. Each actor in the network is interrogated for any initial events it wishes to place on the event queue. These events may be destined for another actor, or they may be destined for the actor itself, in which case they will cause a reaction of the actor to occur at the appropriate time. The execution continues by selecting the earliest event in the event queue and determining which actor should receive that event. The value of that event (if any) is presented as an input to the actor, and the actor reacts ("fires"). The reaction can produce output events, and also events that simply request a later firing of the same actor at some specified time stamp.

At this point, variants of DE MoCs behave differently. Some variants, such as DEVS, require that outputs produced by the actor have a strictly larger time stamp than that of the input just presented. From a modeling perspective, every actor imposes some non-zero delay, in that its reactions (the outputs) become visible to other actors strictly later than the inputs that triggered the reaction. Other variants permit the actor to produce output events with the same time stamp as the input. That is, they can react instantaneously. As with SR models of computation, such instantaneous reactions can create significant subtleties because inputs become simultaneous with outputs.

The subtleties introduced by simultaneous events can be resolved by treating DE as a generalization of SR (Lee and Zheng, 2007). In this variant of a DE semantics, execution proceeds as follows. Again, we use an event queue and interrogate the actors for initial events to place on the queue. We select the event from the queue with the least time stamp, and all other events with the same time stamp, present those events to actors in the model as inputs, and then fire all actors in the manner of a constructive fixed-point iteration, as p.139
normal with SR. In this variant of the semantics, any outputs produced by an actor *must* be simultaneous with the inputs (they have the same time stamp), so they participate in the fixed point. If the actor wishes to produce an output event at a later time, it does so by requesting a firing at a later time (which results in the posting of an event on the event queue).

6.4.3 Continuous-Time Systems

In Chapter 2 we consider models of continuous-time systems based on ordinary differential equations (ODEs). Specifically, we consider equations of the form p.20

$$\dot{\mathbf{x}}(t) = f(\mathbf{x}(t), t),$$

Probing Further: Discrete Event Semantics

Discrete-event models of computation have been a subject of study for many years, with several textbooks available (Zeigler et al., 2000; Cassandras, 1993; Fishman, 2001). The subtleties in the semantics are considerable (see Lee (1999); Cataldo et al. (2006); Liu et al. (2006); Liu and Lee (2008)). Instead of discussing the formal semantics here, we describe how a DE model is executed. Such a description is, in fact, a valid way of giving the semantics of a model. The description is called an **operational semantics** (Scott and Strachey, 1971; Plotkin, 1981).

DE models are often quite large and complex, so execution performance becomes very important. Because of the use of a single event queue, parallelizing or distributing execution of DE models can be challenging (Misra, 1986; Fujimoto, 2000). A recently proposed strategy called **PTIDES** (for programming temporally integrated distributed embedded systems), leverages network time synchronization to provide efficient distributed execution (Zhao et al., 2007; Lee et al., 2009). The claim is that the execution is efficient enough that DE can be used not only as a simulation technology, but also as an implementation technology. That is, the DE event queue and execution engine become part of the deployed embedded software. As of this writing, that claim has not been proven on any practical examples.

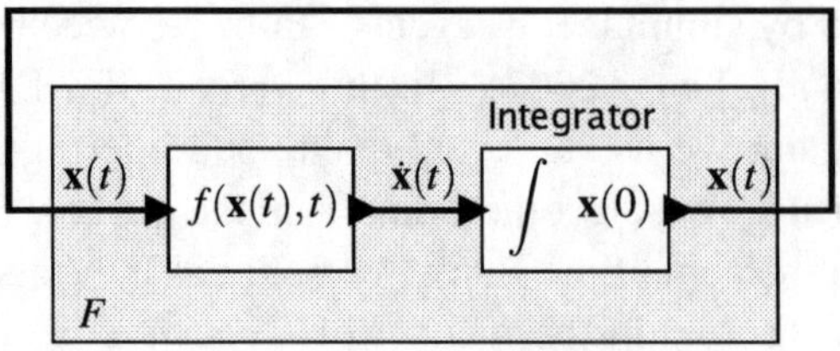

Figure 6.16: Actor model of a system described by equation (6.4).

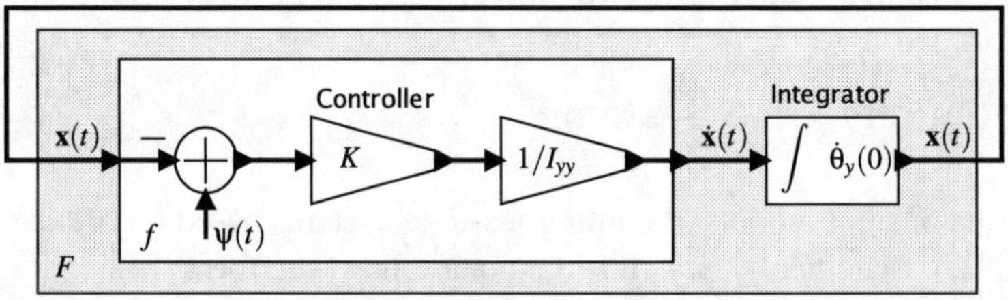

Figure 6.17: The feedback control system of Figure 2.3, using the helicopter model of Example 2.3, redrawn to conform to the pattern of Figure 6.16.

where $\mathbf{x}\colon \mathbb{R} \to \mathbb{R}^n$ is a vector-valued continuous-time function. An equivalent model is an integral equation of the form

$$\mathbf{x}(t) = \mathbf{x}(0) + \int_0^t \dot{\mathbf{x}}(\tau)d\tau \qquad (6.3)$$

$$= \mathbf{x}(0) + \int_0^t f(\mathbf{x}(\tau),\tau)d\tau. \qquad (6.4)$$

In Chapter 2, we show that a model of a system given by such ODEs can be described as an
p.26 interconnection of actors, where the communication between actors is via continuous-time signals. Equation (6.4) can be represented as the interconnection shown in Figure 6.16, which conforms to the feedback pattern of Figure 6.1(d).

Example 6.17: The feedback control system of Figure 2.3, using the helicopter model of Example 2.3, can be redrawn as shown in Figure 6.17, which conforms to the pattern of Figure 6.16. In this case, $\mathbf{x} = \dot{\theta}_y$ is a scalar-valued continuous-time function (or a vector of length one). The function f is defined as follows,

$$f(\mathbf{x}(t),t) = (K/I_{yy})(\psi(t) - \mathbf{x}(t)),$$

and the initial value of the integrator is

$$\mathbf{x}(0) = \dot{\theta}_y(0).$$

Such models, in fact, are actor compositions under a **continuous-time model of computation**, but unlike the previous MoCs, this one cannot strictly be executed on a digital computer. A digital computer cannot directly deal with the time continuum. It can, however, be approximated, often quite accurately.

The approximate execution of a continuous-time model is accomplished by a **solver**, which constructs a numerical approximation to the solution of an ODE. The study of algorithms for solvers is quite old, with the most commonly used techniques dating back to the 19th century. Here, we will consider only one of the simplest of solvers, which is known as a **forward Euler** solver.

A forward Euler solver estimates the value of $\mathbf{x}$ at time points $0, h, 2h, 3h, \cdots$, where h is called the **step size**. The integration is approximated as follows,

$$\begin{aligned}
\mathbf{x}(h) &= \mathbf{x}(0) + hf(\mathbf{x}(0), 0) \\
\mathbf{x}(2h) &= \mathbf{x}(h) + hf(\mathbf{x}(h), h) \\
\mathbf{x}(3h) &= \mathbf{x}(2h) + hf(\mathbf{x}(2h), 2h) \\
&\cdots \\
\mathbf{x}((k+1)h) &= \mathbf{x}(kh) + hf(\mathbf{x}(kh), kh).
\end{aligned}$$

This process is illustrated in Figure 6.18(a), where the "true" value of $\dot{\mathbf{x}}$ is plotted as a function of time. The true value of $\mathbf{x}(t)$ is the area under that curve between 0 and t, plus the initial value $\mathbf{x}(0)$. At the first step of the algorithm, the increment in area is approximated as the area of a rectangle of width h and height $f(\mathbf{x}(0), 0)$. This increment yields an estimate for $\mathbf{x}(h)$, which can be used to calculate $\dot{\mathbf{x}}(h) = f(\mathbf{x}(h), h)$, the height of the second rectangle. And so on.

You can see that the errors in approximation will accumulate over time. The algorithm can be improved considerably by two key techniques. First, a **variable-step solver** will vary the step size based on estimates of the error to keep the error small. Second, a more sophisticated solver will take into account the slope of the curve and use trapezoidal approximations

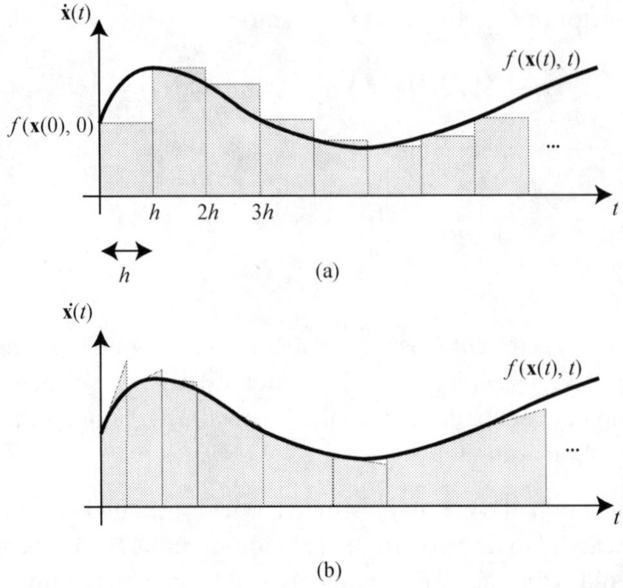

Figure 6.18: (a) Forward Euler approximation to the integration in (6.4), where **x** is assumed to be a scalar. (b) A better approximation that uses a variable step size and takes into account the slope of the curve.

as suggested in Figure 6.18(b). A family of such solvers known as Runge-Kutta solvers are widely used. But for our purposes here, it does not matter what solver is used. All that matters is that (a) the solver determines the step size, and (b) at each step, the solver performs some calculation to update the approximation to the integral.

When using such a solver, we can interpret the model in Figure 6.16 in a manner similar p.29 to SR and DE models. The f actor is memoryless, so it simply performs a calculation to produce an output that depends only on the input and the current time. The integrator is a p.47 state machine whose state is updated at each reaction by the solver, which uses the input to determine what the update should be. The state space of this state machine is infinite, since the state variable $\mathbf{x}(t)$ is a vector of real numbers.

p.132 Hence, a continuous-time model can be viewed as an SR model with a time step between global reactions determined by a solver (Lee and Zheng, 2007). Specifically, a continuous-time model is a network of actors, each of which is a cascade composition of a simple memoryless computation actor and a state machine, and the actor reactions are p.116 simultaneous and instantaneous. The times of the reactions are determined by a solver. The solver will typically consult the actors in determining the time step, so that for example events like level crossings (when a continuous signal crosses a threshold) can be captured precisely. Hence, despite the additional complication of having to provide a solver, the mechanisms required to achieve a continuous-time model of computation are not much different from those required to achieve SR and DE.

A popular software tool that uses a continuous-time MoC is Simulink, from The MathWorks. Simulink represents models similarly as block diagrams, which are interconnections of actors. Continuous-time models can also be simulated using the textual tool MATLAB from the same vendor. MATRIXx, from National Instruments, also supports graphical continuous-time modeling. Continuous-time models can also be integrated within LabVIEW models, either graphically using the Control Design and Simulation Module or textually using the programming language MathScript.

6.5 Summary

This chapter provides a whirlwind tour of a rather large topic, concurrent models of computation. It begins with synchronous-reactive models, which are closest to the synchronous composition of state machines considered in the previous chapter. It then considers dataflow models, where execution can be more loosely coordinated. Only data precedences impose constraints on the order of actor computations. The chapter then concludes with a quick view of a few models of computation that explicitly include a notion of time. Such MoCs are particularly useful for modeling cyber-physical systems.

Exercises

1. Show how each of the following actor models can be transformed into a feedback system by using a reorganization similar to that in Figure 6.1(b). That is, the actors should be aggregated into a single side-by-side composite actor. p.31

   (a)

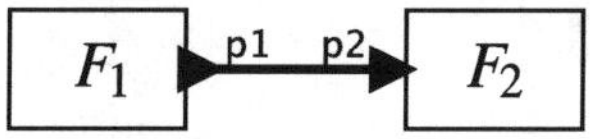

(b)

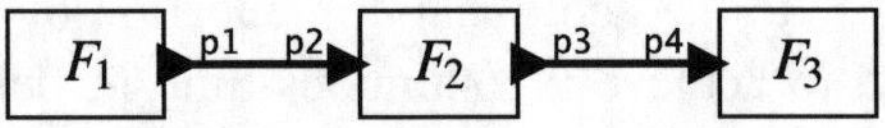

(c)

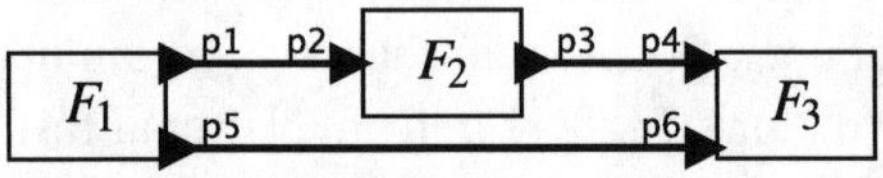

2. Consider the following state machine in a synchronous feedback composition:

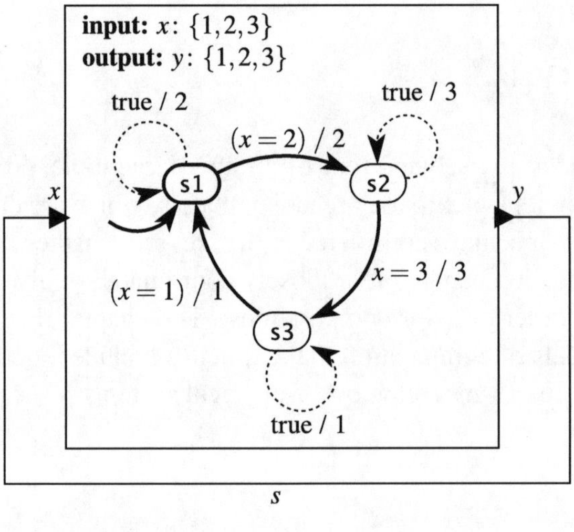

   (a) Is it well-formed? Is it constructive?
   (b) If it is well-formed and constructive, then find the output symbols for the first 10 reactions. If not, explain where the problem is.
   (c) Show the composition machine, assuming that the composition has no input and that the only output is y.

3. For the following synchronous model, determine whether it is well formed and constructive, and if so, determine the sequence of values of the signals s_1 and s_2.

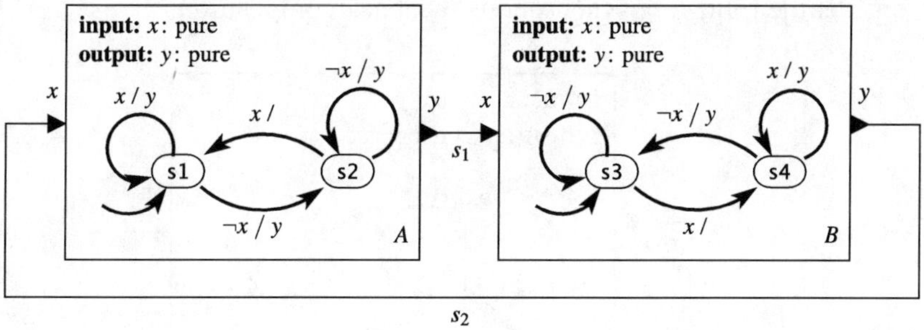

4. For the following synchronous model, determine whether it is well formed and constructive, and if so, determine the possible sequences of values of the signals s_1 and s_2. Note that machine A is nondeterministic.

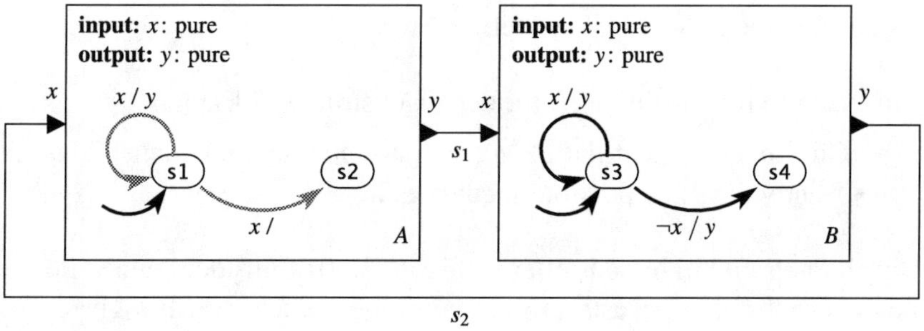

5. (a) Determine whether the following synchronous model is well formed and constructive:

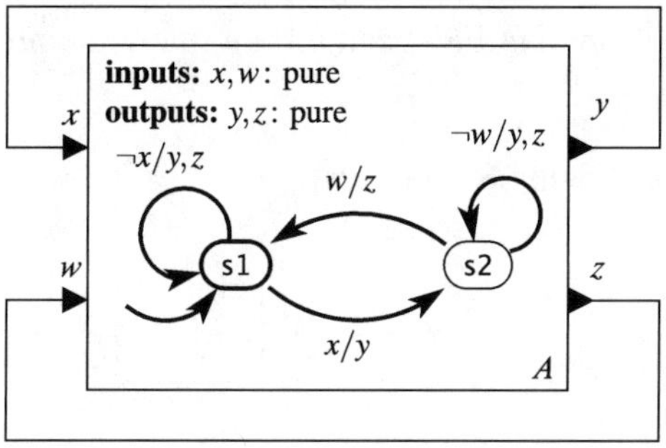

Explain.

(b) For the model in part (a), give the semantics by giving an equivalent flat state machine with no inputs and two outputs.

6. Consider the following synchronous feedback composition:

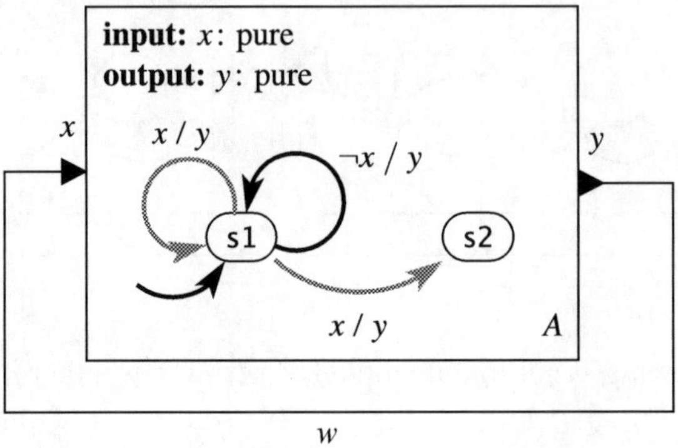

Notice that the FSM A is nondeterministic.

(a) Is this composition well formed? Constructive? Explain.

(b) Give an equivalent flat FSM (with no input and no connections) that produces exactly the same possible sequences w.

7. Recall the traffic light controller of Figure 3.10. Consider connecting the outputs of this controller to a pedestrian light controller, whose FSM is given in Figure 5.10. Using your favorite modeling software that supports state machines (such as Ptolemy II, LabVIEW Statecharts, or Simulink/Stateflow), construct the composition of the
p.56 above two FSMs along with a deterministic extended state machine modeling the environment and generating input symbols *timeR*, *timeG*, *timeY*, and *isCar*. For example, the environment FSM can use an internal counter to decide when to generate these symbols.

p.145 8. Consider the following SDF model:

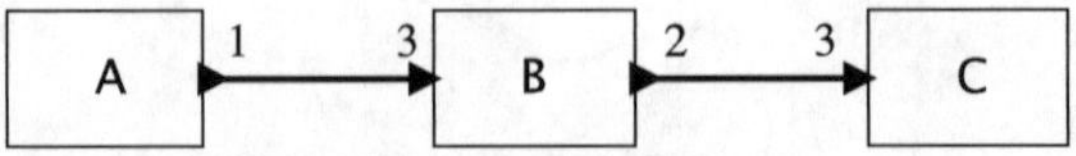

The numbers adjacent to the ports indicate the number of tokens produced or consumed by the actor when it fires. Answer the following questions about this model.

(a) Let q_A, q_B, and q_C denote the number of firings of actors A, B, and C, respectively. Write down the balance equations and find the least positive integer solution.

(b) Find a schedule for an unbounded execution that minimizes the buffer sizes on the two communication channels. What is the resulting size of the buffers?

9. For each of the following dataflow models, determine whether there is an unbounded execution with bounded buffers. If there is, determine the minimum buffer size.

(a)

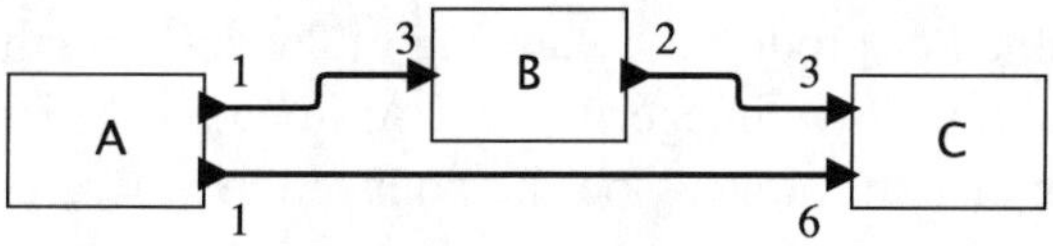

(b)

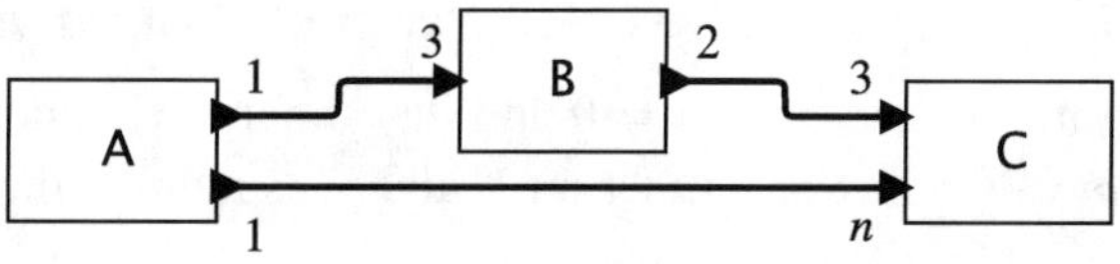

where n is some integer.

(c)

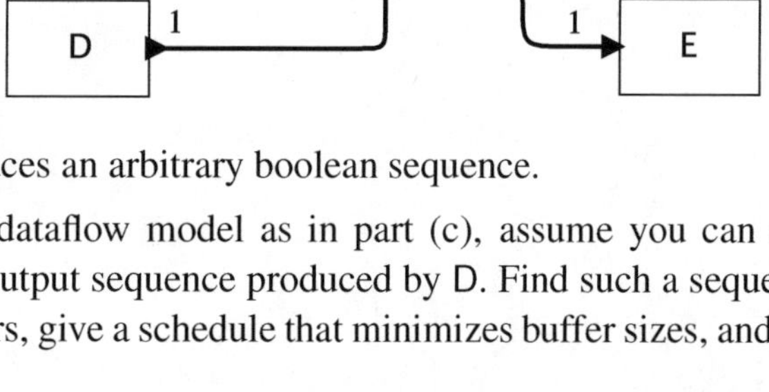

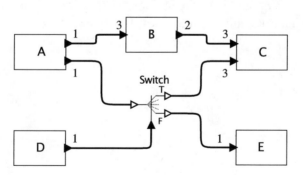

where D produces an arbitrary boolean sequence.

(d) For the same dataflow model as in part (c), assume you can specify a periodic boolean output sequence produced by D. Find such a sequence that yields bounded buffers, give a schedule that minimizes buffer sizes, and give the buffer sizes.

10. Consider the SDF graph shown below:

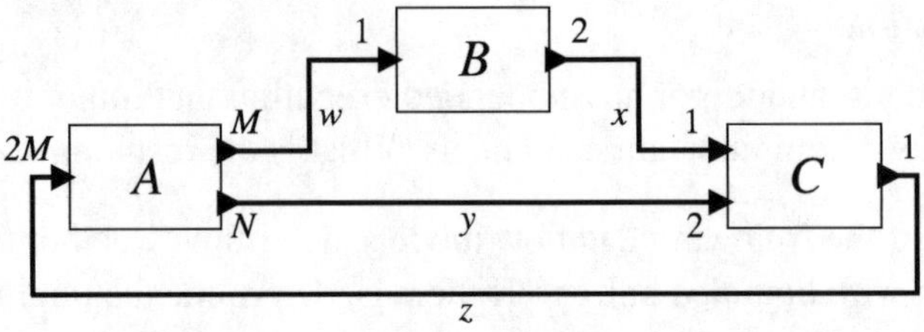

In this figure, A, B, and C are actors. Adjacent to each port is the number of tokens consumed or produced by a firing of the actor on that port, where N and M are variables with positive integer values. Assume the variables w, x, y, and z represent the number of initial tokens on the connection where these variables appear in the diagram. These variables have non-negative integer values.

(a) Derive a simple relationship between N and M such that the model is consistent, or show that no positive integer values of N and M yield a consistent model.

(b) Assume that $w = x = y = 0$ and that the model is consistent and find the minimum value of z (as a function N and M) such that the model does not deadlock.

(c) Assume that $z = 0$ and that the model is consistent. Find values for w, x, and y such that the model does not deadlock and $w + x + y$ is minimized.

(d) Assume that $w = x = y = 0$ and z is whatever value you found in part (b). Let b_w, b_x, b_y, and b_z be the buffer sizes for connections w, x, y, and z, respectively. What is the minimum for these buffer sizes?

Part II

Design of Embedded Systems

This part of this text studies the design p.8 of embedded systems, with emphasis on the techniques used to build concurrent, p.130 real-time embedded software. We proceed bottom up, discussing first in Chapter 7 sensors and actuators, with emphasis on how to model them. Chapter 8 covers the design of embedded processors, with emphasis on parallelism in the hardware and its implications for programmers. Chapter 9 covers memory architectures, with particular emphasis on the effect they have on program timing. Chapter 10 covers the input and output mechanisms that enable programs to interact with the external physical world, with emphasis on how to reconcile the sequential nature of software with the concurrent nature of the physical world. Chapter 11 describes mechanisms for achieving concurrency in software, threads and processes, and synchronization of concurrent software tasks, including semaphores and mutual exclusion. Finally, Chapter 12 covers scheduling, with particular emphasis on controlling timing in concurrent programs.

7 Sensors and Actuators

Contents

A **sensor** is a device that measures a physical quantity. An **actuator** is a device that alters a physical quantity. In electronic systems, sensors often produce a voltage that is proportional to the physical quantity being measured. The voltage may then be converted to a number by an **analog-to-digital converter** (**ADC**). A sensor that is packaged with an ADC is called a **digital sensor**, whereas a sensor without an ADC is called an **analog sensor**. A digital sensor will have a limited precision, determined by the number of bits used to represent the number (this can be as few as one!). Conversely, an actuator is commonly driven by a voltage that may be converted from a number by a **digital-to-analog converter** (**DAC**). An actuator that is packaged with a DAC is called a **digital actuator**.

Today, sensors and actuators are often packaged with microprocessors and network interfaces, enabling them to appear on the Internet as services. The trend is towards a technology that deeply connects our physical world with our information world through such smart sensors and actuators. This integrated world is variously called the **Internet of Things** (**IoT**), **Industry 4.0**, the **Industrial Internet**, **Machine-to-Machine** (**M2M**), the **Internet of Everything**, the **Smarter Planet**, **TSensors** (Trillion Sensors), or **The Fog** (like The Cloud, but closer to the ground).

Some technologies for interfacing to sensors and actuators have emerged that leverage established mechanisms originally developed for ordinary Internet usage. For example, a sensor or actuator may be accessible via a web server using the so-called **Representational State Transfer** (**REST**) architectural style (Fielding and Taylor, 2002). In this style, data may be retrieved from a sensor or commands may be issued to an actuator by constructing a URL (uniform resource locator), as if you were accessing an ordinary web page from a browser, and then transmitting the URL directly to the sensor or actuator device, or to a web server that serves as an intermediary.

In this chapter, we focus not on such high-level interfaces, but rather on foundational properties of sensors and actuators as bridges between the physical and the cyber worlds. Key low-level properties include the rate at which measurements are taken or actuations are performed, the proportionality constant that relates the physical quantity to the measurement or control signal, the offset or bias, and the dynamic range. For many sensors and actuators, it is useful to model the degree to which a sensor or actuator deviates from a proportional measurement (its **nonlinearity**), and the amount of random variation introduced by the p.177 measurement process (its noise).

A key concern for sensors and actuators is that the physical world functions in a multidimensional continuum of time and space. It is an **analog** world. The world of software, however, is **digital**, and strictly quantized. Measurements of physical phenomena must be quantized in both magnitude and time before software can operate on them. And commands to the physical world that originate from software will also be intrinsically quantized. Understanding the effects of this quantization is essential.

This chapter begins in Section 7.1 with an outline of how to construct models of sensors and actuators, specifically focusing on linearity (and nonlinearity), bias, dynamic range, quantization, noise, and sampling. That section concludes with a brief introduction to signal conditioning, a signal processing technique to improve the quality of sensor data and actuator controls. Section 7.2 then discusses a number of common sensing problems, including measuring tilt and acceleration (accelerometers), measuring position and velocity (anemometers, inertial navigation, GPS, and other ranging and triangulation techniques), measuring rotation (gyroscopes), measuring sound (microphones), and measuring distance (rangefinders). The chapter concludes with Section 7.3, which shows how to apply the modeling techniques to actuators, focusing specifically on LEDs and motor controllers.

Higher-level properties that are not addressed in this chapter, but are equally important, include security (specifically access control), privacy (particularly for data flowing over the open Internet), name-space management, and commissioning. The latter is particularly big issue when the number of sensors or actuators get large. Commissioning is the process of associating a sensor or actuator device with a physical location (e.g., a temperature sensor gives the temperature of what?), enabling and configuring network interfaces, and possibly calibrating the device for its particular environment.

7.1 Models of Sensors and Actuators

Sensors and actuators connect the cyber world with the physical world. Numbers in the cyber world bear a relationship with quantities in the physical world. In this section, we provide models of that relationship. Having a good model of a sensor or actuator is essential to effectively using it.

7.1.1 Linear and Affine Models

Many sensors may be approximately modeled by an affine function. Suppose that a physical quantity $x(t)$ at time t is reported by the sensor to have value $f(x(t))$, where $f\colon \mathbb{R} \to \mathbb{R}$ is a function. The function f is **linear** if there exists a **proportionality constant** $a \in \mathbb{R}$ such that for all $x(t) \in \mathbb{R}$

$$f(x(t)) = ax(t).$$

It is an **affine function** if there exists a proportionality constant $a \in \mathbb{R}$ and a **bias** $b \in \mathbb{R}$ such that

$$f(x(t)) = ax(t) + b. \tag{7.1}$$

Clearly, every linear function is an affine function (with $b = 0$), but not vice versa.

Interpreting the readings of such a sensor requires knowledge of the proportionality constant and bias. The proportionality constant represents the **sensitivity** of the sensor, since it specifies the degree to which the measurement changes when the physical quantity changes.

Actuators may also be modeled by affine functions. The affect that a command to the actuator has on the physical environment may be reasonably approximated by a relation like (7.1).

7.1.2 Range

No sensor or actuator truly realizes an affine function. In particular, the **range** of a sensor, the set of values of a physical quantity that it can measure, is always limited. Similarly for an actuator. Outside that range, an affine function model is no longer valid. For example, a thermometer designed for weather monitoring may have a range of $-20°$ to $50°$ Celsius. Physical quantities outside this range will typically **saturate**, meaning that they yield a
p.173 maximum or a minimum reading outside their range. An affine function model of a sensor may be augmented to take this into account as follows,

$$f(x(t)) = \begin{cases} ax(t)+b & \text{if } L \leq x(t) \leq H \\ aH+b & \text{if } x(t) > H \\ aL+b & \text{if } x(t) < L, \end{cases} \tag{7.2}$$

where $L, H \in \mathbb{R}$, $L < H$, are the low and high end of the sensor range, respectively.

A relation between a physical quantity $x(t)$ and a measurement given by (7.2) is not an affine
p.172 relation (it is, however, piecewise affine). In fact, this is a simple form of nonlinearity that is shared by all sensors. The sensor is reasonably modeled by an affine function within an **operating range** (L, H), but outside that operating range, its behavior is distinctly different.

7.1.3 Dynamic Range

Digital sensors are unable to distinguish between two closely-spaced values of the physical quantity. The **precision** p of a sensor is the smallest absolute difference between two values of a physical quantity whose sensor readings are distinguishable. The **dynamic range** $D \in \mathbb{R}_+$ of a digital sensor is the ratio

$$D = \frac{H-L}{p},$$

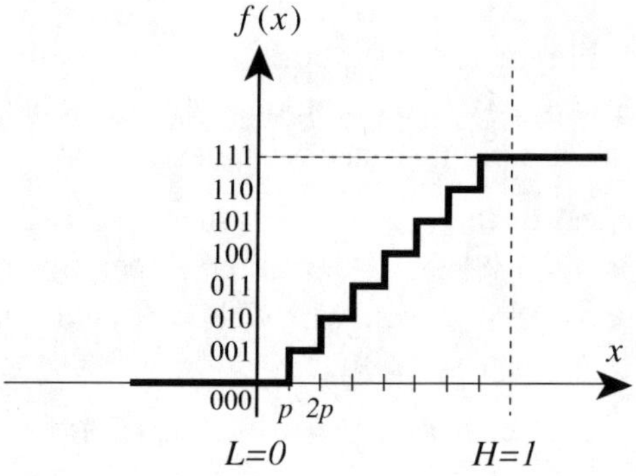

Figure 7.1: Sensor distortion function for a 3-bit digital sensor capable of measuring a range of zero to one volt, where the precision $p = 1/8$.

where H and L are the limits of the range in (7.2). Dynamic range is usually measured in decibels (see sidebar on page 180), as follows: p.180

$$D_{dB} = 20\log_{10}\left(\frac{H-L}{p}\right). \tag{7.3}$$

7.1.4 Quantization

A digital sensor represents a physical quantity using an n-bit number, where n is a small integer. There are only 2^n distinct such numbers, so such a sensor can produce only 2^n distinct measurements. The actual physical quantity may be represented by a real number $x(t) \in \mathbb{R}$, but for each such $x(t)$, the sensor must pick one of the 2^n numbers to represent it. This process is called **quantization**. For an ideal digital sensor, two physical quantities that differ by the precision p will be represented by digital quantities that differ by one bit, so precision and quantization become intertwined.

We can further augment the function f in (7.2) to include quantization, as illustrated in the following example.

Example 7.1: Consider a 3-bit digital sensor that can measure a voltage between zero and one volt. Such a sensor may be modeled by the function $f\colon \mathbb{R} \to$

$\{0,1,\cdots,7\}$ shown in Figure 7.1. The horizontal axis is the input to the sensor (in volts), and the vertical axis is the output, with the value shown in binary to emphasize that this is a 3-bit digital sensor.

In the figure, the low end of the measurable range is $L = 0$, and the high end is $H = 1$. The precision is $p = 1/8$, because within the operating range, any two inputs that differ by more than $1/8$ of a volt will yield different outputs. The dynamic range, therefore, is

$$D_{dB} = 20\log_{10}\left(\frac{H-L}{p}\right) \approx 18dB.$$

A function f like the one in Figure 7.1 that defines the output of a sensor as a function of its input is called the **sensor distortion function**. In general, an ideal n-bit digital sensor with a sensor distortion function like that shown in Figure 7.1 will have a precision given by

$$p = (H-L)/2^n$$

and a dynamic range of

$$D_{dB} = 20\log_{10}\left(\frac{H-L}{p}\right) 20\log_{10}(2^n) = 20n\log_{10}(2) \approx 6n\ dB. \tag{7.4}$$

Each additional bit yields approximately 6 decibels of dynamic range.

Example 7.2: An extreme form of quantization is performed by an **analog comparator**, which compares a signal value against a threshold and produces a binary value, zero or one. Here, the sensor function $f\colon \mathbb{R} \to \{0,1\}$ is given by

$$f(x(t)) = \begin{cases} 0 & \text{if } x(t) \le 0 \\ 1 & \text{otherwise} \end{cases}$$

Such extreme quantization is often useful, because the resulting signal is a very
p.252 simple digital signal that can be connected directly to a GPIO input pin of a microprocessor, as discussed in Chapter 10.

The analog comparator of the previous example is a one-bit ADC. The quantization error is
p.185 high for such a converter, but using signal conditioning, as described below in Section 7.1.8,

if the sample rate is high enough, the noise can be reduced considerably by digital low-pass filtering. Such a process is called **oversampling**; it is commonly used today because processing signals digitally is often less costly than analog processing. p.208

Actuators are also subject to quantization error. A digital actuator takes a digital command and converts it to an analog physical action. A key part of this is the digital to analog converter (DAC). Because the command is digital, it has only a finite number of possible values. The precision with which an analog action can be taken, therefore, will depend on the number of bits of the digital signal and the range of the actuator.

As with ADCs, however, it is possible to trade off precision and speed. A **bang-bang controller**, for example, uses a one-bit digital actuation signal to drive an actuator, but updates that one-bit command very quickly. An actuator with a relatively slow response time, such as a motor, does not have much time to react to each bit, so the reaction to each bit is small. The overall reaction will be an average of the bits over time, much smoother than what you would expect from a one-bit control. This is the mirror image of oversampling.

The design of ADC and DAC hardware is itself quite an art. The effects of choices of sampling interval and number of bits are quite nuanced. Considerable expertise in signal processing is required to fully understand the implications of choices (see Lee and Varaiya (2011)). Below, we give a cursory view of this rather sophisticated topic. Section 7.1.8 discusses how to mitigate the noise in the environment and noise due to quantization, showing the intuitive result that it is beneficial to filter out frequency ranges that are not of interest. These frequency ranges are related to the sample rate. Hence, noise and sampling are the next topics.

7.1.5 Noise

By definition, **noise** is the part of a signal that we do not want. If we want to measure $x(t)$ at time t, but we actually measure $x'(t)$), then the noise is the difference,

$$n(t) = x'(t) - x(t).$$

Equivalently, the actual measurement is

$$x'(t) = x(t) + n(t), \tag{7.5}$$

a sum of what we want plus the noise.

p.186

Example 7.3: Consider using an accelerometer to measure the orientation of a slowly moving object (see Section 7.2.1 below for an explanation of why an accelerometer can measure orientation). The accelerometer is attached to the moving object and reacts to changes in orientation, which change the direction of the gravitational field with respect to the axis of the accelerometer. But it will also report acceleration due to vibration. Let $x(t)$ be the signal due to orientation and $n(t)$ be the signal due to vibration. The accelerometer measures the sum.

In the above example, noise is a side effect of the fact that the sensor is not measuring exactly what we want. We want orientation, but it is measuring acceleration. We can also model sensor imperfections and quantization as noise. In general, a sensor distortion function can be modeled as additive noise,

$$f(x(t)) = x(t) + n(t), \tag{7.6}$$

where $n(t)$ by definition is just $f(x(t)) - x(t)$.

It is useful to be able to characterize how much noise there is in a measurement. The **root mean square** (**RMS**) $N \in \mathbb{R}_+$ of the noise is equal to the square root of the average value of $n(t)^2$. Specifically,

$$N = \lim_{T \to \infty} \sqrt{\frac{1}{2T} \int_{-T}^{T} (n(\tau))^2 d\tau}. \tag{7.7}$$

This is a measure of (the square root of) **noise power**. An alternative (statistical) definition of noise power is the square root of the expected value of the square of $n(t)$. Formula (7.7) defines the noise power as an *average* over time rather than an expected value.

The **signal to noise ratio** (**SNR**, in decibels) is defined in terms of RMS noise,

$$SNR_{dB} = 20 \log_{10} \left(\frac{X}{N} \right),$$

where X is the RMS value of the input signal x (again, defined either as a time average as in (7.7) or using the expected value). In the next example, we illustrate how to calculate SNR using expected values, leveraging elementary probability theory.

Example 7.4: We can find the SNR that results from quantization by using (7.6) as a model of the quantizer. Consider Example 7.1 and Figure 7.1, which show a 3-bit digital sensor with an operating range of zero to one volt. Assume that the input voltage is equally likely to be anywhere in the range of zero to one volt. That is, $x(t)$ is a random variable with uniform distribution ranging from 0 to 1. Then the RMS value of the input x is given by the square root of the expected value of the square of $x(t)$, or

$$X = \sqrt{\int_0^1 x^2 dx} = \frac{1}{\sqrt{3}}.$$

Examining Figure 7.1, we see that if $x(t)$ is a random variable with uniform distribution ranging from 0 to 1, then the error $n(t)$ in the measurement (7.6) is equally likely to be anywhere in the range from $-1/8$ to 0. The RMS noise is therefore given by

$$N = \sqrt{\int_{-1/8}^{0} 8n^2 dn} = \sqrt{\frac{1}{3 \cdot 64}} = \frac{1}{8\sqrt{3}}.$$

The SNR is therefore

$$SNR_{dB} = 20\log_{10}\left(\frac{X}{N}\right) = 20\log_{10}(8) \approx 18dB.$$

Notice that this matches the 6 dB per bit dynamic range predicted by (7.4)!

To calculate the SNR in the previous example, we needed a statistical model of the input x (uniformly distributed from 0 to 1) and the quantization function. In practice, it is difficult to calibrate ADC hardware so that the input x makes full use of its range. That is, the input is likely to be distributed over less than the full range 0 to 1. It is also unlikely to be uniformly distributed. Hence, the actual SNR achieved in a system will likely be considerably less than the 6 dB per bit predicted by (7.4).

7.1.6 Sampling

A physical quantity $x(t)$ is a function of time t. A digital sensor will **sample** the physical quantity at particular points in time to create a discrete signal. In **uniform sampling**, there p.44
is a fixed time interval T between samples; T is called the **sampling interval**. The resulting

signal may be modeled as a function $s\colon \mathbb{Z} \to \mathbb{R}$ defined as follows,

$$\forall\, n \in \mathbb{Z}, \quad s(n) = f(x(nT)), \tag{7.8}$$

Decibels

The term "**decibel**" is literally one tenth of a **bel**, which is named after Alexander Graham Bell. This unit of measure was originally developed by telephone engineers at Bell Telephone Labs to designate the ratio of the **power** of two signals.

Power is a measure of energy dissipation (work done) per unit time. It is measured in **watts** for electronic systems. One bel is defined to be a factor of 10 in power. Thus, a 1000 watt hair dryer dissipates 1 bel, or 10 dB, more power than a 100 watt light bulb. Let $p_1 = 1000$ watts be the power of the hair dryer and $p_2 = 100$ be the power of the light bulb. Then the ratio is

$$\log_{10}(p_1/p_2) = 1 \text{ bel}, \quad \text{or}$$

$$10\log_{10}(p_1/p_2) = 10 \text{ dB}.$$

Comparing against (7.3) we notice a discrepancy. There, the multiplying factor is 20, not 10. That is because the ratio in (7.3) is a ratio of amplitude (magnitude), not powers. In electronic circuits, if an amplitude represents the voltage across a resistor, then the power dissipated by the resistor is proportional to the *square* of the amplitude. Let a_1 and a_2 be two such amplitudes. Then the ratio of their powers is

$$10\log_{10}(a_1^2/a_2^2) = 20\log_{10}(a_1/a_2).$$

Hence the multiplying factor of 20 instead of 10 in (7.3). A 3 dB power ratio amounts to a factor of 2 in power. In amplitudes, this is a ratio of $\sqrt{2}$.

In audio, decibels are used to measure sound pressure. A statement like "a jet engine at 10 meters produces 120 dB of sound," by convention, compares sound pressure to a defined reference of 20 micropascals, where a pascal is a pressure of 1 newton per square meter. For most people, this is approximately the threshold of hearing at 1 kHz. Thus, a sound at 0 dB is barely audible. A sound at 10 dB has 10 times the power. A sound at 100 dB has 10^{10} times the power. The jet engine, therefore, would probably make you deaf without ear protection.

where $\mathbb{Z}$ is the set of integers. That is, the physical quantity $x(t)$ is observed only at times $t = nT$, and the measurement is subjected to the sensor distortion function. The **sampling rate** is $1/T$, which has units of **samples per second**, often given as **Hertz** (written **Hz**, meaning cycles per second).

In practice, the smaller the sampling interval T, the more costly it becomes to provide more bits in an ADC. At the same cost, faster ADCs typically produce fewer bits and hence have either higher quantization error or smaller range.

Example 7.5: The ATSC digital video coding standard includes a format where the frame rate is 30 frames per second and each frame contains $1080 \times 1920 =$ 2,073,600 pixels. An ADC that is converting one color channel to a digital representation must therefore perform 2,073,600 $\times$ 30 = 62,208,000 conversions per second, which yields a sampling interval T of approximately 16 nsec. With such a short sampling interval, increasing the number of bits in the ADC becomes expensive. For video, a choice of $b = 8$ bits is generally adequate to yield good visual fidelity and can be realized at reasonable cost. p.209

An important concern when sampling signals is that there are many distinct functions x that when sampled will yield the same signal s. This phenomenon is known as **aliasing**.

Example 7.6: Consider a sinusoidal sound signal at 1 kHz (kilohertz, or thousands of cycles per second),

$$x(t) = \cos(2000\pi t).$$

Suppose there is no sensor distortion, so the function f in (7.8) is the identity function. If we sample at 8000 samples per second (a rate commonly used in telephony), we get a sampling interval of $T = 1/8000$, which yields the samples

$$s(n) = f(x(nT)) = \cos(\pi n/4).$$

Suppose instead that we are given a sound signal at 9 kHz,

$$x'(t) = \cos(18{,}000\pi t).$$

Sampling at the same 8kHz rate yields

$$s'(n) = \cos(9\pi n/4) = \cos(\pi n/4 + 2\pi n) = \cos(\pi n/4) = s(n).$$

The 1 kHz and 9 kHz sound signals yield exactly the same samples, as illustrated in Figure 7.2. Hence, at this sampling rate, these two signals are aliases of one another. They cannot be distinguished.

Aliasing is a complex and subtle phenomenon (see Lee and Varaiya (2011) for details), but a useful rule of thumb for uniform sampling is provided by the **Nyquist-Shannon sampling theorem**. A full study of the subject requires the machinery of Fourier transforms, and is beyond the scope of this text. Informally, this theorem states that a set of samples at sample rate $R = 1/T$ uniquely defines a continuous-time signal that is a sum of sinusoidal components with frequencies less than $R/2$. That is, among all continuous-time signals that are sums of sinusoids with frequencies less than $R/2$, there is only one that matches any given set of samples taken at rate R. The rule of thumb, therefore, is that if you sample a signal where the most rapid expected variation occurs at frequency $R/2$, then sampling the signal at a rate at least R will result in samples that uniquely represent the signal.

Example 7.7: In traditional telephony, engineers have determined that intelligible human speech signals do not require frequencies higher than 4 kHz. Hence, removing the frequencies above 4 kHz and sampling an audio signal with human speech at 8kHz is sufficient to enable reconstruction of an intelligible audio signal from the samples. The removal of the high frequencies is accomplished by a frequency selective filter called an **anti-aliasing filter**, because it prevents frequency components above 4 kHz from masquerading as frequency components below 4 kHz.

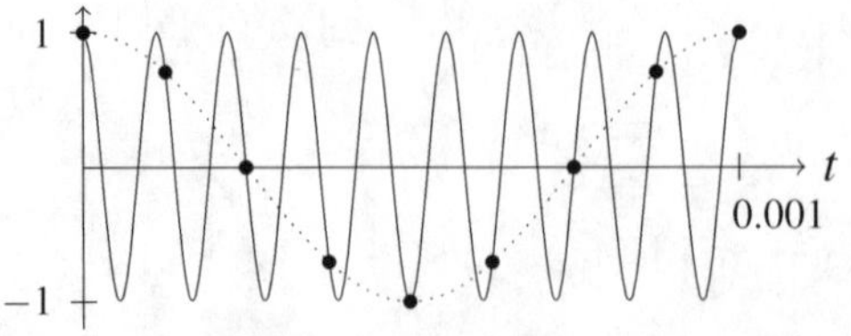

Figure 7.2: Illustration of aliasing, where samples of a 9 kHz sinusoid taken at 8,000 samples per second are the same as samples of a 1 kHz sinusoid taken at 8,000 samples per second.

The human ear, however, can easily discern frequencies up to about 15 kHz, or 20 kHz in young people. Digital audio signals intended for music, therefore, are sampled at frequencies above 40 kHz; 44.1 kHz is a common choice, a rate defined originally for use in compact discs (CDs).

Example 7.8: Air temperature in a room varies quite slowly compared to sound pressure. We might assume, for example, that the most rapid expected air temperature variations occur at rates measured in minutes, not seconds. If we want to capture variations on the scale of a minute or so, then we should take at least two samples per minute of the temperature measurement.

7.1.7 Harmonic Distortion

A form of nonlinearity that occurs even within the operating range of sensors and actuators p.172
is **harmonic distortion**. It typically occurs when the sensitivity of the sensor or actuator is p.174
not constant and depends on the magnitude of the signal. For example, a microphone may be less responsive to high sound pressure than to lower sound pressure.

Harmonic distortion is a nonlinear effect that can be modeled by powers of the physical quantity. Specifically, **second harmonic distortion** is a dependence on the square of the physical quantity. That is, given a physical quantity $x(t)$, the measurement is modeled as

$$f(x(t)) = ax(t) + b + d_2(x(t))^2, \tag{7.9}$$

where d_2 is the amount of second harmonic distortion. If d_2 is small, then the model is nearly affine. If d_2 is large, then it is far from affine. The $d_2(x(t))^2$ term is called second harmonic distortion because of the effect it has the frequency content of a signal $x(t)$ that is varying in time.

p.190

Example 7.9: Suppose that a microphone is stimulated by a purely sinusoidal input sound

$$x(t) = \cos(\omega_0 t),$$

where t is time in seconds and ω_0 is the frequency of the sinusoid in radians per second. If the frequency is within the human auditory range, then this will sound like a pure tone.

A sensor modeled by (7.9) will produce at time t the measurement

$$\begin{aligned} x'(t) &= ax(t) + b + d_2(x(t))^2 \\ &= a\cos(\omega_0 t) + b + d_2\cos^2(\omega_0 t) \\ &= a\cos(\omega_0 t) + b + \frac{d_2}{2} + \frac{d_2}{2}\cos(2\omega_0 t), \end{aligned}$$

where we have used the trigonometric identity

$$\cos^2(\theta) = \frac{1}{2}(1 + \cos(2\theta)).$$

To humans, the bias term $b + d_2/2$ is not audible. Hence, this signal consists of a pure tone, scaled by a, and a distortion term at twice the frequency, scaled by $d_2/2$. This distortion term is audible as harmonic distortion as long as $2\omega_0$ is in the human auditory range.

A cubic term will introduce **third harmonic distortion**, and higher powers will introduce higher harmonics.

The importance of harmonic distortion depends on the application. The human auditory system is very sensitive to harmonic distortion, but the human visual system much less so, for example.

7.1.8 Signal Conditioning[1]

Noise and harmonic distortion often have significant differences from the desired signal. We can exploit those differences to reduce or even eliminate the noise or distortion. The

[1] This section may be skipped on a first reading. It requires a background in signals and systems at the level typically covered in a sophomore or junior engineering course.

easiest way to do this is with **frequency selective filtering**. Such filtering relies on Fourier theory, which states that a signal is an additive composition of sinusoidal signals of different frequencies. While Fourier theory is beyond the scope of this text (see Lee and Varaiya (2003) for details), it may be useful to some readers who have some background to see how to apply that theory in the context of embedded systems. We do that in this section.

Example 7.10: The accelerometer discussed in Example 7.3 is being used to measure the orientation of a slowly moving object. But instead it measures the sum of orientation and vibration. We may be able to reduce the effect of the vibration by **signal conditioning**. If we assume that the vibration $n(t)$ has higher frequency content than the orientation $x(t)$, then frequency-selective filtering will reduce the effects of vibration. Specifically, vibration may be mostly rapidly changing acceleration, whereas orientation changes more slowly, and filtering can remove the rapidly varying components, leaving behind only the slowly varying components. p.186

To understand the degree to which frequency-selective filtering helps, we need to have a model of both the desired signal x and the noise n. Reasonable models are usually statistical, and analysis of the signals requires using the techniques of random processes, estimation, and machine learning. Although such analysis is beyond the scope of this text, we can gain insight that is useful in many practical circumstances through a purely deterministic analysis.

Our approach will be to condition the signal $x' = x + n$ by filtering it with an LTI system S p.30
called a **conditioning filter**. Let the output of the conditioning filter be given by

$$y = S(x') = S(x+n) = S(x) + S(n),$$

where we have used the linearity assumption on S. Let the residual error signal after filtering be defined to be

$$r = y - x = S(x) + S(n) - x. \tag{7.10}$$

This signal tells us how far off the filtered output is from the desired signal. Let R denote the RMS value of r, and X the RMS value of x. Then the SNR after filtering is p.178

$$SNR_{dB} = 20 \log_{10} \left(\frac{X}{R} \right),$$

We would like to design the conditioning filter S to maximize this SNR. Since X does not depend on S, we maximize this SNR if we minimize R. That is, we choose S to minimize the RMS value of r in (7.10).

Although determination of this filter requires statistical methods beyond the scope of this text, we can draw some intuitively appealing conclusions by examining (7.10). It is easy to show that the denominator is bounded as follows,

$$R = RMS(r) \leq RMS(S(x) - x) + RMS(n) \tag{7.11}$$

where RMS is the function defined by (7.7). This suggests that we may be able to minimize R by making $S(x)$ close to x (i.e., make $S(x) \approx x$) while making $RMS(n)$ small. That is, the filter S should do minimal damage to the desired signal x while filtering out as much as possible of the noise.

As illustrated in Example 7.3, x and n often differ in frequency content. In that example, x contains only low frequencies, and n contains only higher frequencies. Therefore, the best choice for S will be a lowpass filter.

7.2 Common Sensors

In this section, we describe a few sensors and show how to obtain and use reasonable models of these sensors.

7.2.1 Measuring Tilt and Acceleration

An **accelerometer** is a sensor that measures **proper acceleration**, which is the acceleration of an object as observed by an observer in free fall. As we explain here, gravitational force is indistinguishable from acceleration, and therefore an accelerometer measures not just acceleration, but also gravitational force. This result is a precursor to Albert Einstein's Theory of General Relativity and is known as Einstein's **equivalence principle** (Einstein, 1907).

A schematic view of an accelerometer is shown in Figure 7.3. A movable mass is attached via a spring to a fixed frame. Assume that the sensor circuitry can measure the position of the movable mass relative to the fixed frame (this can be done, for example, by measuring capacitance). When the frame accelerates in the direction of the double arrow in the figure, the acceleration results in displacement of the movable mass, and hence this acceleration can be measured.

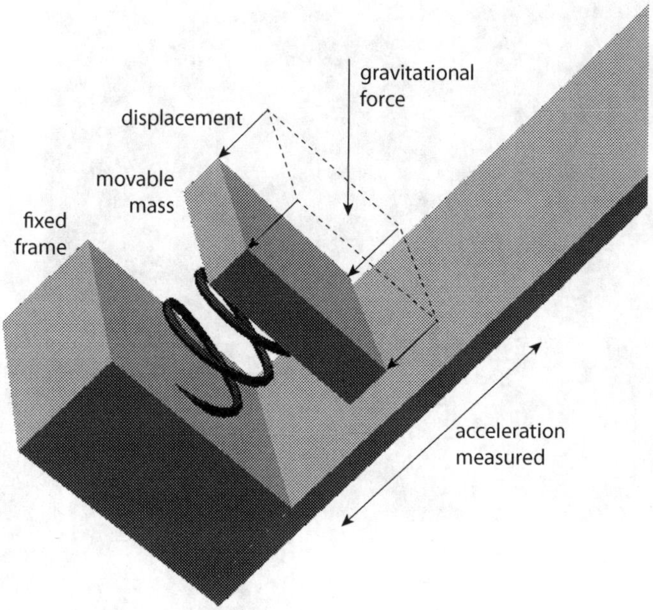

Figure 7.3: A schematic of an accelerometer as a spring-mass system.

The movable mass has a neutral position, which is its position when the spring is not deformed at all. It will occupy this neutral position if the entire assembly is in free fall, or if the assembly is lying horizontally. If the assembly is instead aligned vertically, then gravitational force will compress the spring and displace the mass. To an observer in free fall, this looks exactly as if the assembly were accelerating upwards at the **acceleration of gravity**, which is approximately $g = 9.8$ meters/second^2.

An accelerometer, therefore, can measure the tilt (relative to gravity) of the fixed frame. Any acceleration experienced by the fixed frame will add or subtract from this measurement. It can be challenging to separate these two effects, gravity and acceleration. The combination of the two is what we call proper acceleration. p.186

Assume x is the proper acceleration of the fixed frame of an accelerometer at a particular time. A digital accelerometer will produce a measurement $f(x)$ where

$$f\colon (L,H) \to \{0,\ldots,2^b - 1\}$$

where $L \in \mathbb{R}$ is the minimum measurable proper acceleration and $H \in \mathbb{R}$ is the maximum, and $b \in \mathbb{N}$ is the number of bits of the ADC.

Figure 7.4: A silicon accelerometer consists of flexible silicon fingers that deform under gravitational pull or acceleration (Lemkin and Boser, 1999).

Today, accelerometers are typically implemented in silicon (see Figure 7.4), where silicon fingers deform under gravitational pull or acceleration (see for example Lemkin and Boser (1999)). Circuitry measures the deformation and provides a digital reading. Often, three accelerometers are packaged together, giving a three-axis accelerometer. This can be used to measure orientation of an object relative to gravity, plus acceleration in any direction in three-dimensional space.

7.2.2 Measuring Position and Velocity

In theory, given a measurement x of acceleration over time, it is possible to determine the velocity and location of an object. Consider an object moving in a one-dimensional space. Let the position of the object over time be $p\colon \mathbb{R}_+ \to \mathbb{R}$, with initial position $p(0)$. Let the velocity of the object be $v\colon \mathbb{R}_+ \to \mathbb{R}$, with initial velocity $v(0)$. And let the acceleration be

$x\colon \mathbb{R}_+ \to \mathbb{R}$. Then

$$p(t) = p(0) + \int_0^t v(\tau)d\tau,$$

and

$$v(t) = v(0) + \int_0^t x(\tau)d\tau.$$

Note, however, that if there is a non-zero bias in the measurement of acceleration, then $p(t)$ will have an error that grows proportionally to t^2. Such an error is called **drift**, and it makes using an accelerometer alone to determine position not very useful. However, if the position can be periodically reset to a known-good value, using for example GPS, then an accelerometer becomes useful to approximate the position between such settings. p.173

In some circumstances, we can measure velocity of an object moving through a medium. For example, an **anemometer** (which measures air flow) can estimate the velocity of an aircraft relative to the surrounding air. But using this measurement to estimate position is again subject to drift, particularly since the movement of the surrounding air ensures bias.

Direct measurements of position are difficult. The **global positioning system** (**GPS**) is a sophisticated satellite-based navigation system using triangulation. A GPS receiver listens for signals from four or more GPS satellites that carry extremely precise clocks. The satellites transmit a signal that includes the time of transmission and the location of the satellite at the time of transmission. If the receiver were to have an equally precise clock, then upon receiving such a signal from a satellite, it would be able to calculate its distance from the satellite using the speed of light. Given three such distances, it would be able to calculate its own position. However, such precise clocks are extremely expensive. Hence, the receiver uses a fourth such distance measurement to get a system of four equations with four unknowns, the three dimensions of its location and the error in its own local clock.

The signal from GPS satellites is relatively weak and is easily blocked by buildings and other obstacles. Other mechanisms must be used, therefore, for indoor localization. One such mechanism is **WiFi fingerprinting**, where a device uses the known location of WiFi access points, the signal strength from those access points, and other local information. Another technology that is useful for indoor localization is **bluetooth**, a short-distance wireless communication standard. Bluetooth signals can be used as beacons, and signal strength can give a rough indication of distance to the beacon.

Strength of a radio signal is a notoriously poor measure of distance because it is subject to local diffraction and reflection effects on the radio signal. In an indoor environment, a radio signal is often subject to **multipath**, where it propagates along more than one path to a target, and at the target experiences either constructive or destructive interference. Such interference introduces wide variability in signal strength that can lead to misleading measures of distance. As of this writing, mechanisms for accurate indoor localization are not widely available, in notable contrast to outdoor localization, where GPS is available worldwide.

7.2.3 Measuring Rotation

A **gyroscope** is a device that measures changes in orientation (rotation). Unlike an accelerometer, it is (mostly) unaffected by a gravitational field. Traditional gyroscopes are bulky rotating mechanical devices on a double gimbal mount. Modern gyroscopes are either MEMS devices (microelectromechanical systems) using small resonating structures, or optical devices that measure the difference in distance traveled by a laser beam around a closed path in opposite directions, or (for extremely high precision) devices that leverage quantum effects.

Gyroscopes and accelerometers may be combined to improve the accuracy of **inertial navigation**, where position is estimated by **dead reckoning**. Also called **ded reckoning** (for deduced reckoning), dead reckoning starts from a known initial position and orientation, and then uses measurements of motion to estimate subsequent position and orientation. An **inertial measurement unit (IMU)** or **inertial navigation system (INS)** uses a gyroscope to measure changes in orientation and an accelerometer to measure changes in velocity. Such units are subject to drift, of course, so they are often combined with GPS units, which can periodically provided "known good" location information (though not orientation). IMUs can get quite sophisticated and expensive.

7.2.4 Measuring Sound

A **microphone** measures changes in sound pressure. A number of techniques are used, including electromagnetic induction (where the sound pressure causes a wire to move in a magnetic field), capacitance (where the distance between a plate deformed by the sound pressure and a fixed plate varies, causing a measurable change in capacitance), or the piezoelectric effect (where charge accumulates in a crystal due to mechanical stress).

Microphones for human audio are designed to give low distortion and low noise within the human hearing frequency range, about 20 to 20,000 Hz. But microphones are also used outside this range. For example, an **ultrasonic rangefinder** emits a sound outside the human hearing range and listens for an echo. It can be used to measure the distance to a sound-reflecting surface.

7.2.5 Other Sensors

There are many more types of sensors. For example, measuring temperature is central to HVAC systems, automotive engine controllers, overcurrent protectection, and many industrial chemical processes. Chemical sensors can pick out particular pollutants, measure alcohol concentration, etc. Cameras and photodiodes measure light levels and color. Clocks measure the passage of time. p.49

A switch is a particularly simple sensor. Properly designed, it can sense pressure, tilt, or motion, for example, and it can often be directly connected to the GPIO pins of a microcontroller. One issue with switches, however, is that they may **bounce**. A mechanical switch that is based on closing an electrical contact has metal colliding with metal, and the establishment of the contact may not occur cleanly in one step. As a consequence, system designers need to be careful when reacting to the establishment of an electrical contact or they may inadvertently react several times to a single throwing of the switch. p.252

7.3 Actuators

As with sensors, the variety of available actuators is enormous. Since we cannot provide comprehensive coverage here, we discuss two common examples, LEDs and motor control. Further details may be found in Chapter 10, which discusses particular microcontroller I/O designs.

7.3.1 Light-Emitting Diodes

Very few actuators can be driven directly from the digital I/O pins (GPIO pins) of a microcontroller. These pins can source or sink a limited amount of current, and any attempt to exceed this amount risks damaging the circuits. One exception is **light-emitting diodes (LEDs)**, which when put in series with a resistor, can often be connected directly to a GPIO pin. This provides a convenient way for an embedded system to provide a visual indication of some activity. p.252

Example 7.11: Consider a microcontroller that operates at 3 volts from a coin-cell battery and specifies that its GPIO pins can sink up to 18 mA. Suppose that you wish to turn on and off an LED under software control (see Chapter 10 for how to do this). Suppose you use an LED that, when forward biased (turned on), has a voltage drop of 2 volts. Then what is the smallest resistor you can put in series with the LED to safely keep the current within the 18 mA limit? **Ohm's law** states

$$V_R = IR, \tag{7.12}$$

where V_R is the voltage across the resistor, I is the current, and R is the resistance. The resistor will have a voltage drop of $V_R = 3 - 2 = 1$ volt across it (two of the 3 supply volts drop across the LED), so the current flowing through it will be

$$I = 1/R.$$

To limit this current to 18 mA, we require a resistance

$$R \geq 1/0.018 \approx 56 \text{ ohms}.$$

If you choose a 100 ohm resistor, then the current flowing through the resistor and the LED is

$$I = V_R/100 = 10\text{mA}.$$

If the battery capacity is 200 mAh (milliamp-hours), then driving the LED for 20 hours will completely deplete the battery, not counting any power dissipation in the microcontroller or other circuits. The power dissipated in the resistor will be

$$P_R = V_R I = 10mW.$$

The power dissipated in the LED will be

$$P_L = 2I = 20mW.$$

These numbers give an indication of the heat generated by the LED circuit (which will be modest).

The calculations in the previous example are typical of what you need to do to connect any device to a micro controller.

7.3.2 Motor Control

A **motor** applies a torque (angular force) to a load proportional to the current through the motor windings. It might be tempting, therefore, to apply a voltage to the motor proportional to the desired torque. However, this is rarely a good idea. First, if the voltage is digitally controlled through a DAC, then we have to be very careful to not exceed the current limits of the DAC. Most DACs cannot deliver much power, and require a power amplifier between the DAC and the device being powered. The input to a power amplifier has high impedance, meaning that at a given voltage, it draws very little current, so it can usually be connected directly to a DAC. The output, however, may involve substantial current. p.22 p.172

Example 7.12: An audio amplifier designed to drive 8 ohm speakers can often deliver 100 watts (peak) to the speaker. Power is equal to the product of voltage and current. Combining that with Ohm's law, we get that power is proportional to the square of current,

$$P = RI^2,$$

where R is the resistance. Hence, at 100 watts, the current through an 8 ohm speaker is

$$I = \sqrt{P/R} = \sqrt{100/8} \approx 3.5\text{amps},$$

which is a substantial current. The circuitry in the power amplifier that can deliver such a current without overheating and without introducing distortion is quite sophisticated.

Power amplifiers with good linearity (where the output voltage and current are proportional to the input voltage) can be expensive, bulky, and inefficient (the amplifier itself dissipates significant energy). Fortunately, when driving a motor, we do not usually need such a power amplifier. It is sufficient to use a switch that we can turn on and off with a digital signal from a microcontroller. Making a switch that tolerates high currents is much easier than making a power amplifier.

We use a technique called **pulse width modulation** (**PWM**), which can efficiently deliver large amounts of power under digital control, as long as the device to which the power is being delivered can tolerate rapidly switching on and off its power source. Devices that tolerate this include LEDs, incandescent lamps (this is how dimmers work), and DC motors. A PWM signal, as shown on the bottom of Figure 7.5, switches between a high level and a

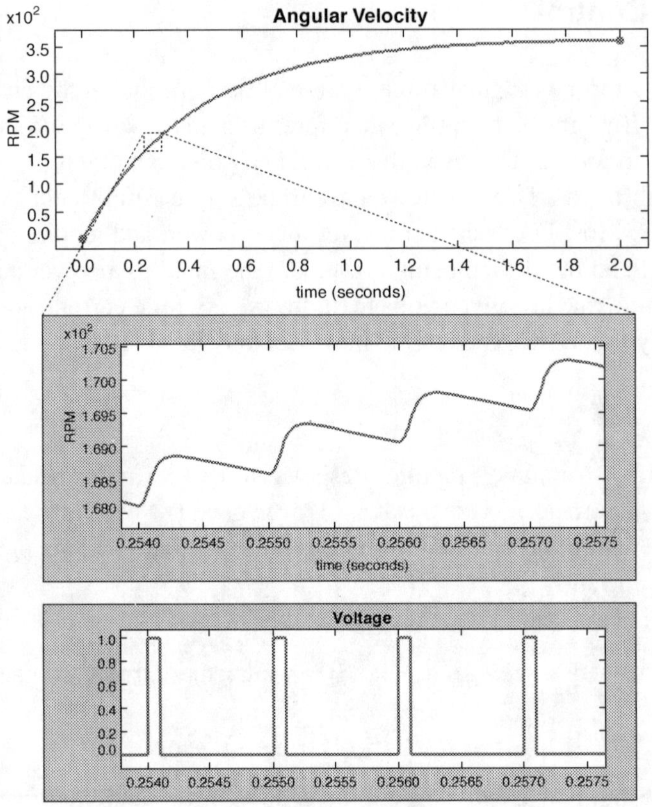

Figure 7.5: PWM control of a DC motor.

low level at a specified frequency. It holds the signal high for a fraction of the cycle period. This fraction is called the **duty cycle**, and in Figure 7.5, is 0.1, or 10%.

A **DC motor** consists of an electromagnet made by winding wires around a core placed in a magnetic field made with permanent magnets or electromagnets. When current flows through the wires, the core spins. Such a motor has both inertia and inductance that smooth its response when the current is abruptly turned on and off, so such motors tolerate PWM signals well.

Let $\omega\colon \mathbb{R} \to \mathbb{R}$ represent the angular velocity of the motor as a function of time. Assume we apply a voltage v to the motor, also a function of time. Then using basic circuit theory, we expect the voltage and current through the motor to satisfy the following equation,

$$v(t) = Ri(t) + L\frac{di(t)}{dt},$$

where R is the resistance and L the inductance of the coils in the motor. That is, the coils of the motor are modeled as series connection of a resistor and an inductor. The voltage drop across the resistor is proportional to current, and the voltage drop across the inductor is proportional to the rate of change of current.

However, motors exhibit a phenomenon that when a coil rotates in a magnetic field, it *generates* a current (and corresponding voltage). In fact, a motor can also function as an electrical generator; if instead of mechanically coupling it to a passive load, you couple it to a source of power that applies a torque to the motor, then the motor will generate electricity. p.22
Even when the motor is being used a motor rather than a generator, there will be some torque resisting the rotation, called the **back electromagnetic force**, due to this tendency to generate electricity when rotated. To account for this, the above equation becomes

$$v(t) = Ri(t) + L\frac{di(t)}{dt} + k_b\omega(t), \tag{7.13}$$

where k_b is an empirically determined **back electromagnetic force constant**, typically expressed in units of volts/RPM (volts per revolutions per minute).

Having described the electrical behavior of the motor in (7.13), we can use the techniques of Section 2.1 to describe the mechanical behavior. We can use the rotational version of Newton's second law, $F = ma$, which replaces the force F with torque, the mass m with p.21
moment of inertia and the acceleration a with angular acceleration. The torque T on the p.24
motor is proportional to the current flowing through the motor, adjusted by friction and any torque that might be applied by the mechanical load,

$$T(t) = k_T i(t) - \eta\omega(t) - \tau(t),$$

where k_T is an empirically determined **motor torque constant**, η is the kinetic friction of the motor, and τ is the torque applied by the load. By Newton's second law, this needs to be equal to the moment of inertia I times the angular acceleration, so

$$I\frac{d\omega(t)}{dt} = k_T i(t) - \eta\omega(t) - \tau(t). \tag{7.14}$$

Together, (7.14) and (7.13) describe how the motor responds to an applied voltage and mechanical torque.

Example 7.13: Consider a particular motor with the following parameters,

$$\begin{aligned} I &= 3.88 \times 10^{-7} \text{ kg}\cdot\text{meters}^2 \\ k_b &= 2.75 \times 10^{-4} \text{ volts/RPM} \\ k_T &= 5.9 \times 10^{-3} \text{ newton}\cdot\text{meters/amp} \\ R &= 1.71 \text{ ohms} \\ L &= 1.1 \times 10^{-4} \text{ henrys} \end{aligned}$$

Assume that there is no additional load on the motor, and we apply a PWM signal with frequency 1 kHz and duty cycle 0.1. Then the response of the motor is as shown in Figure 7.5, which has been calculated by numerically simulating according to equations (7.14) and (7.13). Notice that the motor settles at a bit more than 350 RPM after 2 seconds. As shown in the detailed plot, the angular velocity of the motor jitters at a rate of 1 kHz. It accelerates rapidly when the PWM signal is high, and decelerates when it is low, the latter due to friction and back electromagnetic force. If we increase the frequency of the PWM signal, then we can reduce the magnitude of this jitter.

In a typical use of a PWM controller to drive a motor, we will use the feedback control techniques of Section 2.4 to set the speed of the motor to a desired RPM. To do this, we require a measurement of the speed of the motor. We can use a sensor called a **rotary encoder**, or just **encoder**, which reports either the angular position or velocity (or both) of a rotary shaft. There are many different designs for such encoders. A very simple one provides an electrical pulse each time the shaft rotates by a certain angle, so that counting pulses per unit time will provide a measurement of the angular velocity.

7.4 Summary

The variety of sensors and actuators that are available to engineers is enormous. In this chapter, we emphasize *models* of these sensors and actuators. Such models are an essential part of the toolkit of embedded systems designers. Without such models, engineers would be stuck with guesswork and experimentation.

Exercises

1. Show that the composition $f \circ g$ of two affine functions f and g is affine. p.173

2. The dynamic range of human hearing is approximately 100 decibels. Assume that the smallest difference in sound levels that humans can effectively discern is a sound pressure of about 20 μPa (micropascals).

   (a) Assuming a dynamic range of 100 decibels, what is the sound pressure of the loudest sound that humans can effectively discriminate?

   (b) Assume a perfect microphone with a range that matches the human hearing range. What is the minimum number of bits that an ADC should have to match the dynamic range of human hearing? p.171

3. The following questions are about how to determine the function

$$f\colon (L,H) \to \{0,\ldots,2^B-1\},$$

   for an accelerometer, which given a proper acceleration x yields a digital number $f(x)$. We will assume that x has units of "g's," where 1g is the acceleration of gravity, approximately $g = 9.8\text{meters}/\text{second}^2$. p.186 p.187

   (a) Let the bias $b \in \{0,\ldots,2^B-1\}$ be the output of the ADC when the accelerometer measures no proper acceleration. How can you measure b? p.173 p.186

   (b) Let $a \in \{0,\ldots,2^B-1\}$ be the *difference* in output of the ADC when the accelerometer measures 0g and 1g of acceleration. This is the ADC conversion of the sensitivity of the accelerometer. How can you measure a? p.174

   (c) Suppose you have measurements of a and b from parts (3b) and (3a). Give an affine function model for the accelerometer, assuming the proper acceleration is x in units of g's. Discuss how accurate this model is. p.173

   (d) Given a measurement $f(x)$ (under the affine model), find x, the proper acceleration in g's.

   (e) The process of determining a and b by measurement is called **calibration** of the sensor. Discuss why it might be useful to individually calibrate each particular accelerometer, rather than assume fixed calibration parameters a and b for a collection of accelerometers.

   (f) Suppose you have an ideal 8-bit digital accelerometer that produces the value $f(x) = 128$ when the proper acceleration is 0g, value $f(x) = 1$ when the proper acceleration is 3g to the right, and value $f(x) = 255$ when the proper acceleration is 3g to the left. Find the sensitivity a and bias b. What is the dynamic range (in decibels) of this accelerometer? Assume the accelerometer never yields $f(x) = 0$.

p.174

4. (this problem is due to Eric Kim)

You are a Rebel Alliance fighter pilot evading pursuit from the Galactic Empire by hovering your space ship beneath the clouds of the planet Cory. Let the positive z direction point upwards and be your ship's position relative to the ground and v be your vertical velocity. The gravitational force is strong with this planet and induces an acceleration (in a vacuum) with absolute value g. The force from air resistance is linear with respect to velocity and is equal to rv, where the drag coefficient $r \leq 0$ is a constant parameter of the model. The ship has mass M. Your engines provide a vertical force.

(a) Let $L(t)$ be the input be the vertical lift force provided from your engines. Write down the dynamics for your ship for the position $z(t)$ and velocity $v(t)$. Ignore the scenario when your ship crashes. The right hand sides should contain $v(t)$ and $L(t)$.

(b) Given your answer to the previous problem, write down the explicit solution to $z(t)$ and $v(t)$ when the air resistance force is negligible and $r = 0$. At initial time $t = 0$, you are $30m$ above the ground and have an initial velocity of $-10\frac{m}{s}$. *Hint: Write $v(t)$ first then write $z(t)$ in terms of $v(t)$.*

(c) Draw an actor model using integrators, adders, etc. for the system that generates your vertical position and velocity. Make sure to label all variables in your actor model.

(d) Your engine is slightly damaged and you can only control it by giving a pure input, switch, that when present instantaneously switches the state of the engine from on to off and vice versa. When on, the engine creates a positive lift force L and when off $L = 0$. Your instrumentation panel contains an accelerometer. Assume your spaceship is level (i.e. zero pitch angle) and the accelerometer's positive z axis points upwards. Let the input sequence of engine switch commands be

$$\text{switch}(t) = \left\{ \begin{array}{ll} \textit{present} & \text{if } t \in \{.5, 1.5, 2.5, \ldots\} \\ \textit{absent} & \text{otherwise} \end{array} \right\}.$$

To resolve ambiguity at switching times $t = .5, 1.5, 2.5, \ldots$, at the moment of transition the engine's force takes on the new value instantaneously. Assume that air resistance is negligible (i.e. $r = 0$), ignore a crashed state, and the engine is on at $t = 0$.

Sketch the vertical component of the accelerometer reading as a function of time $t \in \mathbb{R}$. Label important values on the axes. *Hint: Sketching the graph for force first would be helpful.*

(e) If the spaceship is flying at a constant height, what is the value read by the accelerometer?

8

Embedded Processors

Contents

In **general-purpose computing**, the variety of instruction set architectures today is limited, with the Intel x86 architecture overwhelmingly dominating all. There is no such dominance in embedded computing. On the contrary, the variety of processors can be daunting to a system designer. Our goal in this chapter is to give the reader the tools and vocabulary to understand the options and to critically evaluate the properties of processors. We particularly focus on the mechanisms that provide concurrency and control over timing, because these issues loom large in the design of cyber-physical systems.

When deployed in a product, embedded processors typically have a dedicated function. They control an automotive engine or measure ice thickness in the Arctic. They are not asked to perform arbitrary functions with user-defined software. Consequently, the processors can be more specialized. Making them more specialized can bring enormous benefits. For example, they may consume far less energy, and consequently be usable with small batteries for long periods of time. Or they may include specialized hardware to perform operations that would be costly to perform on general-purpose hardware, such as image analysis.

When evaluating processors, it is important to understand the difference between an **instruction set architecture** (**ISA**) and a **processor realization** or a **chip**. The latter is a piece of silicon sold by a semiconductor vendor. The former is a definition of the instructions that the processor can execute and certain structural constraints (such as word size) that realizations must share. x86 is an ISA. There are many realizations. An ISA is an abstraction shared by many realizations. A single ISA may appear in many different chips, often made by different manufacturers, and often having widely varying performance profiles.

The advantage of sharing an ISA in a family of processors is that software tools, which are costly to develop, may be shared, and (sometimes) the same programs may run correctly on multiple realizations. This latter property, however, is rather treacherous, since an ISA does not normally include any constraints on timing. Hence, although a program may execute logically the same way on multiple chips, the system behavior may be radically different when the processor is embedded in a cyber-physical system.

8.1 Types of Processors

As a consequence of the huge variety of embedded applications, there is a huge variety of processors that are used. They range from very small, slow, inexpensive, low-power devices, to high-performance, special-purpose devices. This section gives an overview of some of the available types of processors.

8.1.1 Microcontrollers

A **microcontroller** (μC) is a small computer on a single integrated circuit consisting of a relatively simple **central processing unit** (**CPU**) combined with peripheral devices such as memories, I/O devices, and timers. By some accounts, more than half of all CPUs sold worldwide are microcontrollers, although such a claim is hard to substantiate because the difference between microcontrollers and general-purpose processors is indistinct. The simplest microcontrollers operate on 8-bit words and are suitable for applications that require small amounts of memory and simple logical functions (vs. performance-intensive arithmetic functions). They may consume extremely small amounts of energy, and often include a **sleep mode** that reduces the power consumption to nanowatts. Embedded components such as sensor network nodes and surveillance devices have been demonstrated that can operate on a small battery for several years.

Microcontrollers can get quite elaborate. Distinguishing them from general-purpose processors can get difficult. The Intel Atom, for example, is a family of x86 CPUs used mainly in netbooks and other small mobile computers. Because these processors are designed to use relatively little energy without losing too much performance relative to processors used in higher-end computers, they are suitable for some embedded applications and in servers where cooling is problematic. AMD's Geode is another example of a processor near the blurry boundary between general-purpose processors and microcontrollers.

8.1.2 DSP Processors

Many embedded applications do quite a bit of signal processing. A signal is a collection of sampled measurements of the physical world, typically taken at a regular rate called the sample rate. A motion control application, for example, may read position or location information from sensors at sample rates ranging from a few Hertz (Hz, or samples per second) to a few hundred Hertz. Audio signals are sampled at rates ranging from 8,000 Hz (or 8 kHz, the sample rate used in telephony for voice signals) to 44.1 kHz (the sample rate of CDs). Ultrasonic applications (such as medical imaging) and high-performance music applications may sample sound signals at much higher rates. Video typically uses sample rates of 25 or 30 Hz for consumer devices to much higher rates for specialty measurement applications. Each sample, of course, contains an entire image (called a frame), which itself has many samples (called pixels) distributed in space rather than time. Software-defined radio applications have sample rates that can range from hundreds of kHz (for baseband processing) to several GHz (billions of Hertz). Other embedded applications that make heavy use of signal processing include interactive games; radar, sonar, and LIDAR (light detection and ranging) imaging systems; video analytics (the extraction of information from video, for example for surveillance); driver-assist systems for cars; medical electronics; and scientific instrumentation.

Signal processing applications all share certain characteristics. First, they deal with large amounts of data. The data may represent samples in time of a physical processor (such

Microcontrollers

Most semiconductor vendors include one or more families of microcontrollers in their product line. Some of the architectures are quite old. The **Motorola 6800** and **Intel 8080** are 8-bit microcontrollers that appeared on the market in 1974. Descendants of these architectures survive today, for example in the form of the **Freescale 6811**. The **Zilog Z80** is a fully-compatible descendant of the 8080 that became one of the most widely manufactured and widely used microcontrollers of all time. A derivative of the Z80 is the Rabbit 2000 designed by Rabbit Semiconductor.

Another very popular and durable architecture is the **Intel 8051**, an 8-bit microcontroller developed by Intel in 1980. The 8051 ISA p.202 is today supported by many vendors, including Atmel, Infineon Technologies, Dallas Semiconductor, NXP, ST Microelectronics, Texas Instruments, and Cypress Semiconductor. The **Atmel AVR** 8-bit microcontroller, developed by Atmel in 1996, was one of the first microcontrollers to use on-chip flash memory p.231 for program storage. Although Atmel says AVR is not an acronym, it is believed that the architecture was conceived by two students at the Norwegian Institute of Technology, Alf-Egil Bogen and Vegard Wollan, so it may have originated as Alf and Vegard's RISC p.218.

Many 32-bit microcontrollers implement some variant of an **ARM** instruction set, developed by ARM Limited. ARM originally stood for Advanced RISC Machine, and before that Acorn RISC Machine, but today it is simply ARM. Processors that implement the ARM ISA are widely used in mobile phones to realize the user interface functions, as well as in many other embedded systems. Semiconductor vendors license the instruction set from ARM Limited and produce their own chips. ARM processors are currently made by Alcatel, Atmel, Broadcom, Cirrus Logic, Freescale, LG, Marvell Technology Group, NEC, NVIDIA, NXP, Samsung, Sharp, ST Microelectronics, Texas Instruments, VLSI Technology, Yamaha, and others.

Other notable embedded microcontroller architectures include the **Motorola ColdFire** (later the Freescale ColdFire), the **Hitachi H8** and SuperH, the **MIPS** (originally developed by a team led by John Hennessy at Stanford University), the **PIC** (originally Programmable Interface Controller, from Microchip Technology), and the **PowerPC** (created in 1991 by an alliance of Apple, IBM, and Motorola).

Programmable Logic Controllers

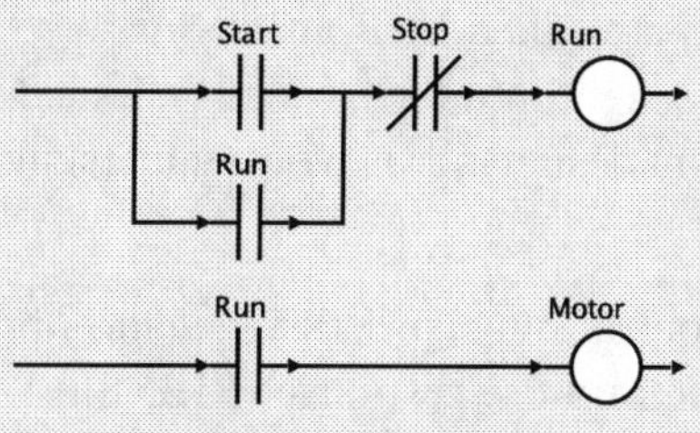

Ladder Logic Diagram

A **programmable logic controller** (**PLC**) is a specialized form of a microcontroller for industrial automation. PLCs originated as replacements for control circuits using electrical relays to control machinery. They are typically designed for continuous operation in hostile environments (high temperature, humidity, dust, etc.).

PLCs are often programmed using **ladder logic**, a notation originally used to specify logic constructed with relays and switches. A **relay** is a switch where the contact is controlled by coil. When a voltage is applied to the coil, the contact closes, enabling current to flow through the relay. By interconnecting contacts and coils, relays can be used to build digital controllers that follow specified patterns.

In common notation, a contact is represented by two vertical bars, and a coil by a circle, as shown in the diagram above. The above diagram has two **rungs**. The Motor coil on the lower rung turns a motor on or off. The Start and Stop contacts represent pushbutton switches. When an operator pushes the Start button, the contact is closed, and current can flow from the left (the power rail) to the right (ground). Start is a **normally open** contact. The Stop contact is **normally closed**, indicated by the slash, meaning that it becomes open when the operator pushes the switch. The logic in the upper rung is interesting. When the operator pushes Start, current flows to the Run coil, causing both Run contacts to close. The motor will run, even after the Start button is released. When the operator pushes Stop, current is interrupted, and both Run contacts become open, causing the motor to stop. Contacts wired in parallel perform a logical OR function, and contacts wired in series perform a logical AND. The upper rung has feedback; the meaning of the rung is a fixed point solution to the logic equation implied by the diagram. p.137

Today, PLCs are just microcontrollers in rugged packages with I/O interfaces suitable for industrial control, and ladder logic is a graphical programming notation for programs. These diagrams can get quite elaborate, with thousands of rungs. For details, we recommend Kamen (1999).

as samples of a wireless radio signal), samples in space (such as images), or both (such as video and radar). Second, they typically perform sophisticated mathematical operations on the data, including filtering, system identification, frequency analysis, machine learning, and feature extraction. These operations are mathematically intensive.

Processors designed specifically to support numerically intensive signal processing applications are called **DSP processors**, or **DSP**s (**digital signal processor**s), for short. To get some insight into the structure of such processors and the implications for the embedded software designer, it is worth understanding the structure of typical signal processing algorithms.

A canonical signal processing algorithm, used in some form in all of the above applications, is **finite impulse response (FIR)** filtering. The simplest form of this algorithm is straightforward, but has profound implications for hardware. In this simplest form, an input signal x consists of a very long sequence of numerical values, so long that for design purposes it should be considered infinite. Such an input can be modeled as a function $x\colon \mathbb{N} \to D$, where D is a set of values in some data type.[1] For example, D could be the set of all 16-bit integers, in which case, $x(0)$ is the first input value (a 16-bit integer), $x(1)$ is the second input value, etc. For mathematical convenience, we can augment this to $x\colon \mathbb{Z} \to D$ by defining $x(n) = 0$ for all $n < 0$. For each input value $x(n)$, an FIR filter must compute an output value $y(n)$

[1] For a review of this notation, see Appendix A on page 471.

The x86 Architecture

p.202 The dominant ISA for desktop and portable computers is known as the **x86**. This term originates with the Intel 8086, a 16-bit microprocessor chip designed by Intel in 1978. A variant of the 8086, designated the 8088, was used in the original IBM PC, and the processor family has dominated the PC market ever since. Subsequent processors in this family were given names ending in "86," and generally maintained backward compatibility. The Intel 80386 was the first 32-bit version of this instruction set, introduced in 1985. Today, the term "x86" usually refers to the 32-bit version, with 64-bit versions designated "x86-64." The **Intel Atom**, introduced in 2008, is an x86 processor with significantly reduced energy consumption. Although it is aimed primarily at netbooks and other small mobile computers, it is also an attractive option for some embedded applications. The x86 architecture has also been implemented in processors from AMD, Cyrix, and several other manufacturers.

according to the formula,

$$y(n) = \sum_{i=0}^{N-1} a_i x(n-i) , \quad (8.1)$$

where N is the length of the FIR filter, and the coefficients a_i are called its **tap values**. You can see from this formula why it is useful to augment the domain of the function x, since the computation of $y(0)$, for example, involves values $x(-1)$, $x(-2)$, etc.

Example 8.1: Suppose $N = 4$ and $a_0 = a_1 = a_2 = a_3 = 1/4$. Then for all $n \in \mathbb{N}$,

$$y(n) = (x(n) + x(n-1) + x(n-2) + x(n-3))/4 .$$

DSP Processors

Specialized computer architectures for signal processing have been around for quite some time (Allen, 1975). Single-chip DSP microprocessors first appeared in the early 1980s, beginning with the Western Electric DSP1 from Bell Labs, the S28211 from AMI, the TMS32010 from Texas Instruments, the uPD7720 from NEC, and a few others. Early applications of these devices included voiceband data modems, speech synthesis, consumer audio, graphics, and disk drive controllers. A comprehensive overview of DSP processor generations through the mid-1990s can be found in Lapsley et al. (1997).

Central characteristics of DSPs include a hardware multiply-accumulate unit; several variants of the Harvard architecture (to support multiple simultaneous data and program fetches); and addressing modes supporting auto increment, circular buffers, and bit-reversed addressing (the latter to support FFT calculation). Most support fixed-point data precisions of 16-24 bits, typically with much wider accumulators (40-56 bits) so that a large number of successive multiply-accumulate instructions can be executed without overflow. A few DSPs have appeared with floating point hardware, but these have not dominated the marketplace. p.233

DSPs are difficult to program compared to RISC architectures, primarily because of complex specialized instructions, a pipeline that is exposed to the programmer, and asymmetric memory architectures. Until the late 1990s, these devices were almost always programmed in assembly language. Even today, C programs make extensive use of libraries that are hand-coded in assembly language to take advantage of the most esoteric features of the architectures. p.218

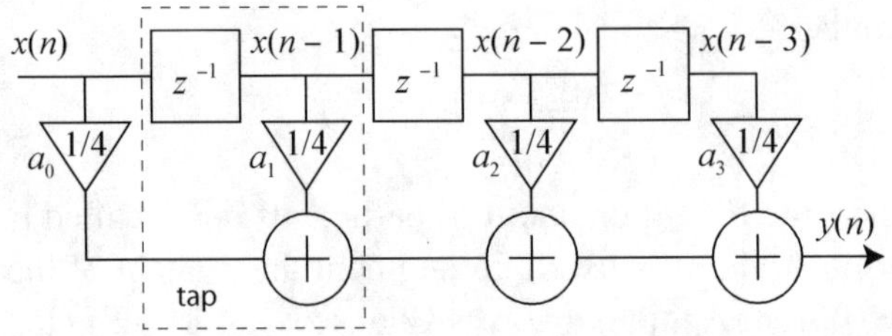

Figure 8.1: Structure of a tapped delay line implementation of the FIR filter of example 8.1. This diagram can be read as a dataflow diagram. For each $n \in \mathbb{N}$, each component in the diagram consumes one input value from each input path and produces one output value on each output path. The boxes labeled z^{-1} are unit delays. Their task is to produce on the output path the previous value of the input (or an initial value if there was no previous input). The triangles multiply their input by a constant, and the circles add their inputs.

Each output sample is the average of the most recent four input samples. The structure of this computation is shown in Figure 8.1. In that figure, input values come in from the left and propagate down the **delay line**, which is tapped after each delay element. This structure is called a **tapped delay line**.

The rate at which the input values $x(n)$ are provided and must be processed is called the **sample rate**. If you know the sample rate and N, you can determine the number of arithmetic operations that must be computed per second.

Example 8.2: Suppose that an FIR filter is provided with samples at a rate of 1 MHz (one million samples per second), and that $N = 32$. Then outputs must be computed at a rate of 1 MHz, and each output requires 32 multiplications and 31 additions. A processor must be capable of sustaining a computation rate of 63 million arithmetic operations per second to implement this application. Of course, to sustain the computation rate, it is necessary not only that the arithmetic hardware be fast enough, but also that the mechanisms for getting data in and out of memory and on and off chip be fast enough.

An image can be similarly modeled as a function $x: H \times V \to D$, where $H \subset \mathbb{N}$ represents the horizontal index, $V \subset \mathbb{N}$ represents the vertical index, and D is the set of all possible pixel values. A **pixel** (or picture element) is a sample representing the color and intensity of a point in an image. There are many ways to do this, but all use one or more numerical values for each pixel. The sets H and V depend on the **resolution** of the image.

Example 8.3: Analog television is steadily being replaced by digital formats such as **ATSC**, a set of standards developed by the Advanced Television Systems Committee. In the US, the vast majority of over-the-air **NTSC** transmissions (National Television System Committee) were replaced with ATSC on June 12, 2009. ATSC supports a number of frame rates ranging from just below 24 Hz to 60 Hz and a number of resolutions. High-definition video under the ATSC standard supports, for example, a resolution of 1080 by 1920 pixels at a frame rate of 30 Hz. Hence, $H = \{0, \cdots, 1919\}$ and $V = \{0, \cdots, 1079\}$. This resolution is called 1080p in the industry. Professional video equipment today goes up to four times this resolution (4320 by 7680). Frame rates can also be much higher than 30 Hz. Very high frame rates are useful for capturing extremely fast phenomena in slow motion.

For a grayscale image, a typical filtering operation will construct a new image y from an original image x according to the following formula,

$$\forall\, i \in H, j \in V, \quad y(i,j) = \sum_{n=-N}^{N} \sum_{m=-M}^{M} a_{n,m} x(i-n, j-m) , \tag{8.2}$$

where $a_{n,m}$ are the filter coefficients. This is a two-dimensional FIR filter. Such a calculation requires defining x outside the region $H \times V$. There is quite an art to this (to avoid edge effects), but for our purposes here, it suffices to get a sense of the structure of the computation without being concerned for this detail.

A color image will have multiple **color channels**. These may represent luminance (how bright the pixel is) and chrominance (what the color of the pixel is), or they may represent colors that can be composed to get an arbitrary color. In the latter case, a common choice is an **RGBA** format, which has four channels representing red, green, blue, and the alpha channel, which represents transparency. For example, a value of zero for R, G, and B represents the color black. A value of zero for A represents fully transparent (invisible). Each channel also has a maximum value, say 1.0. If all four channels are at the maximum, the resulting color is a fully opaque white.

The computational load of the filtering operation in (8.2) depends on the number of channels, the number of filter coefficients (the values of N and M), the resolution (the sizes of the sets H and V), and the frame rate.

Example 8.4: Suppose that a filtering operation like (8.2) with $N = 1$ and $M = 1$ (minimal values for useful filters) is to be performed on a high-definition video signal as in Example 8.3. Then each pixel of the output image y requires performing 9 multiplications and 8 additions. Suppose we have a color image with three channels (say, RGB, without transparency), then this will need to performed 3 times for each pixel. Thus, each frame of the resulting image will require $1080 \times 1920 \times 3 \times 9 = 55,987,200$ multiplications, and a similar number of additions. At 30 frames per second, this translates into $1,679,616,000$ multiplications per second, and a similar number of additions. Since this is about the simplest operation one may perform on a high-definition video signal, we can see that processor architectures handling such video signals must be quite fast indeed.

In addition to the large number of arithmetic operations, the processor has to handle the movement of data down the delay line, as shown in Figure 8.1 (see box on page 211). By providing support for delay lines and multiply-accumulate instructions, as shown in Example 8.6, DSP processors can realize one tap of an FIR filter in one cycle. In that cycle, they multiply two numbers, add the result to an accumulator, and increment or decrement two pointers using modulo arithmetic.

8.1.3 Graphics Processors

A **graphics processing unit** (**GPU**) is a specialized processor designed especially to perform the calculations required in graphics rendering. Such processors date back to the 1970s, when they were used to render text and graphics, to combine multiple graphic patterns, and to draw rectangles, triangles, circles, and arcs. Modern GPUs support 3D graphics, shading, and digital video. Dominant providers of GPUs today are Intel, NVIDIA and AMD.

Some embedded applications, particularly games, are a good match for GPUs. Moreover, GPUs have evolved towards more general programming models, and hence have started to appear in other compute-intensive applications, such as instrumentation. GPUs are typically quite power hungry, and therefore today are not a good match for energy constrained embedded applications.

Circular Buffers

An FIR filter requires a delay-line like that shown in Figure 8.1. A naive implementation p.206 would allocate an array in memory, and each time an input sample arrives, move each element in the array to the next higher location to make room for the new element in the first location. This would be enormously wasteful of memory bandwidth. A better approach is to use a **circular buffer**, where an array in memory is interpreted as having a ring-like structure, as shown below for a length-8 delay line:

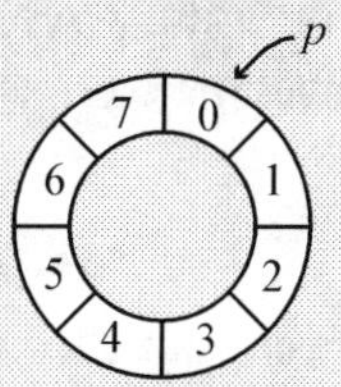

Here, 8 successive memory locations, labeled 0 to 7, store the values in the delay line. A pointer p, initialized to location 0, provides access.

An FIR filter can use this circular buffer to implement the summation of (8.1). One implementation first accepts a new input value $x(n)$, and then calculates the summation backwards, beginning with the $i = N-1$ term, where in our example, $N = 8$. Suppose that when the nth input arrives, the value of p is some number $p_i \in \{0, \cdots, 7\}$ (for the first input $x(0)$, $p_i = 0$). The program writes the new input $x(n)$ into the location given by p and then increments p, setting $p = p_i + 1$. All arithmetic on p is done modulo 8, so for example, if $p_i = 7$, then $p_i + 1 = 0$. The FIR filter calculation then reads $x(n-7)$ from location $p = p_i + 1$ and multiplies it by a_7. The result is stored in an **accumulator** register. It again increments p by one, setting it to $p = p_i + 2$. It next reads $x(n-6)$ from location $p = p_i + 2$, multiplies it by a_6, and adds the result to the accumulator (this explains the name "accumulator" for the register, since it accumulates the products in the tapped delay line). It continues until it reads $x(n)$ from location $p = p_i + 8$, which because of the modulo operation is the same location that the latest input $x(n)$ was written to, and multiplies that value by a_0. It again increments p, getting $p = p_i + 9 = p_i + 1$. Hence, at the conclusion of this operation, the value of p is $p_i + 1$, which gives the location into which the next input $x(n+1)$ should be written.

8.2 Parallelism

Most processors today provide various forms of parallelism. These mechanisms strongly affect the timing of the execution of a program, so embedded system designers have to understand them. This section provides an overview of the several forms and their consequences for system designers.

8.2.1 Parallelism vs. Concurrency

p.130 Concurrency is central to embedded systems. A computer program is said to be concurrent if different parts of the program *conceptually* execute simultaneously. A program is said to be **parallel** if different parts of the program *physically* execute simultaneously on distinct hardware (such as on multicore machines, on servers in a server farm, or on distinct microprocessors).

Non-concurrent programs specify a *sequence* of instructions to execute. A programming language that expresses a computation as a sequence of operations is called an **imperative** language. C is an imperative language. When using C to write concurrent programs, we must step outside the language itself, typically using a **thread library**. A thread library uses facilities provided not by C, but rather provided by the operating system and/or the hardware. Java is a mostly imperative language extended with constructs that directly support threads. Thus, one can write concurrent programs in Java without stepping outside the language.

Every (correct) execution of a program in an imperative language must behave as if the instructions were executed exactly in the specified sequence. It is often possible, however, to execute instructions in parallel or in an order different from that specified by the program and still get behavior that matches what would have happened had they been executed in sequence.

Example 8.5: Consider the following C statements:

```
double pi, piSquared, piCubed;
pi = 3.14159;
piSquared = pi * pi ;
piCubed = pi * pi * pi;
```

The last two assignment statements are independent, and hence can be executed in parallel or in reverse order without changing the behavior of the program. Had we written them as follows, however, they would no longer be independent:

```
double pi, piSquared, piCubed;
pi = 3.14159;
piSquared = pi * pi ;
piCubed = piSquared * pi;
```

In this case, the last statement depends on the third statement in the sense that the third statement must complete execution before the last statement starts.

A compiler may analyze the dependencies between operations in a program and produce parallel code, if the target machine supports it. This analysis is called **dataflow analysis**. Many microprocessors today support parallel execution, using multi-issue instruction streams or VLIW (very large instruction word) architectures. Processors with multi-issue p.221 instruction streams can execute independent instructions simultaneously. The hardware analyzes instructions on-the-fly for dependencies, and when there is no dependency, executes more than one instruction at a time. In the latter, VLIW machines have assembly-level instructions that specify multiple operations to be performed together. In this case, the compiler is usually required to produce the appropriate parallel instructions. In these cases, the dependency analysis is done at the level of assembly language or at the level of individual operations, not at the level of lines of C. A line of C may specify multiple operations, or even complex operations like procedure calls. In both cases (multi-issue and VLIW), an imperative program is analyzed for concurrency in order to enable parallel execution. The overall objective is to speed up execution of the program. The goal is improved **performance**, where the presumption is that finishing a task earlier is always better than finishing it later.

In the context of embedded systems, however, concurrency plays a part that is much more central than merely improving performance. Embedded programs interact with physical processes, and in the physical world, many activities progress at the same time. An embedded program often needs to monitor and react to multiple concurrent sources of stimulus, and simultaneously control multiple output devices that affect the physical world. Embedded programs are almost always concurrent programs, and concurrency is an intrinsic part of the logic of the programs. It is not just a way to get improved performance. Indeed, finishing a task earlier is not necessarily better than finishing it later. *Timeliness* matters, of course; actions performed in the physical world often need to be done at the *right time* (neither early nor late). Picture for example an engine controller for a gasoline engine. Firing the spark plugs earlier is most certainly not better than firing them later. They must be fired at the *right* time.

Just as imperative programs can be executed sequentially or in parallel, concurrent programs can be executed sequentially or in parallel. Sequential execution of a concurrent program is done typically today by a **multitasking operating system**, which interleaves the execution of multiple tasks in a single sequential stream of instructions. Of course, the hardware may parallelize that execution if the processor has a multi-issue or VLIW architecture. Hence, a concurrent program may be converted to a sequential stream by an operating system and back to concurrent program by the hardware, where the latter translation is done to improve performance. These multiple translations greatly complicate the problem of ensuring that things occur at the *right* time. This problem is addressed in Chapter 12.

Parallelism in the hardware, the main subject of this chapter, exists to improve performance for computation-intensive applications. From the programmer's perspective, concurrency arises as a consequence of the hardware designed to improve performance, not as a consequence of the application problem being solved. In other words, the application does not (necessarily) demand that multiple activities proceed simultaneously, it just demands that things be done very quickly. Of course, many interesting applications will combine both forms of concurrency, arising from parallelism and from application requirements.

The sorts of algorithms found in compute-intensive embedded programs has a profound affect on the design of the hardware. In this section, we focus on hardware approaches that deliver parallelism, namely pipelining, instruction-level parallelism, and multicore architectures. All have a strong influence on the programming models for embedded software. In Chapter 9, we give an overview of memory systems, which strongly influence how parallelism is handled.

8.2.2 Pipelining

Most modern processors are **pipelined**. A simple five-stage pipeline for a 32-bit machine is shown in Figure 8.2. In the figure, the shaded rectangles are latches, which are clocked at processor clock rate. On each edge of the clock, the value at the input is stored in the latch register. The output is then held constant until the next edge of the clock, allowing the p.132 circuits between the latches to settle. This diagram can be viewed as a synchronous-reactive model of the behavior of the processor.

In the fetch (leftmost) stage of the pipeline, a **program counter** (**PC**) provides an address to the instruction memory. The instruction memory provides encoded instructions, which in the figure are assumed to be 32 bits wide. In the fetch stage, the PC is incremented by 4 (bytes), to become the address of the next instruction, unless a conditional branch instruction is providing an entirely new address for the PC. The decode pipeline stage extracts register addresses from the 32-bit instruction and fetches the data in the specified registers from the register bank. The execute pipeline stage operates on the data fetched from the registers or on the PC (for a computed branch) using an **arithmetic logic unit** (**ALU**), which performs arithmetic and logical operations. The memory pipeline stage reads or writes to a memory location given by a register. The writeback pipeline stage stores results in the register file.

DSP processors normally add an extra stage or two that performs a multiplication, provide separate ALUs for address calculation, and provide a dual data memory for simultaneous access to two operands (this latter design is known as a Harvard architecture). But the simple version without the separate ALUs suffices to illustrate the issues that an embedded system designer faces. p.233

The portions of the pipeline between the latches operate in parallel. Hence, we can see immediately that there are simultaneously five instructions being executed, each at a different stage of execution. This is easily visualized with a **reservation table** like that in Figure 8.3. The table shows hardware resources that may be simultaneously used on the left. In this

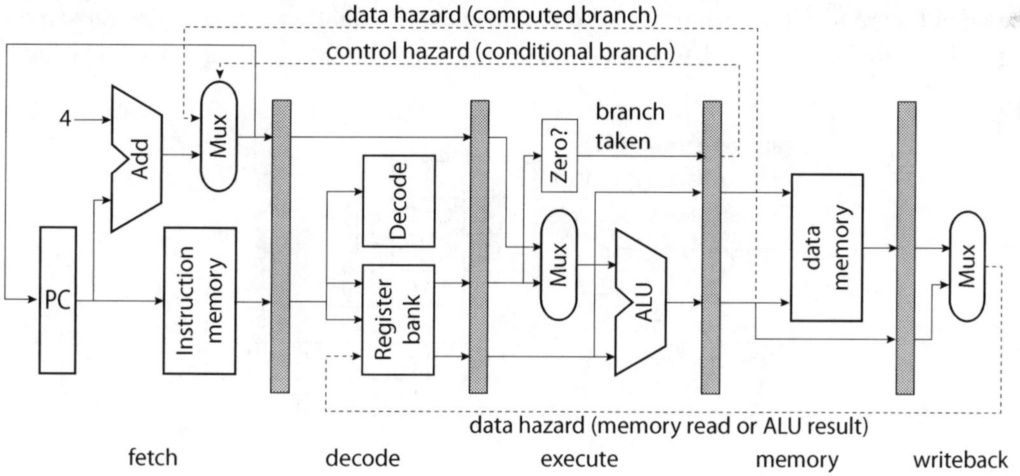

Figure 8.2: Simple pipeline (after Patterson and Hennessy (1996)).

case, the register bank appears three times because the pipeline of Figure 8.2 assumes that two reads and write of the register file can occur in each cycle.

The reservation table in Figure 8.3 shows a sequence A, B, C, D, E of instructions in a program. In cycle 5, E is being fetched while D is reading from the register bank, while C is using the ALU, while B is reading from or writing to data memory, while A is writing results to the register bank. The write by A occurs in cycle 5, but the read by B occurs in cycle 3. Thus, the value that B reads will not be the value that A writes. This phenomenon is known as a **data hazard**, one form of **pipeline hazard**. Pipeline hazards are caused by the dashed lines in Figure 8.2. Programmers normally expect that if instruction A is before instruction B, then any results computed by A will be available to B, so this behavior may not be acceptable.

Computer architects have tackled the problem of pipeline hazards in a variety of ways. The simplest technique is known as an **explicit pipeline**. In this technique, the pipeline hazard is simply documented, and the programmer (or compiler) must deal with it. For the example where B reads a register written by A, the compiler may insert three **no-op** instructions (which do nothing) between A and B to ensure that the write occurs before the read. These no-op instructions form a **pipeline bubble** that propagates down the pipeline.

A more elaborate technique is to provide **interlocks**. In this technique, the instruction decode hardware, upon encountering instruction B that reads a register written by A, will detect the hazard and delay the execution of B until A has completed the writeback stage. For this pipeline, B should be delayed by three clock cycles to permit A to complete, as shown in Figure 8.4. This can be reduced to two cycles if slightly more complex **forwarding** logic is provided, which detects that A is writing the same location that B is reading, and

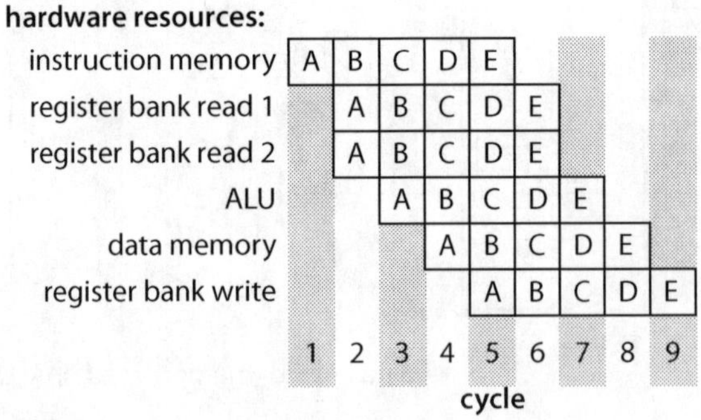

Figure 8.3: Reservation table for the pipeline shown in Figure 8.2.

directly provides the data rather than requiring the write to occur before the read. Interlocks therefore provide hardware that automatically inserts pipeline bubbles.

A still more elaborate technique is **out-of-order execution**, where hardware is provided that detects a hazard, but instead of simply delaying execution of B, proceeds to fetch C, and if C does not read registers written by either A or B, and does not write registers read by B, then proceeds to execute C before B. This further reduces the number of pipeline bubbles.

Another form of pipeline hazard illustrated in Figure 8.2 is a **control hazard**. In the figure, a conditional branch instruction changes the value of the PC if a specified register has value zero. The new value of the PC is provided (optionally) by the result of an ALU operation. In this case, if A is a conditional branch instruction, then A has to have reached the memory stage before the PC can be updated. The instructions that follow A in memory will have been fetched and will be at the decode and execute stages already by the time it is determined that those instructions should not in fact be executed.

Like data hazards, there are multiple techniques for dealing with control hazards. A **delayed branch** simply documents the fact that the branch will be taken some number of cycles after it is encountered, and leaves it up to the programmer (or compiler) to ensure that the instructions that follow the conditional branch instruction are either harmless (like no-ops) or do useful work that does not depend on whether the branch is taken. An interlock provides hardware to insert pipeline bubbles as needed, just as with data hazards. In the most elaborate technique, **speculative execution**, hardware estimates whether the branch is likely to be taken, and begins executing the instructions it expects to execute. If its

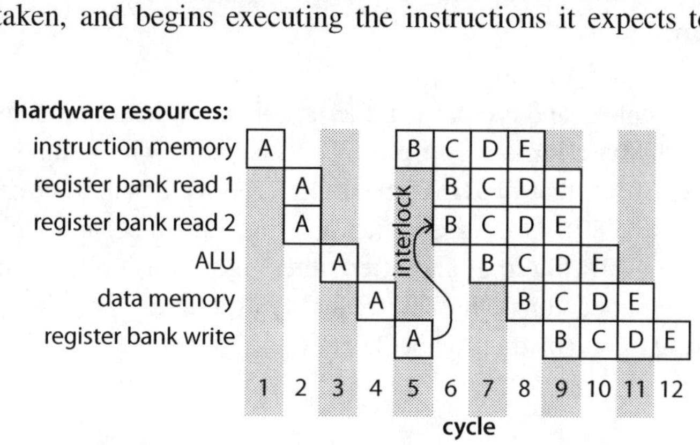

Figure 8.4: Reservation table for the pipeline shown in Figure 8.2 with interlocks, assuming that instruction B reads a register that is written by instruction A.

expectation is not met, then it undoes any side effects (such as register writes) that the speculatively executed instructions caused.

Except for explicit pipelines and delayed branches, all of these techniques introduce variability in the timing of execution of an instruction sequence. Analysis of the timing of a program can become extremely difficult when there is a deep pipeline with elaborate forwarding and speculation. Explicit pipelines are relatively common in DSP processors, which are often applied in contexts where precise timing is essential. Out-of-order and speculative execution are common in general-purpose processors, where timing matters only in an aggregate sense. An embedded system designer needs to understand the requirements of the application and avoid processors where the requisite level of timing precision is unachievable.

8.2.3 Instruction-Level Parallelism

Achieving high performance demands parallelism in the hardware. Such parallelism can take two broad forms, multicore architectures, described later in Section 8.2.4, or **instruction-level parallelism** (**ILP**), which is the subject of this section. A processor supporting ILP is able to perform multiple independent operations in each instruction cycle. We discuss four major forms of ILP: CISC instructions, subword parallelism, superscalar, and VLIW.

CISC Instructions

A processor with complex (and typically, rather specialized) instructions is called a **CISC** machine (**complex instruction set computer**). The philosophy behind such processors is distinctly different from that of **RISC** machines (**reduced instruction set computers**)
p.206 (Patterson and Ditzel, 1980). DSPs are typically CISC machines, and include instructions
p.206 specifically supporting FIR filtering (and often other algorithms such as FFTs (fast Fourier transforms) and Viterbi decoding). In fact, to qualify as a DSP, a processor must be able to perform FIR filtering in one instruction cycle per tap.

Example 8.6: The Texas Instruments TMS320c54x family of DSP processors is intended to be used in power-constrained embedded applications that demand high signal processing performance, such as wireless communication systems and personal digital assistants (**PDA**s). The inner loop of an FIR computation (8.1) is

```
RPT numberOfTaps - 1
MAC *AR2+, *AR3+, A
```

The first instruction illustrates the **zero-overhead loops** commonly found in DSPs. The instruction that comes after it will execute a number of times equal to one plus the argument of the `RPT` instruction. The `MAC` instruction is a **multiply-accumulate instruction**, also prevalent in DSP architectures. It has three arguments specifying the following calculation,

$$a := a + x * y \,,$$

where a is the contents of an accumulator register named `A`, and x and y are values p.211 found in memory. The addresses of these values are contained by auxiliary registers `AR2` and `AR3`. These registers are incremented automatically after the access. Moreover, these registers can be set up to implement circular buffers, as described p.211 in the box on page 211. The c54x processor includes a section of on-chip memory that supports two accesses in a single cycle, and as long as the addresses refer to this section of the memory, the `MAC` instruction will execute in a single cycle. Thus, each cycle, the processor performs two memory fetches, one multiplication, one ordinary addition, and two (possibly modulo) address increments. All DSPs have similar capabilities.

CISC instructions can get quite esoteric.

Example 8.7: The coefficients of the FIR filter in (8.1) are often symmetric, meaning that N is even and

$$a_i = a_{N-i-1} \,.$$

The reason for this is that such filters have linear phase (intuitively, this means that symmetric input signals result in symmetric output signals, or that all frequency components are delayed by the same amount). In this case, we can reduce the number of multiplications by rewriting (8.1) as

$$y(n) = \sum_{i=0}^{(N/2)-1} a_i (x(n-i) + x(n-N+i+1)) \,.$$

The Texas Instruments TMS320c54x instruction set includes a `FIRS` instruction that functions similarly to the `MAC` in Example 8.6, but using this calculation rather than that of (8.1). This takes advantage of the fact that the c54x has two ALUs, and hence can do twice as many additions as multiplications. The time to execute an FIR filter now reduces to 1/2 cycle per tap.

CISC instruction sets have their disadvantages. For one, it is extremely challenging (perhaps impossible) for a compiler to make optimal use of such an instruction set. As a consequence, DSP processors are commonly used with code libraries written and optimized in assembly language.

In addition, CISC instruction sets can have subtle timing issues that can interfere with p.313 achieving hard real-time scheduling. In the above examples, the layout of data in memory strongly affects execution times. Even more subtle, the use of zero-overhead loops (the `RPT` instruction above) can introduce some subtle problems. On the TI c54x, interrupts are disabled during repeated execution of the instruction following the `RPT`. This can result in unexpectedly long latencies in responding to interrupts.

Subword Parallelism

Many embedded applications operate on data types that are considerably smaller than the word size of the processor.

Example 8.8: In Examples 8.3 and 8.4, the data types are typically 8-bit integers, each representing a color intensity. The color of a pixel may be represented by three bytes in the RGB format. Each of the RGB bytes has a value ranging from 0 to 255 representing the intensity of the corresponding color. It would be wasteful of resources to use, say, a 64-bit ALU to process a single 8-bit number.

To support such data types, some processors support **subword parallelism**, where a wide ALU is divided into narrower slices enabling simultaneous arithmetic or logical operations on smaller words.

Example 8.9: Intel introduced subword parallelism into the widely used general purpose Pentium processor and called the technology MMX (Eden and Kagan, 1997). MMX instructions divide the 64-bit datapath into slices as small as 8 bits, supporting simultaneous identical operations on multiple bytes of image pixel data. The technology has been used to enhance the performance of image manipulation applications as well as applications supporting video streaming. Similar techniques were introduced by Sun Microsystems for Sparc™processors (Tremblay et al., 1996) and by Hewlett Packard for the PA RISC processor (Lee, 1996). Many processor architectures designed for embedded applications, including many DSP processors, also support subword parallelism. p.206

A **vector processor** is one where the instruction set includes operations on multiple data elements simultaneously. Subword parallelism is a particular form of vector processing.

Superscalar

Superscalar processors use fairly conventional sequential instruction sets, but the hardware can simultaneously dispatch multiple instructions to distinct hardware units when it detects that such simultaneous dispatch will not change the behavior of the program. That is, the execution of the program is identical to what it would have been if it had been executed in sequence. Such processors even support out-of-order execution, where instructions later in p.217 the stream are executed before earlier instructions. Superscalar processors have a significant disadvantage for embedded systems, which is that execution times may be extremely difficult to predict, and in the context of multitasking (interrupts and threads), may not even p.261 be repeatable. The execution times may be very sensitive to the exact timing of interrupts, in that small variations in such timing may have big effects on the execution times of programs.

VLIW

Processors intended for embedded applications often use VLIW architectures instead of superscalar in order to get more repeatable and predictable timing. **VLIW (very large instruction word)** processors include multiple function units, like superscalar processors, but instead of dynamically determining which instructions can be executed simultaneously, each instruction specifies what each function unit should do in a particular cycle. That is, a VLIW instruction set combines multiple independent operations into a single instruction. Like superscalar architectures, these multiple operations are executed simultaneously on distinct hardware. Unlike superscalar, however, the order and simultaneity of the execution is fixed in the program rather than being decided on-the-fly. It is up to the programmer (working at assembly language level) or the compiler to ensure that the simultaneous operations are indeed independent. In exchange for this additional complexity in programming, execution times become repeatable and (often) predictable.

Example 8.10: In Example 8.7, we saw the specialized instruction `FIRS` of the c54x architecture that specifies operations for two ALUs and one multiplier. This can be thought of as a primitive form of VLIW, but subsequent generations of processors are much more explicit about their VLIW nature. The Texas Instruments TMS320c55x, the next generation beyond the c54x, includes two multiply-accumulate units, and can support instructions that look like this:

```
MAC      *AR2+, *CDP+, AC0
:: MAC   *AR3+, *CDP+, AC1
```

Here, `AC0` and `AC1` are two accumulator registers and `CDP` is a specialized register for pointing to filter coefficients. The notation `::` means that these two instructions should be issued and executed in the same cycle. It is up to the programmer or compiler to determine whether these instructions can in fact be executed simultaneously. Assuming the memory addresses are such that the fetches can occur simultaneously, these two `MAC` instructions execute in a single cycle, effectively dividing in half the time required to execute an FIR filter.

For applications demanding higher performance still, VLIW architectures can get quite elaborate.

Example 8.11: The Texas Instruments c6000 family of processors have a VLIW instruction set. Included in this family are three subfamilies of processors, the c62x and c64x fixed-point processors and the c67x floating-point processors. These processors are designed for use in wireless infrastructure (such as cellular base stations and adaptive antennas), telecommunications infrastructure (such as voice over

IP and video conferencing), and imaging applications (such as medical imaging, surveillance, machine vision or inspection, and radar).

Example 8.12: The **TriMedia** processor family, from NXP, is aimed at digital television, and can perform operations like that in (8.2) very efficiently. NXP Semiconductors used to be part of Philips, a diversified consumer electronics company that, among many other products, makes flat-screen TVs. The strategy in the TriMedia architecture is to make it easier for a compiler to generate efficient code, reducing the need for assembly-level programming (though it includes specialized CISC instructions that are difficult for a compiler to exploit). It makes things easier for the compiler by having a larger register set than is typical (128 registers), a RISC-like instruction set, where several instructions can be issued simultaneously, and hardware supporting IEEE 754 floating point operations. p.226

8.2.4 Multicore Architectures

A **multicore** machine is a combination of several processors on a single chip. Although multicore machines have existed since the early 1990s, they have only recently penetrated into general-purpose computing. This penetration accounts for much of the interest in them today. **Heterogeneous multicore** machines combine a variety of processor types on a single chip, vs. multiple instances of the same processor type.

Example 8.13: Texas Instruments **OMAP** (open multimedia application platform) architectures are widely used in cell phones, which normally combine one or more DSP processors with one or more processors that are closer in style to p.206 general-purpose processors. The DSP processors handle the radio, speech, and media processing (audio, images, and video). The other processors handle the user interface, database functions, networking, and downloadable applications. Specifically, the OMAP4440 includes a 1 GHz dual-core ARM Cortex processor, a c64x DSP, a GPU, and an image signal processor. p.210

For embedded applications, multicore architectures have a significant potential advantage over single-core architectures because real-time and safety-critical tasks can have a dedicated processor. This is the reason for the heterogeneous architectures used for cell phones, since the radio and speech processing functions are hard real-time functions with considerable computational load. In such architectures, user applications cannot interfere with real-time functions. p.310

Fixed-Point Numbers

Many embedded processors provide hardware for integer arithmetic only. Integer arithmetic, however, can be used for non-whole numbers, with some care. Given, say, a 16-bit integer, a programmer can *imagine* a **binary point**, which is like a decimal point, except that it separates bits rather than digits of the number. For example, a 16-bit integer can be used to represent numbers in the range -1.0 to 1.0 (roughly) by placing a (conceptual) binary point just below the high-order bit of the number, as shown below:

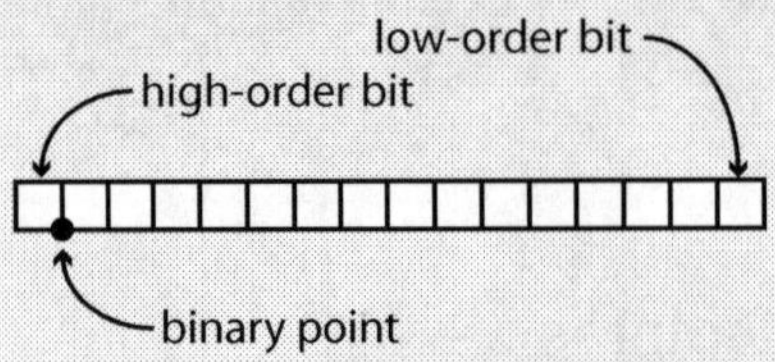

Without the binary point, a number represented by the 16 bits is a whole number $x \in \{-2^{15}, \cdots, 2^{15}-1\}$ (assuming the twos-complement binary representation, which has become nearly universal for signed integers). With the binary point, we *interpret* the 16 bits to represent a number $y = x/2^{15}$. Hence, y ranges from -1 to $1-2^{-15}$. This is known as a **fixed-point number**. The format of this fixed-point number can be written 1.15, indicating that there is one bit to the left of the binary point and 15 to the right. When two such numbers are multiplied at full precision, the result is a 32-bit number. The binary point is located as follows:

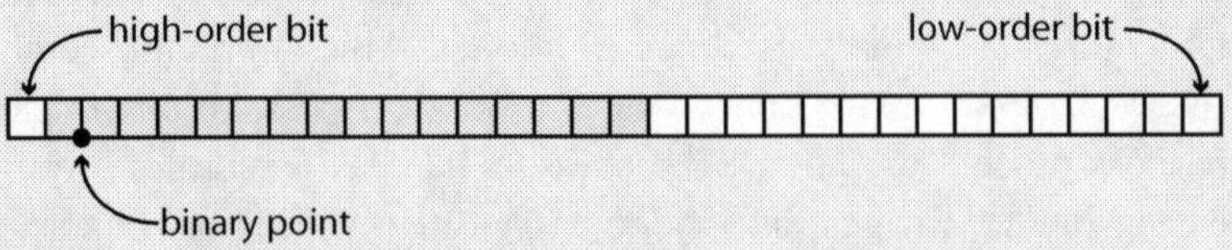

... **Continued on page 225.**

Fixed-Point Numbers (continued)

The location of the binary point follows from the **law of conservation of bits**. When multiplying two numbers with formats $n.m$ and $p.q$, the result has format $(n+p).(m+q)$. Processors often support such full-precision multiplications, where the result goes into an accumulator register that has at least twice as many bits as the ordinary data registers. To write the result back to a data register, however, we have to extract 16 bits from the 32 bit result. If we extract the shaded bits on page 225, then we preserve the position of the binary point, and the result still represents a number roughly in the range -1 to 1.

There is a loss of information, however, when we extract 16 bits from a 32-bit result. First, there is a possibility of **overflow**, because we are discarding the high-order bit. Suppose the two numbers being multiplied are both -1, which has binary representation in twos complement as follows:

```
1.000000000000000
```

When these two number are multiplied, the result has the following bit pattern:

```
01.000000000000000000000000000000
```

which in twos complement, represents 1, the correct result. However, when we extract the shaded 16 bits, the result is now -1! Indeed, 1 is not representable in the fixed-point format 1.15, so overflow has occurred. Programmers must guard against this, for example by ensuring that all numbers are strictly less than 1 in magnitude, prohibiting -1.

A second problem is that when we extract the shaded 16 bits from a 32-bit result, we discard 15 low-order bits. There is a loss of information here. If we simply discard the low-order 15 bits, the strategy is known as **truncation**. If instead we first add the following bit pattern the 32-bit result, then the result is known as **rounding**:

```
00.000000000000000100000000000000
```

Rounding chooses the result that is closest to the full-precision result, while truncation chooses the closest result that is smaller in magnitude.

DSP processors typically perform the above extraction with either rounding or truncation in hardware when data is moved from an accumulator to a general-purpose register or to memory.

Fixed-Point Arithmetic in C

Most C programmers will use `float` or `double` data types when performing arithmetic on non-whole numbers. However, many embedded processors lack hardware for floating-point arithmetic. Thus, C programs that use the `float` or `double` data types often result in unacceptably slow execution, since floating point must be emulated in software. Programmers are forced to use integer arithmetic to implement operations on numbers that are not whole numbers. How can they do that?

p.224 First, a programmer can *interpret* a 32-bit `int` differently from the standard representation, using the notion of a binary point, explained in the boxes on pages 224 and 225. However, when a C program specifies that two `ints` be multiplied, the result is an `int`, not the full precision 64-bit result that we need. In fact, the strategy outlined on page 224, of putting one bit to the left of the binary point and extracting the shaded bits from the result, will not work, because most of the shaded bits will be missing from the result. For example, suppose we want to multiply 0.5 by 0.5. This number can be represented in 32-bit `ints` as follows:

0.1000000000000000000000000000000

Without the binary point (which is invisible to C and to the hardware, residing only in the programmer's mind), this bit pattern represents the integer 2^{30}, a large number indeed. When multiplying these two numbers, the result is 2^{60}, which is not representable in an `int`. Typical processors will set an overflow bit in the processor status register (which the programmer must check) and deliver as a result the number 0, which is the low-order 32 bits of the product. To guard against this, a programmer can shift each 32 bit integer to the right by 16 bits before multiplying. In that case, the result of the multiply 0.5×0.5 is the following bit pattern:

0.0100000000000000000000000000000

With the binary point as shown, this result is interpreted as 0.25, the correct answer. Of course, shifting data to the right by 16 bits discards the 16 low-order bits in the `int`. p.225 There is a loss of precision that amounts to truncation. The programmer may wish to round instead, adding the `int` 2^{15} to the numbers before shifting to the right 16 times. Floating-point data types make things easier. The hardware (or software) keeps track of the amount of shifting required and preserves precision when possible. However, not all embedded processors with floating-point hardware conform with the **IEEE 754** standard. This can complicate the design process for the programmer, because numerical results will not match those produced by a desktop computer.

This lack of interference is more problematic in general-purpose multicore architectures. It is common, for example, to use multi-level caches, where the second or higher level cache p.236 is shared among the cores. Unfortunately, such sharing makes it very difficult to isolate the real-time behavior of the programs on separate cores, since each program can trigger cache misses in another core. Such multi-level caches are not suitable for real-time applications.

A very different type of multicore architecture that is sometimes used in embedded applications uses one or more **soft core**s together with custom hardware on a **field-programmable gate array (FPGA)**. FPGAs are chips whose hardware function is programmable using hardware design tools. Soft cores are processors implemented on FPGAs. The advantage of soft cores is that they can be tightly coupled to custom hardware more easily than off-the-shelf processors.

8.3 Summary

The choice of processor architecture for an embedded system has important consequences for the programmer. Programmers may need to use assembly language to take advantage of esoteric architectural features. For applications that require precise timing, it may be difficult to control the timing of a program because of techniques in the hardware for dealing with pipeline hazards and parallel resources.

Exercises

1. Consider the reservation table in Figure 8.4. Suppose that the processor includes forwarding logic that is able to tell that instruction *A* is writing to the same register that instruction *B* is reading from, and that therefore the result written by *A* can be forwarded directly to the ALU before the write is done. Assume the forwarding logic itself takes no time. Give the revised reservation table. How many cycles are lost to the pipeline bubble?

2. Consider the following instruction, discussed in Example 8.6:

```
MAC *AR2+, *AR3+, A
```

Suppose the processor has three ALUs, one for each arithmetic operation on the addresses contained in registers `AR2` and `AR3` and one to perform the addition in the `MAC` multiply-accumulate instruction. Assume these ALUs each require one clock cycle to execute. Assume that a multiplier also requires one clock cycle to execute.

Assume further that the register bank supports two reads and two writes per cycle, and that the accumulator register `A` can be written separately and takes no time to write. Give a reservation table showing the execution of a sequence of such instructions.

3. Assuming fixed-point numbers with format 1.15 as described in the boxes on pages 224 and 225, show that the *only* two numbers that cause overflow when multiplied are -1 and -1. That is, if either number is anything other than -1 in the 1.15 format, then extracting the 16 shaded bits in the boxes does not result in overflow.

9

Memory Architectures

Contents

Many processor architects argue that memory systems have more impact on overall system performance than data pipelines. This depends, of course, on the application, but for many applications it is true. There are three main sources of complexity in memory. First, it is commonly necessary to mix a variety of memory technologies in the same embedded system. Many memory technologies are **volatile**, meaning that the contents of the memory is lost if power is lost. Most embedded systems need at least some non-volatile memory and some volatile memory. Moreover, within these categories, there are several choices, and the choices have significant consequences for the system designer. Second, memory hierarchy is often needed because memories with larger capacity and/or lower power consumption are slower. To achieve reasonable performance at reasonable cost, faster memories must be mixed with slower memories. Third, the address space of a processor architecture is divided up to provide access to the various kinds of memory, to provide support for common programming models, and to designate addresses for interaction with devices other than memories, such as I/O devices. In this chapter, we discuss these three issues in order.

9.1 Memory Technologies

In embedded systems, memory issues loom large. The choices of memory technologies have important consequences for the system designer. For example, a programmer may need to worry about whether data will persist when the power is turned off or a power-saving standby mode is entered. A memory whose contents are lost when the power is cut off is called a **volatile memory**. In this section, we discuss some of the available technologies and their tradeoffs.

9.1.1 RAM

In addition to the register file, a microcomputer typically includes some amount of **RAM** (random access memory), which is a memory where individual items (bytes or words) can be written and read one at a time relatively quickly. **SRAM** (static RAM) is faster than **DRAM** (dynamic RAM), but it is also larger (each bit takes up more silicon area). DRAM holds data for only a short time, so each memory location must be periodically refreshed. SRAM holds data for as long as power is maintained. Both types of memories lose their contents if power is lost, so both are volatile memory, although arguably DRAM is more volatile than SRAM because it loses its contents even if power is maintained.

Most embedded computer systems include an SRAM memory. Many also include DRAM because it can be impractical to provide enough memory with SRAM technology alone. A programmer that is concerned about the time it takes a program to execute must be aware of whether memory addresses being accessed are mapped to SRAM or DRAM. Moreover, the refresh cycle of DRAM can introduce variability to the access times because the DRAM may be busy with a refresh at the time that access is requested. In addition, the access history can affect access times. The time it takes to access one memory address may depend on what memory address was last accessed.

A manufacturer of a DRAM memory chip will specify that each memory location must be refreshed, say, every 64 ms, and that a number of locations (a "row") are refreshed together. The mere act of reading the memory will refresh the locations that are read (and locations on the same row), but since applications may not access all rows within the specified time interval, DRAM has to be used with a controller that ensures that all locations are refreshed sufficiently often to retain the data. The memory controller will stall accesses if the memory is busy with a refresh when the access is initiated. This introduces variability in the timing of the program.

9.1.2 Non-Volatile Memory

Embedded systems invariably need to store data even when the power is turned off. There are several options for this. One, of course, is to provide battery backup so that power is never lost. Batteries, however, wear out, and there are better options available, known collectively as **non-volatile memories**. An early form of non-volatile memory was **magnetic core memory** or just **core**, where a ferromagnetic ring was magnetized to store data. The term "core" persists in computing to refer to computer memories, although this may change as multicore machines become ubiquitous. p.223

The most basic non-volatile memory today is **ROM** (read-only memory) or **mask ROM**, the contents of which is fixed at the chip factory. This can be useful for mass produced products that only need to have a program and constant data stored, and these data never change. Such programs are known as **firmware**, suggesting that they are not as "soft" as software. There are several variants of ROM that can be programmed in the field, and the technology has gotten good enough that these are almost always used today over mask ROM. **EEPROM**, electrically-erasable programmable ROM, comes in several forms, but it is possible to write to all of these. The write time is typically much longer than the read time, and the number of writes is limited during the lifetime of the device. A particularly useful form of EEPROM is flash memory. Flash is commonly used to store firmware and user data that needs to persist when the power is turned off.

Flash memory, invented by Dr. Fujio Masuoka at Toshiba around 1980, is a particularly convenient form of non-volatile memory, but it presents some interesting challenges for embedded systems designers. Typically, flash memories have reasonably fast read times, but not as fast as SRAM and DRAM, so frequently accessed data will typically have to be moved from the flash to RAM before being used by a program. The write times are much longer than the read times, and the total number of writes are limited, so these memories are not a substitute for working memory. p.231

There are two types of flash memories, known as NOR and NAND flash. NOR flash has longer erase and write times, but it can be accessed like a RAM. NAND flash is less expensive and has faster erase and write times, but data must be read a block at a time, where a block is hundreds to thousands of bits. This means that from a system perspective it behaves more like a secondary storage device like a hard disk or optical media like CD or DVD. Both types of flash can only be erased and rewritten a bounded number of times, typically under 1,000,000 for NOR flash and under 10,000,000 for NAND flash, as of this writing.

The longer access times, limited number of writes, and block-wise accesses (for NAND flash), all complicate the problem for embedded system designers. These properties must be taken into account not only while designing hardware, but also software.

Disk memories are also non-volatile. They can store very large amounts of data, but access times can become quite large. In particular, the mechanics of a spinning disk and a read-write head require that the controller wait until the head is positioned over the requested location before the data at that location can be read. The time this takes is highly variable. Disks are also more vulnerable to vibration than the solid-state memories discussed above, and hence are more difficult to use in many embedded applications.

9.2 Memory Hierarchy

Many applications require substantial amounts of memory, more than what is available on-chip in a microcomputer. Many processors use a **memory hierarchy**, which combines different memory technologies to increase the overall memory capacity while optimizing cost, latency, and energy consumption. Typically, a relatively small amount of on-chip p.230 SRAM will be used with a larger amount of off-chip DRAM. These can be further combined with a third level, such as disk drives, which have very large capacity, but lack random access and hence can be quite slow to read and write.

The application programmer may not be aware that memory is fragmented across these technologies. A commonly used scheme called **virtual memory** makes the diverse technologies look to the programmer like a contiguous **address space**. The operating system and/or the hardware provides **address translation**, which converts logical addresses in the address space to physical locations in one of the available memory technologies. This translation is often assisted by a specialized piece of hardware called a **translation lookaside buffer** (**TLB**), which can speed up some address translations. For an embedded system designer, these techniques can create serious problems because they make it very difficult to predict or understand how long memory accesses will take. Thus, embedded system designers typically need to understand the memory system more deeply than general-purpose programmers.

9.2.1 Memory Maps

A **memory map** for a processor defines how addresses get mapped to hardware. The total size of the address space is constrained by the address width of the processor. A 32-bit processor, for example, can address 2^{32} locations, or 4 gigabytes (GB), assuming each address refers to one byte. The address width typically matches the word width, except for 8-bit processors, where the address width is typically higher (often 16 bits). An ARM Cortex^{TM} - M3 architecture, for example, has the memory map shown in Figure 9.1. Other architectures will have other layouts, but the pattern is similar.

Notice that this architecture separates addresses used for program memory (labeled A in the figure) from those used for data memory (B and D). This (typical) pattern allows these memories to be accessed via separate buses, permitting instructions and data to be fetched simultaneously. This effectively doubles the memory bandwidth. Such a separation of program memory from data memory is known as a **Harvard architecture**. It contrasts with the classical **von Neumann architecture**, which stores program and data in the same memory.

Any particular realization in silicon of this architecture is constrained by this memory map. For example, the Luminary Micro[1] LM3S8962 controller, which includes an ARM Cortex^{TM} - M3 core, has 256 KB of on-chip flash memory, nowhere near the total of 0.5 GB that the architecture allows. This memory is mapped to addresses `0x00000000` through `0x0003FFFF`. The remaining addresses that the architecture allows for program memory, which are `0x00040000` through `0x1FFFFFFF`, are "reserved addresses," meaning that they should not be used by a compiler targeting this particular device.

The LM3S8962 has 64 KB of SRAM, mapped to addresses `0x20000000` through `0x2000FFFF`, a small portion of area B in the figure. It also includes a number of on-chip **peripherals**, which are devices that are accessed by the processor using some of the memory addresses in the range from `0x40000000` to `0x5FFFFFFF` (area C in the figure). These include timers, ADCs, GPIO, UARTs, and other I/O devices. Each of these devices p.263
occupies a few of the memory addresses by providing **memory-mapped registers**. The processor may write to some of these registers to configure and/or control the peripheral, or to provide data to be produced on an output. Some of the registers may be read to retrieve

[1] Luminary Micro was acquired by Texas Instruments in 2009.

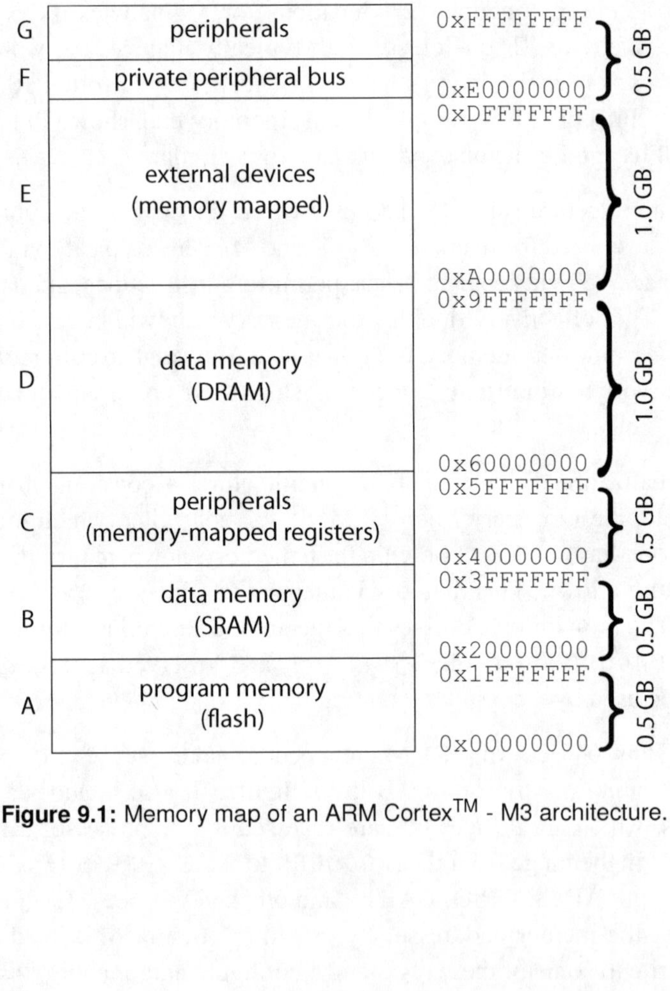

Figure 9.1: Memory map of an ARM Cortex^{TM} - M3 architecture.

input data obtained by the peripheral. A few of the addresses in the private peripheral bus region are used to access the interrupt controller. p.265

The LM3S8962 is mounted on a printed circuit board that will provide additional devices such as DRAM data memory and additional external devices. As shown in Figure 9.1, these p.230 will be mapped to memory addresses in the range from `0xA0000000` to `0xDFFFFFFF` (area E). For example, the Stellaris® LM3S8962 evaluation board from Luminary Micro includes no additional external memory, but does add a few external devices such as an LCD display, a MicroSD slot for additional flash memory, and a USB interface.

This leaves many memory addresses unused. ARM has introduced a clever way to take advantage of these unused addresses called **bit banding**, where some of the unused addresses can be used to access individual bits rather than entire bytes or words in the memory and peripherals. This makes certain operations more efficient, since extra instructions to mask the desired bits become unnecessary.

9.2.2 Register Files

The most tightly integrated memory in a processor is the **register file**. Each register in the file stores a **word**. The size of a word is a key property of a processor architecture. It is one byte on an 8-bit architecture, four bytes on a 32-bit architecture, and eight bytes on a 64-bit architecture. The register file may be implemented directly using flip flops in the processor circuitry, or the registers may be collected into a single memory bank, typically using the same SRAM technology discussed above. p.230

Harvard Architecture

The term “Harvard architecture” comes from the Mark I computer, which used distinct memories for program and data. The Mark I was made with electro-mechanical relays by IBM and shipped to Harvard in 1944. The machine stored instructions on punched tape and data in electro-mechanical counters. It was called the Automatic Sequence Controlled Calculator (ASCC) by IBM, and was devised by Howard H. Aiken to numerically solve differential equations. Rear Admiral Grace Murray Hopper of the p.19 United States Navy and funding from IBM were instrumental in making the machine a reality.

The number of registers in a processor is usually small. The reason for this is not so much the cost of the register file hardware, but rather the cost of bits in an instruction word. An p.202 instruction set architecture (ISA) typically provides instructions that can access one, two, or three registers. To efficiently store programs in memory, these instructions cannot require too many bits to encode them, and hence they cannot devote too many bits to identifying the registers. If the register file has 16 registers, then each reference to a register requires 4 bits. If an instruction can refer to 3 registers, that requires a total of 12 bits. If an instruction word is 16 bits, say, then this leaves only 4 bits for other information in the instruction, such as the identity of the instruction itself, which also must be encoded in the instruction. This identifies, for example, whether the instruction specifies that two registers should be added or subtracted, with the result stored in the third register.

9.2.3 Scratchpads and Caches

Many embedded applications mix memory technologies. Some memories are accessed before others; we say that the former are "closer" to the processor than the latter. For example, a close memory (SRAM) is typically used to store working data temporarily while the program operates on it. If the close memory has a distinct set of addresses and the program is responsible for moving data into it or out of it to the distant memory, then it is called a **scratchpad**. If the close memory duplicates data in the distant memory with the hardware automatically handling the copying to and from, then it is called a **cache**. For embedded applications with tight real-time constraints, cache memories present some formidable obstacles because their timing behavior can vary substantially in ways that are difficult to predict. On the other hand, manually managing the data in a scratchpad memory can be quite tedious for a programmer, and automatic compiler-driven methods for doing so are in their infancy.

As explained in Section 9.2.1, an architecture will typically support a much larger address space than what can actually be stored in the physical memory of the processor, with a p.232 virtual memory system used to present the programmer with the view of a contiguous address space. If the processor is equipped with a **memory management unit** (**MMU**), then programs reference **logical addresses** and the MMU translates these to **physical addresses**. p.298 For example, using the memory map in Figure 9.1, a process might be allowed to use logical addresses `0x60000000` to `0x9FFFFFFF` (area D in the figure), for a total of 1 GB of addressable data memory. The MMU may implement a cache that uses however much physical memory is present in area B. When the program provides a memory address, the MMU determines whether that location is cached in area B, and if it is, translates the address and completes the fetch. If it is not, then we have a **cache miss**, and the MMU handles fetching data from the secondary memory (in area D) into the cache (area B). If the location is also not present in area D, then the MMU triggers a **page fault**, which can result in soft-

Parameter	Description
m	Number of physical address bits
$S = 2^s$	Number of (cache) sets
E	Number of lines per set
$B = 2^b$	Block size in bytes
$t = m - s - b$	Number of tag bits
C	Overall cache size in bytes

Table 9.1: Summary of cache parameters.

ware handling movement of data from disk into the memory. Thus, the program is given the illusion of a vast amount of memory, with the cost that memory access times become quite difficult to predict. It is not uncommon for memory access times to vary by a factor of 1000 or more, depending on how the logical addresses happen to be disbursed across the physical memories.

Given this sensitivity of execution time to the memory architecture, it is important to understand the organization and operation of caches. That is the focus of this section.

Basic Cache Organization

Suppose that each address in a memory system comprises m bits, for a maximum of $M = 2^m$ unique addresses. A cache memory is organized as an array of $S = 2^s$ **cache sets**. Each cache set in turn comprises E **cache lines**. A cache line stores a single **block** of $B = 2^b$ bytes of data, along with **valid** and **tag** bits. The valid bit indicates whether the cache line stores meaningful information, while the tag (comprising $t = m - s - b$ bits) uniquely identifies the block that is stored in the cache line. Figure 9.2 depicts the basic cache organization and address format.

Thus, a cache can be characterized by the tuple (m, S, E, B). These parameters are summarized in Table 9.1. The overall cache size C is given as $C = S \times E \times B$ bytes.

Suppose a program reads the value stored at address a. Let us assume for the rest of this section that this value is a single data word w. The CPU first sends address a to the cache to determine if it is present there. The address a can be viewed as divided into three segments of bits: the top t bits encode the tag, the next s bits encode the set index, and the last b bits encode the position of the word within a block. If w is present in the cache, the memory access is a **cache hit**; otherwise, it is a **cache miss**.

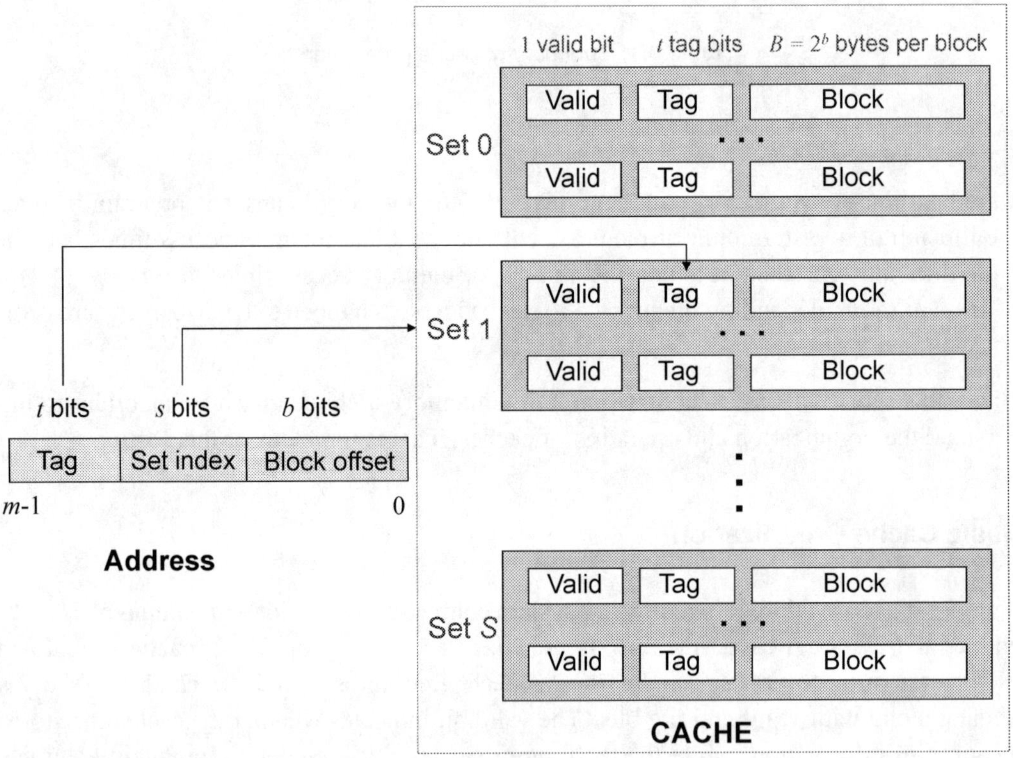

Figure 9.2: Cache Organization and Address Format. A cache can be viewed as an array of sets, where each set comprises of one or more cache lines. Each cache line includes a valid bit, tag bits, and a cache block.

Caches are categorized into classes based on the value of E. We next review these categories of cache memories, and describe briefly how they operate.

Direct-Mapped Caches

A cache with exactly one line per set ($E = 1$) is called a **direct-mapped cache**. For such a cache, given a word w requested from memory, where w is stored at address a, there are three steps in determining whether w is a cache hit or a miss:

1. *Set Selection:* The s bits encoding the set are extracted from address a and used as an index to select the corresponding cache set.
2. *Line Matching:* The next step is to check whether a copy of w is present in the unique cache line for this set. This is done by checking the valid and tag bits for that cache line. If the valid bit is set and the tag bits of the line match those of the address a, then the word is present in the line and we have a cache hit. If not, we have a cache miss.
3. *Word Selection:* Once the word is known to be present in the cache block, we use the b bits of the address a encoding the word's position within the block to read that data word.

On a cache miss, the word w must be requested from the next level in the memory hierarchy. Once this block has been fetched, it will replace the block that currently occupies the cache line for w.

While a direct-mapped cache is simple to understand and to implement, it can suffer from **conflict misses**. A conflict miss occurs when words in two or more blocks that map to the same cache line are repeatedly accessed so that accesses to one block evict the other, resulting in a string of cache misses. Set-associative caches can help to resolve this problem.

Set-Associative Caches

A **set-associative cache** can store more than one cache line per set. If each set in a cache can store E lines, where $1 < E < C/B$, then the cache is called an *E-way* set-associative cache. The word "associative" comes from **associative memory**, which is a memory that is addressed by its contents. That is, each word in the memory is stored along with a unique key and is retrieved using the key rather than the physical address indicating where it is stored. An associative memory is also called a **content-addressable memory**.

For a set-associative cache, accessing a word w at address a consists of the following steps:

1. *Set Selection:* This step is identical to a direct-mapped cache.
2. *Line Matching:* This step is more complicated than for a direct-mapped cache because there could be multiple lines that w might lie in; i.e., the tag bits of a could match the tag bits of any of the lines in its cache set. Operationally, each set in a set-associative cache can be viewed as an associative memory, where the keys are the concatenation of the tag and valid bits, and the data values are the contents of the corresponding block.
3. *Word Selection:* Once the cache line is matched, the word selection is performed just as for a direct-mapped cache.

In the case of a miss, cache line replacement can be more involved than it is for a direct-mapped cache. For the latter, there is no choice in replacement since the new block will displace the block currently present in the cache line. However, in the case of a set-associative cache, we have an option to select the cache line from which to evict a block. A common policy is **least-recently used** (**LRU**), in which the cache line whose most recent access occurred the furthest in the past is evicted. Another common policy is **first-in, first-out** (**FIFO**), where the cache line that is evicted is the one that has been in the cache for the longest, regardless of when it was last accessed. Good cache replacement policies are essential for good cache performance. Note also that implementing these cache replacement policies requires additional memory to remember the access order, with the amount of additional memory differing from policy to policy and implementation to implementation.

A **fully-associative cache** is one where $E = C/B$, i.e., there is only one set. For such a cache, line matching can be quite expensive for a large cache size because an
p.239 associative memory is expensive. Hence, fully-associative caches are typically only used
p.232 for small caches, such as the translation lookaside buffers (TLBs) mentioned earlier.

9.3 Memory Models

A **memory model** defines how memory is used by programs. The hardware, the operating system (if any), and the programming language and its compiler all contribute to the memory model. This section discusses a few of the common issues that arise with memory models.

9.3.1 Memory Addresses

At a minimum, a memory model defines a range of **memory addresses** accessible to the program. In C, these addresses are stored in **pointers**. In a **32-bit architecture**, memory addresses are 32-bit unsigned integers, capable of representing addresses 0 to $2^{32}-1$, which is about four billion addresses. Each address refers to a byte (eight bits) in memory. The C `char` data type references a byte. The C `int` data type references a sequence of at least two bytes. In a 32-bit architecture, it will typically reference four bytes, able to represent integers from -2^{31} to $2^{31}-1$. The `double` data type in C refers to a sequence of eight bytes encoded according to the IEEE floating point standard (IEEE 754).

Since a memory address refers to a byte, when writing a program that directly manipulates memory addresses, there are two critical compatibility concerns. The first is the **alignment** of the data. An `int` will typically occupy four consecutive bytes starting at an address that is a multiple of four. In hexadecimal notation these addresses always end in 0, 4, 8, or c.

The second concern is the byte order. The first byte (at an address ending in 0, 4, 8, or c), may represent the eight low order bits of the int (a representation called **little endian**), or it may represent the eight high order bits of the int (a representation called **big endian**). Unfortunately, although many data representation questions have become universal standards (such as the bit order in a byte), the byte order is not one those questions. Intel's x86 architectures and ARM processors, by default, use a little-endian representation, whereas IBM's PowerPC uses big endian. Some processors support both. Byte order also matters in network protocols, which generally use big endian.

The terminology comes from Gulliver's Travels, by Jonathan Swift, where a royal edict in Lilliput requires cracking open one's soft-boiled egg at the small end, while in the rival kingdom of Blefuscu, inhabitants crack theirs at the big end.

9.3.2 Stacks

A **stack** is a region of memory that is dynamically allocated to the program in a last-in, first-out (**LIFO**) pattern. A **stack pointer** (typically a register) contains the memory address of the top of the stack. When an item is pushed onto the stack, the stack pointer is incremented and the item is stored at the new location referenced by the stack pointer. When an item is popped off the stack, the memory location referenced by the stack pointer is (typically) copied somewhere else (e.g., into a register) and the stack pointer is decremented.

Stacks are typically used to implement procedure calls. Given a procedure call in C, for example, the compiler produces code that pushes onto the stack the location of the instruction to execute upon returning from the procedure, the current value of some or all of the machine registers, and the arguments to the procedure, and then sets the program counter equal to the location of the procedure code. The data for a procedure that is pushed onto the stack is known as the **stack frame** of that procedure. When a procedure returns, the compiler pops its stack frame, retrieving finally the program location at which to resume execution.

For embedded software, it can be disastrous if the stack pointer is incremented beyond the memory allocated for the stack. Such a **stack overflow** can result in overwriting memory that is being used for other purposes, leading to unpredictable results. Bounding the stack usage, therefore, is an important goal. This becomes particularly difficult with **recursive programs**, where a procedure calls itself. Embedded software designers often avoid using recursion to circumvent this difficulty.

More subtle errors can arise as a result of misuse or misunderstanding of the stack. Consider the following C program:

```
1   int* foo(int a) {
2     int b;
3     b = a * 10;
4     return &b;
5   }
6   int main(void) {
7     int* c;
8     c = foo(10);
9     ...
10  }
```

p.244 The variable b is a local variable, with its memory on the stack. When the procedure returns, the variable c will contain a pointer to a memory location *above the stack pointer*. The contents of that memory location will be overwritten when items are next pushed onto the stack. It is therefore incorrect for the procedure foo to return a pointer to b. By the time that pointer is de-referenced (i.e., if a line in main refers to *c after line 8), the memory location may contain something entirely different from what was assigned in foo. Unfortunately, C provides no protection against such errors.

9.3.3 Memory Protection Units

A key issue in systems that support multiple simultaneous tasks is preventing one task from disrupting the execution of another. This is particularly important in embedded applications that permit downloads of third party software, but it can also provide an important defense against software bugs in safety-critical applications.

Many processors provide **memory protection** in hardware. Tasks are assigned their own address space, and if a task attempts to access memory outside its own address space, a **segmentation fault** or other exception results. This will typically result in termination of the offending application. p.232

9.3.4 Dynamic Memory Allocation

General-purpose software applications often have indeterminate requirements for memory, depending on parameters and/or user input. To support such applications, computer scientists have developed dynamic memory allocation schemes, where a program can at any time request that the operating system allocate additional memory. The memory is allocated from a data structure known as a **heap**, which facilitates keeping track of which portions of memory are in use by which application. Memory allocation occurs via an operating system call (such as `malloc` in C). When the program no longer needs access to memory that has been so allocated, it deallocates the memory (by calling `free` in C).

Support for memory allocation often (but not always) includes garbage collection. For example, garbage collection is intrinsic in the Java programming language. A **garbage collector** is a task that runs either periodically or when memory gets tight that analyzes the data structures that a program has allocated and automatically frees any portions of memory that are no longer referenced within the program. When using a garbage collector, in principle, a programmer does not need to worry about explicitly freeing memory.

With or without garbage collection, it is possible for a program to inadvertently accumulate memory that is never freed. This is known as a memory leak, and for embedded applications, which typically must continue to execute for a long time, it can be disastrous. The program will eventually fail when physical memory is exhausted.

Another problem that arises with memory allocation schemes is memory fragmentation. This occurs when a program chaotically allocates and deallocates memory in varying sizes. A fragmented memory has allocated and free memory chunks interspersed, and often the free memory chunks become too small to use. In this case, defragmentation is required.

Defragmentation and garbage collection are both very problematic for real-time systems. Straightforward implementations of these tasks require all other executing tasks to be stopped while the defragmentation or garbage collection is performed. Implementations using such "stop the world" techniques can have substantial pause times, running sometimes for many milliseconds. Other tasks cannot execute during this time because references to data within data structures (pointers) are inconsistent during the task. A technique that can reduce pause times is incremental garbage collection, which isolates sections of memory and garbage collects them separately. As of this writing, such techniques are experimental and not widely deployed.

9.3.5 Memory Model of C

C programs store data on the stack, on the heap, and in memory locations fixed by by the compiler. Consider the following C program:

```
1   int a = 2;
2   void foo(int b, int* c) {
3     ...
4   }
5   int main(void) {
6     int d;
7     int* e;
8     d = ...;                  // Assign some value to d.
9     e = malloc(sizeInBytes);  // Allocate memory for e.
10    *e = ...;                 // Assign some value to e.
11    foo(d, e);
12    ...
13  }
```

In this program, the variable a is a **global variable** because it is declared outside any procedure definition. The compiler will assign it a fixed memory location. The variables b and p.241 c are **parameters**, which are allocated locations on the stack when the procedure foo is called (a compiler could also put them in registers rather than on the stack). The variables d and e are **automatic variables** or **local variables**. They are declared within the body of a procedure (in this case, main). The compiler will allocate space on the stack for them.

When the procedure foo is called on line 11, the stack location for b will acquire a *copy* of the value of variable d assigned on line 8. This is an example of **pass by value**, where a parameter's value is copied onto the stack for use by the called procedure. The data referred p.243 to by the pointer e, on the other hand, is stored in memory allocated on the heap, and then it is **passed by reference** (the pointer to it, e, is passed by value). The *address* is stored in the stack location for c. If foo includes an assignment to *c, then then after foo returns, that value can be read by dereferencing e.

The global variable a is assigned an initial value on line 1. There is a subtle pitfall here, however. The memory location storing a will be initialized with value 2 *when the program is loaded*. This means that if the program is run a second time without reloading, then the initial value of a will not necessarily be 2! Its value will be whatever it was when the first invocation of the program ended. In most desktop operating systems, the program is reloaded on each run, so this problem does not show up. But in many embedded systems, the program is not necessarily reloaded for each run. The program may be run from the beginning, for example, each time the system is reset. To guard against this problem, it is safer to initialize global variables in the body of main, rather than on the declaration line, as done above.

9.4 Summary

An embedded system designer needs to understand the memory architecture of the target computer and the memory model of the programming language. Incorrect uses of memory can lead to extremely subtle errors, some of which will not show up in testing. Errors that only show up in a fielded product can be disastrous, for both the user of the system and the technology provider.

Specifically, a designer needs to understand which portions of the address space refer to volatile and non-volatile memory. For time-sensitive applications (which is most embedded systems), the designer also needs to be aware of the memory technology and cache architecture (if any) in order to understand execution times of the program. In addition, the programmer needs to understand the memory model of the programming language in order to avoid reading data that may be invalid. In addition, the programmer needs to be very careful with dynamic memory allocation, particularly for embedded systems that are expected to run for a very long time. Exhausting the available memory can cause system crashes or other undesired behavior.

Exercises

1. Consider the function compute_variance listed below, which computes the variance of integer numbers stored in the array data.

```
1  int data[N];
2
3  int compute_variance() {
4    int sum1 = 0, sum2 = 0, result;
5    int i;
6
7    for(i=0; i < N; i++) {
```

```
      sum1 += data[i];
    }
    sum1 /= N;

    for(i=0; i < N; i++) {
      sum2 += data[i] * data[i];
    }
    sum2 /= N;

    result = (sum2 - sum1*sum1);

    return result;
}
```

Suppose this program is executing on a 32-bit processor with a direct-mapped cache with parameters $(m,S,E,B) = (32,8,1,8)$. We make the following additional assumptions:

- An `int` is 4 bytes wide.
- `sum1`, `sum2`, `result`, and `i` are all stored in registers.
- `data` is stored in memory starting at address `0x0`.

Answer the following questions:

(a) Consider the case where `N` is 16. How many cache misses will there be?

(b) Now suppose that `N` is 32. Recompute the number of cache misses.

(c) Now consider executing for `N` = 16 on a 2-way set-associative cache with parameters $(m,S,E,B) = (32,8,2,4)$. In other words, the block size is halved, while there are two cache lines per set. How many cache misses would the code suffer?

2. Recall from Section 9.2.3 that caches use the middle range of address bits as the set index and the high order bits as the tag. Why is this done? How might cache performance be affected if the middle bits were used as the tag and the high order bits were used as the set index?

3. Consider the C program and simplified memory map for a 16-bit microcontroller shown below. Assume that the stack grows from the top (area D) and that the program and static variables are stored in the bottom (area C) of the data and program memory region. Also, assume that the entire address space has physical memory associated with it.

```
#include <stdio.h>
#define FOO 0x0010
int n;
int* m;
void foo(int a) {
    if (a > 0) {
      n = n + 1;
      foo(n);
    }
}
int main() {
    n = 0;
    m = (int*)FOO;
    foo(*m);
    printf("n = %d\n", n);
}
```

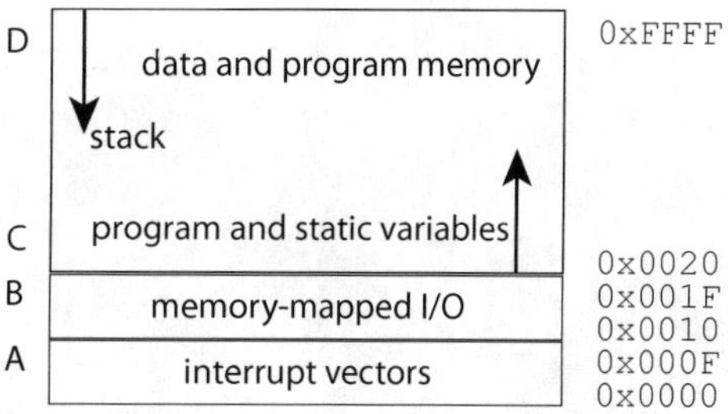

You may assume that in this system, an `int` is a 16-bit number, that there is no operating system and no memory protection, and that the program has been compiled and loaded into area C of the memory.

(a) For each of the variables `n`, `m`, and `a`, indicate where in memory (region A, B, C, or D) the variable will be stored.

(b) Determine what the program will do if the contents at address 0x0010 is 0 upon entry.

(c) Determine what the program will do if the contents of memory location 0x0010 is 1 upon entry.

4. Consider the following program:

```
int a = 2;
void foo(int b) {
  printf("%d", b);
}
int main(void) {
  foo(a);
  a = 1;
}
```

Is it true or false that the value of `a` passed to `foo` will always be 2? Explain. Assume that this is the entire program, that this program is stored in persistent memory, and that the program is executed on a bare-iron microcontroller each time a reset button is pushed. p.282

10
Input and Output

Contents

Because cyber-physical systems integrate computing and physical dynamics, the mechanisms in processors that support interaction with the outside world are central to any design. A system designer has to confront a number of issues. Among these, the mechanical and electrical properties of the interfaces are important. Incorrect use of parts, such as drawing too much current from a pin, may cause a system to malfunction or may reduce its useful lifetime. In addition, in the physical world, many things happen at once. Software, by contrast, is mostly sequential. Reconciling these two disparate properties is a major challenge, and is often the biggest risk factor in the design of embedded systems. Incorrect p.1

interactions between sequential code and concurrent events in the physical world can cause dramatic system failures. In this chapter, we deal with issues.

10.1 I/O Hardware

p.202 Embedded processors, be they microcontrollers, DSP processors, or general-purpose processors, typically include a number of input and output (**I/O**) mechanisms on chip, exposed to designers as pins of the chip. In this section, we review some of the more common interfaces provided, illustrating their properties through the following running example.

Example 10.1: Figure 10.1 shows an evaluation board for the Luminary Micro Stellaris® microcontroller, which is an ARM Cortex^{TM} - M3 32-bit processor. The microcontroller itself is in the center below the graphics display. Many of the pins of the microcontroller are available at the connectors shown on either side of the microcontroller and at the top and bottom of the board. Such a board would typically be used to prototype an embedded application, and in the final product it would be replaced with a custom circuit board that includes only the hardware required by the application. An engineer will develop software for the board using an integrated development environment (**IDE**) provided by the vendor and load the p.231 software onto flash memory to be inserted into the slot at the bottom of the board. p.258 Alternatively, software might be loaded onto the board through the USB interface at the top from the development computer.

The evaluation board in the above example is more than a processor since it includes a display and various hardware interfaces (switches and a speaker, for example). Such a board is often called a **single-board computer** or a **microcomputer board**. We next discuss a few of the interfaces provided by a microcontroller or single-board computer. For a more comprehensive description of the many kinds of I/O interfaces in use, we recommend Valvano (2007) and Derenzo (2003).

10.1.1 Pulse Width Modulation

p.193 Pulse width modulation (PWM) is a technique for delivering a variable amount of power efficiently to external hardware devices. It can be used to control for example the speed of

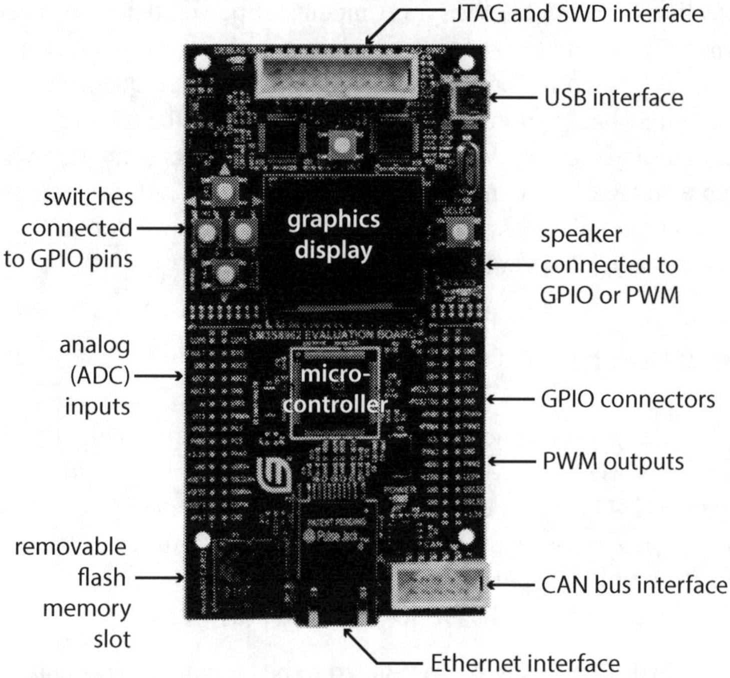

Figure 10.1: Stellaris® LM3S8962 evaluation board (Luminary Micro®, 2008a). (Luminary Micro was acquired by Texas Instruments in 2009.)

electric motors, the brightness of an LED light, and the temperature of a heating element. In general, it can deliver varying amounts of power to devices that tolerate rapid and abrupt changes in voltage and current.

PWM hardware uses only digital circuits, and hence is easy to integrate on the same chip with a microcontroller. Digital circuits, by design, produce only two voltage levels, high and low. A PWM signal rapidly switches between high and low at some fixed frequency, varying the amount of time that it holds the signal high. The duty cycle is the proportion of p.194 time that the voltage is high. If the duty cycle is 100%, then the voltage is always high. If the duty cycle is 0%, then the voltage is always low.

Many microcontrollers provide PWM peripheral devices (see Figure 10.1). To use these, a programmer typically writes a value to a memory-mapped register to set the duty cycle (the p.233 frequency may also be settable). The device then delivers power to external hardware in proportion to the specified duty cycle.

PWM is an effective way to deliver varying amounts of power, but only to certain devices. A heating element, for example, is a resistor whose temperature increases as more current passes through it. Temperature varies slowly, compared to the frequency of a PWM signal, so the rapidly varying voltage of the signal is averaged out by the resistor, and the temperature will be very close to constant for a fixed duty cycle. Motors similarly average out rapid variations in input voltage. So do incandescent and LED lights. Any device whose response to changes in current or voltage is slow compared to the frequency of the PWM signal is a candidate for being controlled via PWM.

10.1.2 General-Purpose Digital I/O

Embedded system designers frequently need to connect specialized or custom digital hardware to embedded processors. Many embedded processors have a number of **general-purpose I/O** pins (**GPIO**), which enable the software to either read or write voltage levels representing a logical zero or one. If the processor **supply voltage** is V_{DD}, in **active high logic** a voltage close to V_{DD} represents a logical one, and a voltage near zero represents a logical zero. In **active low logic**, these interpretations are reversed.

In many designs, a GPIO pin may be configured to be an output. This enables software to
p.233 then write to a memory-mapped register to set the output voltage to be either high or low. By this mechanism, software can directly control external physical devices.

However, caution is in order. When interfacing hardware to GPIO pins, a designer needs to understand the specifications of the device. In particular, the voltage and current levels vary by device. If a GPIO pin produces an output voltage of V_{DD} when given a logical one, then the designer needs to know the current limitations before connecting a device to it. If
p.192 a device with a resistance of R ohms is connected to it, for example, then Ohm's law tells us that the output current will be

$$I = V_{DD}/R\ .$$

It is essential to keep this current within specified tolerances. Going outside these tolerances could cause the device to overheat and fail. A **power amplifier** may be needed to deliver adequate current. An amplifier may also be needed to change voltage levels.

Example 10.2: The GPIO pins of the Luminary Micro Stellaris® microcontroller shown in Figure 10.1 may be configured to source or sink varying amounts of current up to 18 mA. There are restrictions on what combinations of pins can handle such relatively high currents. For example, Luminary Micro® (2008b) states "The high-current GPIO package pins must be selected such that there are only a maximum of two per side of the physical package ... with the total number of high-current GPIO outputs not exceeding four for the entire package." Such constraints are designed to prevent overheating of the device.

In addition, it may be important to maintain **electrical isolation** between processor circuits and external devices. The external devices may have messy (noisy) electrical characteristics that will make the processor unreliable if the noise spills over into the power or ground lines of the processor. Or the external device may operate in a very different voltage or power regime compared to the processor. A useful strategy is to divide a circuit into **electrical domains**, possibly with separate power supplies, that have relatively little influence on one another. Isolation devices such as opto-isolators and transformers may be used to enable communication across electrical domains. The former convert an electrical signal in one electrical domain into light, and detect the light in the other electrical domain and convert it back to an electrical signal. The latter use inductive coupling between electrical domains.

GPIO pins can also be configured as inputs, in which case software will be able to react to externally provided voltage levels. An input pin may be **Schmitt triggered**, in which case they have hysteresis, similar to the thermostat of Example 3.5. A Schmitt triggered input p.52
pin is less vulnerable to noise. It is named after Otto H. Schmitt, who invented it in 1934 while he was a graduate student studying the neural impulse propagation in squid nerves.

Example 10.3: The GPIO pins of the microcontroller shown in Figure 10.1, when configured as inputs, are Schmitt triggered.

In many applications, several devices may share a single electrical connection. The designer must take care to ensure that these devices do not simultaneously drive the voltage of this single electrical connection to different values, resulting in a short circuit that can cause overheating and device failure.

Example 10.4: Consider a factory floor where several independent microcontrollers are all able to turn off a piece of machinery by asserting a logical zero on

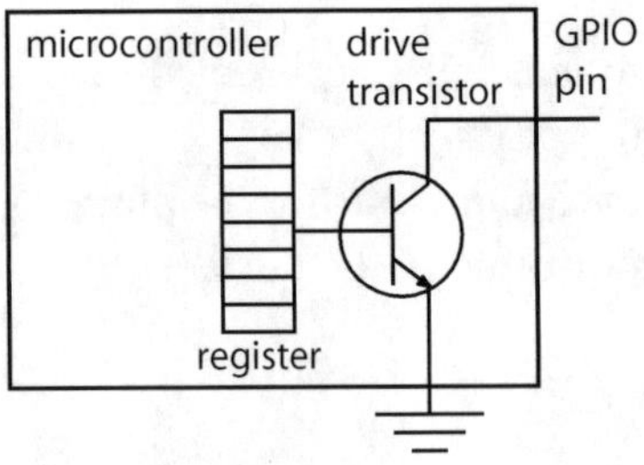

Figure 10.2: An open collector circuit for a GPIO pin.

an output GPIO line. Such a design may provide additional safety because the microcontrollers may be redundant, so that failure of one does not prevent a safety-related shutdown from occurring. If all of these GPIO lines are wired together to a single control input of the piece of machinery, then we have to take precautions to ensure that the microcontrollers do not short each other out. This would occur if one microcontroller attempts to drive the shared line to a high voltage while another attempts to drive the same line to a low voltage.

GPIO outputs may use **open collector** circuits, as shown in Figure 10.2. In such a circuit, writing a logical one into the (memory mapped) register turns on the transistor, which pulls the voltage on the output pin down to (near) zero. Writing a logical zero into the register turns off the transistor, which leaves the output pin unconnected, or "open."

A number of open collector interfaces may be connected as shown in Figure 10.3. The shared line is connected to a **pull-up resistor**, which brings the voltage of the line up to V_{DD} when all the transistors are turned off. If any one transistor is turned on, then it will bring the voltage of the entire line down to (near) zero without creating a short circuit with the other GPIO pins. Logically, all registers must have zeros in them for the output to be high. If any one of the registers has a one in it, then the output will be low. Assuming p.252 active high logic, the logical function being performed is NOR, so such a circuit is called a **wired NOR**. By varying the configuration, one can similarly create wired OR or wired AND.

The term "open collector" comes from the name for the terminal of a bipolar transistor. In CMOS technologies, this type of interface will typically be called an **open drain** interface. It functions essentially in the same way.

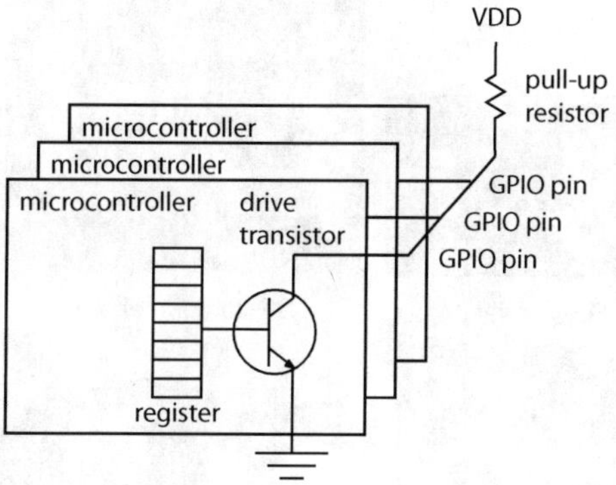

Figure 10.3: A number of open collector circuits wired together.

Example 10.5: The GPIO pins of the microcontroller shown in Figure 10.1, when configured as outputs, may be specified to be open drain circuits. They may also optionally provide the pull-up resistor, which conveniently reduces the number of external discrete components required on a printed circuit board.

GPIO outputs may also be realized with **tristate** logic, which means that in addition to producing an output high or low voltage, the pin may be simply turned off. Like an open-collector interface, this can facilitate sharing the same external circuits among multiple devices. Unlike an open-collector interface, a tristate design can assert both high and low voltages, rather than just one of the two.

10.1.3 Serial Interfaces

One of the key constraints faced by embedded processor designers is the need to have physically small packages and low power consumption. A consequence is that the number of pins on the processor integrated circuit is limited. Thus, each pin must be used efficiently. In addition, when wiring together subsystems, the number of wires needs to be limited to keep the overall bulk and cost of the product in check. Hence, wires must also be used efficiently. One way to use pins and wires efficiently is to send information over them serially as sequences of bits. Such an interface is called a **serial interface**. A number of

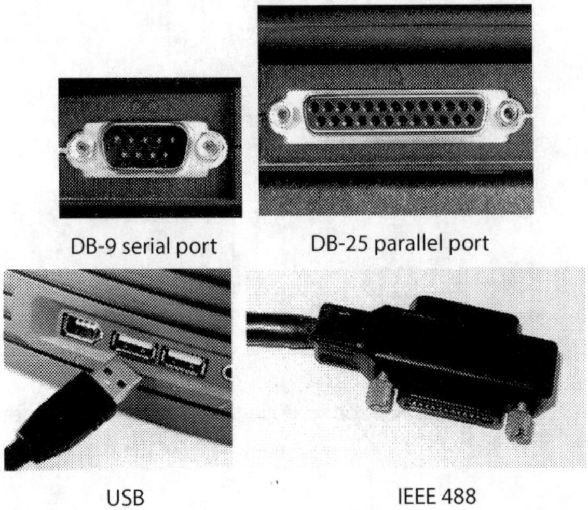

Figure 10.4: Connectors for serial and parallel interfaces.

standards have evolved for serial interfaces so that devices from different manufacturers can (usually) be connected.

An old but persistent standard, **RS-232**, standardized by the Electronics Industries Association (EIA), was first introduced in 1962 to connect teletypes to modems. This standard defines electrical signals and connector types; it persists because of its simplicity and because of continued prevalence of aging industrial equipment that uses it. The standard defines how one device can transmit a byte to another device asynchronously (meaning that the devices do not share a clock signal). On older PCs, an RS-232 connection may be provided via a DB-9 connector, as shown in Figure 10.4. A microcontroller will typically use a **universal asynchronous receiver/transmitter** (**UART**) to convert the contents of an 8-bit register into a sequence of bits for transmission over an RS-232 serial link.

For an embedded system designer, a major issue to consider is that RS-232 interfaces can be quite slow and may slow down the application software, if the programmer is not very careful.

p.204 **Example 10.6:** All variants of the Atmel AVR microcontroller include a UART that can be used to provide an RS-232 serial interface. To send a byte over the serial port, an application program may include the lines

```
while(!(UCSR0A & 0x20));
UDR0 = x;
```

where x is a variable of type uint8_t (a C data type specifying an 8-bit unsigned integer). The symbols UCSR0A and UDR0 are defined in header files provided in the AVR IDE. They are defined to refer to memory locations corresponding to memory-mapped registers in the AVR architecture. p.250 p.233

The first line above executes an empty while loop until the serial transmit buffer is empty. The AVR architecture indicates that the transmit buffer is empty by setting the sixth bit of the memory mapped register UCSR0A to 1. When that bit becomes 1, the expression !(UCSR0A & 0x20) becomes 0 and the while loop stops looping. The second line loads the value to be sent, which is whatever the variable x contains, into the memory-mapped register UDR0.

Suppose you wish to send a sequence of 8 bytes stored in an array x. You could do this with the C code

```
for(i = 0; i < 8; i++) {
  while(!(UCSR0A & 0x20));
  UDR0 = x[i];
}
```

How long would it take to execute this code? Suppose that the serial port is set to operate at 57600 baud, or bits per second (this is quite fast for an RS-232 interface). Then after loading UDR0 with an 8-bit value, it will take $8/57600$ seconds or about 139 microseconds for the 8-bit value to be sent. Suppose that the frequency of the processor is operating at 18 MHz (relatively slow for a microcontroller). Then except for the first time through the for loop, each while loop will need to consume approximately 2500 cycles, during which time the processor is doing no useful work.

To receive a byte over the serial port, a programmer may use the following C code:

```
while(!(UCSR0A & 0x80));
return UDR0;
```

In this case, the while loop waits until the UART has received an incoming byte. The programmer must ensure that there will be an incoming byte, or this code will execute forever. If this code is again enclosed in a loop to receive a sequence of bytes, then the while loop will need to consume a considerable number of cycles each time it executes.

p.261

For both sending and receiving bytes over a serial port, a programmer may use an interrupt instead to avoid having an idle processor that is waiting for the serial communication to occur. Interrupts will be discussed below.

The RS-232 mechanism is very simple. The sender and receiver first must agree on a transmission rate (which is slow by modern standards). The sender initiates transmission of a byte with a **start bit**, which alerts the receiver that a byte is coming. The sender then clocks out the sequence of bits at the agreed-upon rate, following them by one or two **stop bits**. The receiver's clock resets upon receiving the start bit and is expected to track the sender's clock closely enough to be able to sample the incoming signal sequentially and recover the sequence of bits. There are many descendants of the standard that support higher rate communication, such as **RS-422**, **RS-423**, and more.

Newer devices designed to connect to personal computers typically use **universal serial bus** (**USB**) interfaces, standardized by a consortium of vendors. USB 1.0 appeared in 1996 and supports a data rate of 12 Mbits/sec. USB 2.0 appeared in 2000 and supports data rates up to 480 Mbits/sec. USB 3.0 appeared in 2008 and supports data rates up to 4.8 Gbits/sec.

USB is electrically simpler than RS-232 and uses simpler, more robust connectors, as shown in Figure 10.4. But the USB standard defines much more than electrical transport of bytes, and more complicated control logic is required to support it. Since modern peripheral devices such as printers, disk drives, and audio and video devices all include microcontrollers, supporting the more complex USB protocol is reasonable for these devices.

Another serial interface that is widely implemented in embedded processors is known as **JTAG** (Joint Test Action Group), or more formally as the IEEE 1149.1 standard test access port and boundary-scan architecture. This interface appeared in the mid 1980s to solve the problem that integrated circuit packages and printed circuit board technology had evolved to the point that testing circuits using electrical probes had become difficult or impossible. Points in the circuit that needed to be accessed became inaccessible to probes. The notion of a **boundary scan** allows the state of a logical boundary of a circuit (what would traditionally have been pins accessible to probes) to be read or written serially through pins that are made accessible. Today, JTAG ports are widely used to provide a debug interface to embedded processors, enabling a PC-hosted debugging environment to examine and control the state of an embedded processor. The JTAG port is used, for example, to read out the state of processor registers, to set breakpoints in a program, and to single step through a program. A newer variant is **serial wire debug** (**SWD**), which provides similar functionality with fewer pins.

There are several other serial interfaces in use today, including for example **I^{2}C** (inter-integrated circuit), **SPI** (serial peripheral interface bus), **PCI Express** (peripheral component interconnect express), **FireWire**, **MIDI** (musical instrument digital interface), and serial versions of SCSI (described below). Each of these has its use. Also, network interfaces are typically serial.

10.1.4 Parallel Interfaces

A serial interface sends or receives a sequence of bits sequentially over a single line. A **parallel interface** uses multiple lines to simultaneously send bits. Of course, each line of a parallel interface is also a serial interface, but the logical grouping and coordinated action of these lines is what makes the interface a parallel interface.

Historically, one of the most widely used parallel interfaces is the IEEE-1284 printer port, which on the IBM PC used a DB-25 connector, as shown in Figure 10.4. This interface originated in 1970 with the Centronics model 101 printer, and hence is sometimes called a Centronics printer port. Today, printers are typically connected using USB or wireless networks.

With careful programming, a group of GPIO pins can be used together to realize a parallel p.252 interface. In fact, embedded system designers sometimes find themselves using GPIO pins to emulate an interface not supported directly by their hardware.

It seems intuitive that parallel interfaces should deliver higher performance than serial interfaces, because more wires are used for the interconnection. However, this is not necessarily the case. A significant challenge with parallel interfaces is maintaining synchrony across the multiple wires. This becomes more difficult as the physical length of the interconnection increases. This fact, combined with the requirement for bulkier cables and more I/O pins has resulted in many traditionally parallel interfaces being replaced by serial interfaces.

10.1.5 Buses

A **bus** is an interface shared among multiple devices, in contrast to a point-to-point interconnection linking exactly two devices. Busses can be serial interfaces (such as USB) or parallel interfaces. A widespread parallel bus is **SCSI** (pronounced scuzzy, for small computer system interface), commonly used to connect hard drives and tape drives to computers. Recent variants of SCSI interfaces, however, depart from the traditional parallel interface to become serial interfaces. SCSI is an example of a **peripheral bus** architecture, used to connect computers to peripherals such as sound cards and disk drives.

Other widely used peripheral bus standards include the **ISA bus** (industry standard architecture, used in the ubiquitous IBM PC architecture), **PCI** (peripheral component interface), and **Parallel ATA** (advanced technology attachment). A somewhat different kind of peripheral bus standard is **IEEE-488**, originally developed more than 30 years ago to connect automated test equipment to controlling computers. This interface was designed at Hewlett Packard and is also widely known as **HP-IB** (Hewlett Packard interface bus) and **GPIB** (general purpose interface bus). Many networks also use a bus architecture.

Because a bus is shared among several devices, any bus architecture must include a **media-access control** (**MAC**) protocol to arbitrate competing accesses. A simple MAC protocol has a single bus master that interrogates bus slaves. USB uses such a mechanism. An alternative is a **time-triggered bus**, where devices are assigned time slots during which they can transmit (or not, if they have nothing to send). A third alternative is a **token ring**, where devices on the bus must acquire a token before they can use the shared medium, and the token is passed around the devices according to some pattern. A fourth alternative is to use a bus arbiter, which is a circuit that handles requests for the bus according to some priorities. A fifth alternative is **carrier sense multiple access** (**CSMA**), where devices sense the carrier to determine whether the medium is in use before beginning to use it, detect collisions that might occur when they begin to use it, and try again later when a collision occurs.

In all cases, sharing of the physical medium has implications on the timing of applications.

Example 10.7: A peripheral bus provides a mechanism for external devices to communicate with a CPU. If an external device needs to transfer a large amount of data to the main memory, it may be inefficient and/or disruptive to require the CPU to perform each transfer. An alternative is **direct memory access** (**DMA**). In the DMA scheme used on the ISA bus, the transfer is performed by a separate device called a **DMA controller** which takes control of the bus and transfers the data. In some more recent designs, such as PCI, the external device directly takes control of the bus and performs the transfer without the help of a dedicated DMA controller. In both cases, the CPU is free to execute software while the transfer is occurring, but if the executed code needs access to the memory or the peripheral bus, then the timing of the program is disrupted by the DMA. Such timing effects can be difficult to analyze.

10.2 Sequential Software in a Concurrent World

As we saw in Example 10.6, when software interacts with the external world, the timing of the execution of the software may be strongly affected. Software is intrinsically sequential, typically executing as fast as possible. The physical world, however, is concurrent, with many things happening at once, and with the pace at which they happen determined by their physical properties. Bridging this mismatch in semantics is one of the major challenges that an embedded system designer faces. In this section, we discuss some of the key mechanisms for accomplishing this.

10.2.1 Interrupts and Exceptions

An **interrupt** is a mechanism for pausing execution of whatever a processor is currently doing and executing a pre-defined code sequence called an **interrupt service routine** (**ISR**) or **interrupt handler**. Three kinds of events may trigger an interrupt. One is a **hardware interrupt**, where some external hardware changes the voltage level on an interrupt request line. In the case of a **software interrupt**, the program that is executing triggers the interrupt by executing a special instruction or by writing to a memory-mapped register. A third variant is called an **exception**, where the interrupt is triggered by internal hardware that detects a fault, such as a segmentation fault.

p.233

p.243

For the first two variants, once the ISR completes, the program that was interrupted resumes where it left off. In the case of an exception, once the ISR has completed, the program that triggered the exception is not normally resumed. Instead, the program counter is set to some fixed location where, for example, the operating system may terminate the offending program.

Upon occurrence of an interrupt trigger, the hardware must first decide whether to respond. If interrupts are disabled, it will not respond. The mechanism for enabling or disabling interrupts varies by processor. Moreover, it may be that some interrupts are enabled and others are not. Interrupts and exceptions generally have priorities, and an interrupt will be serviced only if the processor is not already in the middle of servicing an interrupt with a higher priority. Typically, exceptions have the highest priority and are always serviced.

When the hardware decides to service an interrupt, it will usually first disable interrupts, push the current program counter and processor status register(s) onto the stack, and branch to a designated address that will normally contain a jump to an ISR. The ISR must store on the stack the values currently in any registers that it will use, and restore their values before returning from the interrupt, so that the interrupted program can resume where it left off.

p.241

Either the interrupt service routine or the hardware must also re-enable interrupts before returning from the interrupt.

Example 10.8: The ARM Cortex™ - M3 is a 32-bit microcontroller used in p.263 industrial automation and other applications. It includes a system timer called SysTick. This timer can be used to trigger an ISR to execute every 1ms. Suppose for example that every 1ms we would like to count down from some initial count until the count reaches zero, and then stop counting down. The following C code defines an ISR that does this:

```
volatile uint timerCount = 0;
void countDown(void) {
    if (timerCount != 0) {
        timerCount--;
    }
}
```

p.244 Here, the variable `timerCount` is a global variable, and it is decremented each time `countDown()` is invoked, until it reaches zero. We will specify below that this is to occur once per millisecond by registering `countDown()` as an ISR. The variable `timerCount` is marked with the C **volatile keyword**, which tells the compiler that the value of the variable will change at unpredictable times during execution of the program. This prevents the compiler from performing certain optimizations, such as caching the value of the variable in a register and reading it p.286 repeatedly. Using a C API provided by Luminary Micro® (2008c), we can specify that `countDown()` should be invoked as an interrupt service routine once per millisecond as follows:

```
SysTickPeriodSet(SysCtlClockGet() / 1000);
SysTickIntRegister(&countDown);
SysTickEnable();
SysTickIntEnable();
```

The first line sets the number of clock cycles between "ticks" of the SysTick timer. The timer will request an interrupt on each tick. `SysCtlClockGet()` is a library procedure that returns the number of cycles per second of the target platform's clock (e.g., 50,000,000 for a 50 MHz part). The second line registers the ISR by providing a **function pointer** for the ISR (the address of the `countDown()` procedure).

(Note: Some configurations do not support run-time registration of ISRs, as shown in this code. See the documentation for your particular system.) The third line starts the clock, enabling ticks to occur. The fourth line enables interrupts.

The timer service we have set up can be used, for example, to perform some function for two seconds and then stop. A program to do that is:

```
int main(void) {
    timerCount = 2000;
    ... initialization code from above ...
    while(timerCount != 0) {
        ... code to run for 2 seconds ...
    }
}
```

Processor vendors provide many variants of the mechanisms used in the previous example, so you will need to consult the vendor's documentation for the particular processor you are using. Since the code is not **portable** (it will not run correctly on a different processor), it is wise to isolate such code from your application logic and document carefully what needs to be re-implemented to target a new processor.

Basics: Timers

Microcontrollers almost always include some number of peripheral devices called **timer**s. A **programmable interval timer** (**PIT**), the most common type, simply counts down from some value to zero. The initial value is set by writing to a memory-mapped register, and when the value hits zero, the PIT raises an interrupt request. By writing to a memory-mapped control register, a timer might be set up to trigger repeatedly without having to be reset by the software. Such repeated triggers will be more precisely periodic than what you would get if the ISR restarts the timer each time it gets invoked. This is because the time between when the count reaches zero in the timer hardware and the time when the counter gets restarted by the ISR is difficult to control and variable. For example, if the timer reaches zero at a time when interrupts happen to be disabled, then there will be a delay before the ISR gets invoked. It cannot be invoked before interrupts are re-enabled.

p.233

10.2.2 Atomicity

An interrupt service routine can be invoked between any two instructions of the main program (or between any two instructions of a lower priority ISR). One of the major challenges for embedded software designers is that reasoning about the possible interleavings of instructions can become extremely difficult. In the previous example, the interrupt service routine and the main program are interacting through a **shared variable**, namely `timerCount`. The value of that variable can change between any two **atomic operation**s of the main program. Unfortunately, it can be quite difficult to know what operations are atomic. The term "atomic" comes from the Greek work for "indivisible," and it is far from obvious to a programmer what operations are indivisible. If the programmer is writing assembly code, then it may be safe to assume that each assembly language instruction is p.202 atomic, but many ISAs include assembly level instructions that are not atomic.

Example 10.9: The ARM instruction set includes a LDM instruction, which loads multiple registers from consecutive memory locations. It can be interrupted part way through the loads (ARM Limited, 2006).

At the level of a C program, it can be even more difficult to know what operations are atomic. Consider a single, innocent looking statement

```
timerCount = 2000;
```

On an 8-bit microcontroller, this statement may take more than one instruction cycle to execute (an 8-bit word cannot store both the instruction and the constant 2000; in fact, the constant alone does not fit in an 8-bit word). An interrupt could occur part way through the execution of those cycles. Suppose that the ISR also writes to the variable `timerCount`. In this case, the final value of the `timerCount` variable may be composed of 8 bits set in the ISR and the remaining bits set by the above line of C, for example. The final value could be very different from 2000, and also different from the value specified in the interrupt service routine. Will this bug occur on a 32-bit microcontroller? The only way to know for sure is to fully understand the ISA and the compiler. In such circumstances, there is no advantage to having written the code in C instead of assembly language.

Bugs like this in a program are extremely difficult to identify and correct. Worse, the problematic interleavings are quite unlikely to occur, and hence may not show up in testing. For safety-critical systems, programmers have to make every effort to avoid such bugs. One way to do this is to build programs using higher-level concurrent models of computation, as discussed in Chapter 6. Of course, the implementation of those models of computation needs to be correct, but presumably, that implementation is constructed by experts in concurrency, rather than by application engineers.

When working at the level of C and ISRs, a programmer must carefully reason about the *order* of operations. Although many interleavings are possible, operations given as a sequence of C statements must execute in order (more precisely, they must behave as if they had executed in order, even if out-of-order execution is used). p.217

Example 10.10: In example 10.8, the programmer can rely on the statements within `main()` executing in order. Notice that in that example, the statement

```
timerCount = 2000;
```

appears before

```
SysTickIntEnable();
```

The latter statement enables the SysTick interrupt. Hence, the former statement cannot be interrupted by the SysTick interrupt.

10.2.3 Interrupt Controllers

An **interrupt controller** is the logic in the processor that handles interrupts. It supports some number of interrupts and some number of priority levels. Each interrupt has an **interrupt vector**, which is the address of an ISR or an index into an array called the **interrupt vector table** that contains the addresses of all the ISRs.

Example 10.11: The Luminary Micro LM3S8962 controller, shown in Figure 10.1, includes an ARM Cortex™ - M3 core microcontroller that supports 36 interrupts with eight priority levels. If two interrupts are assigned the same priority number, then the one with the lower vector will have priority over the one with the higher vector.

When an interrupt is asserted by changing the voltage on a pin, the response may be either **level triggered** or **edge triggered**. For level-triggered interrupts, the hardware asserting the interrupt will typically hold the voltage on the line until it gets an acknowledgement, which indicates that the interrupt is being handled. For edge-triggered interrupts, the hardware as-
p.254 serting the interrupt changes the voltage for only a short time. In both cases, open collector lines can be used so that the same physical line can be shared among several devices (of course, the ISR will require some mechanism to determine which device asserted the in-
p.233 terrupt, for example by reading a memory-mapped register in each device that could have asserted the interrupt).

Sharing interrupts among devices can be tricky, and careful consideration must be given to prevent low priority interrupts from blocking high priority interrupts. Asserting interrupts by writing to a designated address on a bus has the advantage that the same hardware can support many more distinct interrupts, but the disadvantage that peripheral devices get more complex. The peripheral devices have to include an interface to the memory bus.

10.2.4 Modeling Interrupts

The behavior of interrupts can be quite difficult to fully understand, and many catastrophic system failures are caused by unexpected behaviors. Unfortunately, the logic of interrupt controllers is often described in processor documentation very imprecisely, leaving many possible behaviors unspecified. One way to make this logic more precise is to model it as
p.47 an FSM.

Example 10.12: The program of Example 10.8, which performs some action for two seconds, is shown in Figure 10.5 together with two finite state machines that model the ISR and the main program. The states of the FSMs correspond to positions in the execution labeled A through E, as shown in the program listing. These positions are between C statements, so we are assuming here that these statements
p.264 are atomic operations (a questionable assumption in general).

We may wish to determine whether the program is assured of always reaching position C. In other words, can we assert with confidence that the program will

```
    volatile uint timerCount = 0;
    void ISR(void) {
       ... disable interrupts
D-->   if(timerCount != 0) {
E-->      timerCount--;
       }
       ... enable interrupts
    }
    int main(void) {
       // initialization code
       SysTickIntRegister(&ISR);
       ... // other init
       timerCount = 2000;
A-->   while(timerCount != 0) {
B-->    ... code to run for 2 seconds
       }
    }
C--> ... whatever comes next
```

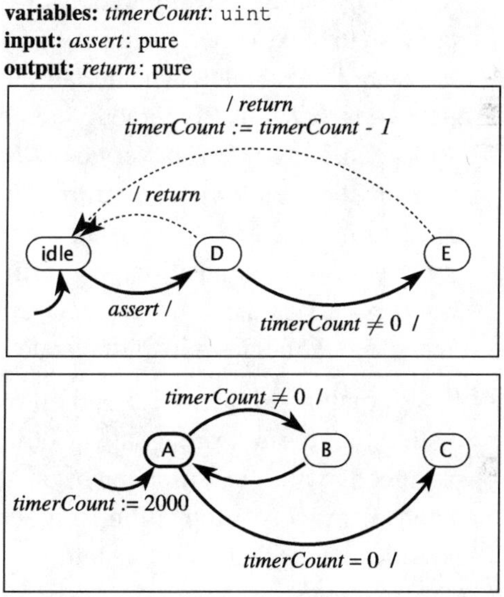

Figure 10.5: State machine models and main program for a program that does something for two seconds and then continues to do something else.

eventually move beyond whatever computation it was to perform for two seconds? A state machine model will help us answer that question.

The key question now becomes how to compose these state machines to correctly model the interaction between the two pieces of sequential code in the procedures `ISR` and `main`. It is easy to see that asynchronous composition is not the right choice because the interleavings are not arbitrary. In particular, `main` can be interrupted by `ISR`, but `ISR` cannot be interrupted by `main`. Asynchronous composition would fail to capture this asymmetry.

p.110

Assuming that the interrupt is always serviced immediately upon being requested, we wish to have a model something like that shown in Figure 10.6. In that figure, a two-state FSM models whether an interrupt is being serviced. The transition from Inactive to Active is triggered by a pure input *assert*, which models the timer hardware requesting interrupt service. When the ISR completes its execution, another pure input *return* triggers a return to the Inactive state. Notice here that the transition from Inactive to Active is a preemptive transition, indicated by the small circle at the start of the transition, suggesting that it should be taken immediately when *assert* occurs, and that it is a reset transition, suggesting that the state refinement of Active should begin in its initial state upon entry.

p.122
p.122

If we combine Figures 10.5 and 10.6 we get the hierarchical FSM in Figure 10.7. Notice that the *return* signal is both an input and an output now. It is an output produced by the state refinement of Active, and it is an input to the top-level FSM, where it triggers a transition to Inactive. Having an output that is also an input provides a mechanism for a state refinement to trigger a transition in its container state machine.

p.119

To determine whether the program reaches state C, we can study the flattened state machine shown in Figure 10.8. Studying that machine carefully, we see that in fact there is no assurance that state C will be reached! If, for example, *assert* is present on every reaction, then C is never reached.

Could this happen in practice? With this program, it is improbable, but not impossible. It could happen if the ISR itself takes longer to execute than the time between interrupts. Is there any assurance that this will not happen? Unfortunately, our only assurance is a vague notion that processors are faster than that. There is no guarantee.

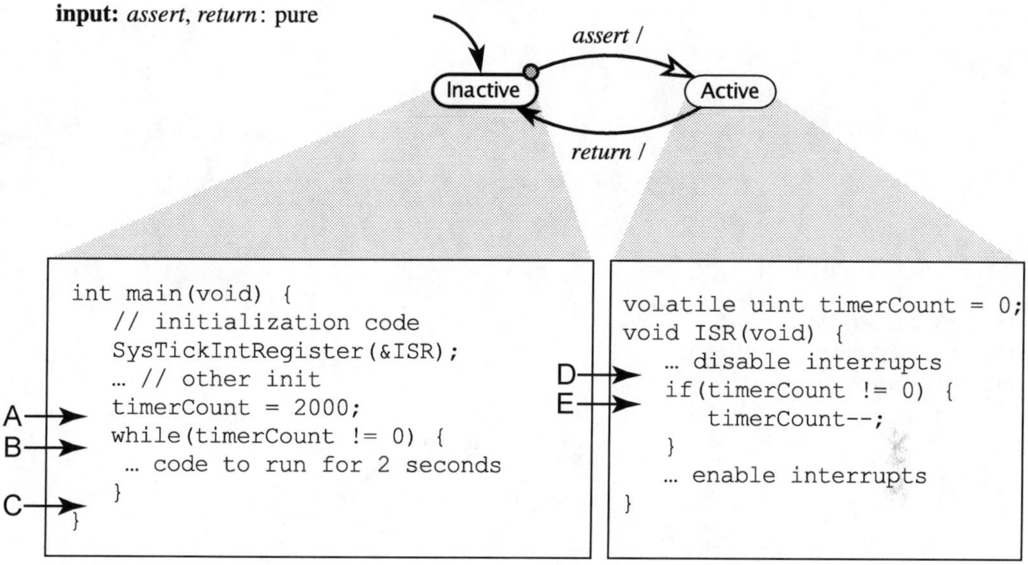
input: assert, return: pure
assert /
Inactive
Active
return /
int main(void) {
// initialization code
SysTickIntRegister(&ISR);
... // other init
timerCount = 2000;
while(timerCount != 0) {
... code to run for 2 seconds
}
}
A
B
C
volatile uint timerCount = 0;
void ISR(void) {
... disable interrupts
if(timerCount != 0) {
timerCount--;
}
... enable interrupts
}
D
E

Figure 10.6: Sketch of a state machine model for the interaction between an ISR and the main program.

In the above example, modeling the interaction between a main program and an interrupt service routine exposes a potential flaw in the program. Although the flaw may be unlikely to occur in practice in this example, the fact that the flaw is present at all is disturbing. In any case, it is better to know that the flaw is present, and to decide that the risk is acceptable, than to not know it is present.

Interrupt mechanisms can be quite complex. Software that uses these mechanisms to provide I/O to an external device is called a **device driver**. Writing device drivers that are correct and robust is a challenging engineering task requiring a deep understanding of the architecture and considerable skill reasoning about concurrency. Many failures in computer systems are caused by unexpected interactions between device drivers and other programs.

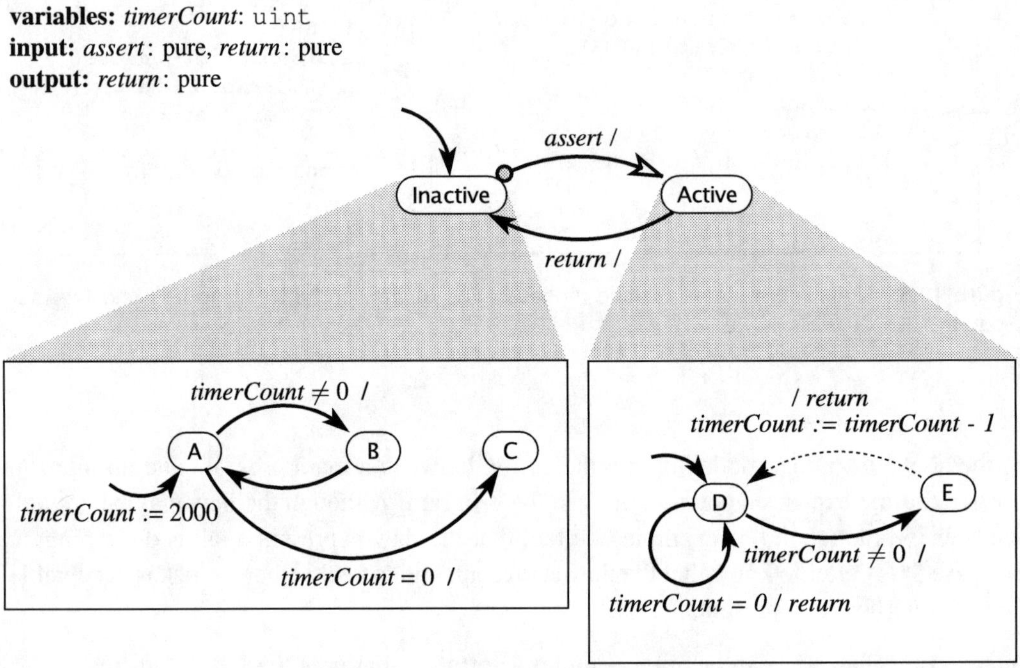

Figure 10.7: Hierarchical state machine model for the interaction between an ISR and the main program.

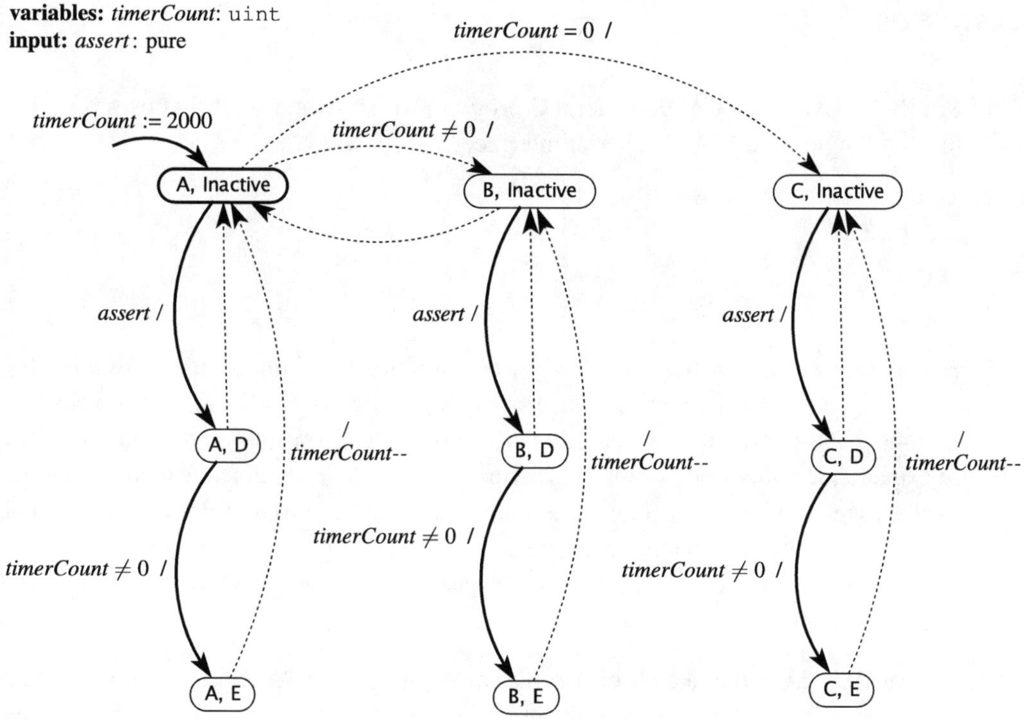

Figure 10.8: Flattened version of the hierarchical state machine in Figure 10.7.

10.3 Summary

This chapter has reviewed hardware and software mechanisms used to get sensor data into processors and commands from the processor to actuators. The emphasis is on understanding the principles behind the mechanisms, with a particular focus on the bridging between the sequential world of software and the parallel physical world.

Exercises

p.204

1. Similar to Example 10.6, consider a C program for an Atmel AVR that uses a UART to send 8 bytes to an RS-232 serial interface, as follows:

```
for(i = 0; i < 8; i++) {
  while(!(UCSR0A & 0x20));
  UDR0 = x[i];
}
```

   Assume the processor runs at 50 MHz; also assume that initially the UART is idle, so when the code begins executing, `UCSR0A & 0x20 == 0x20` is true; further, assume that the serial port is operating at 19,200 baud. How many cycles are required to execute the above code? You may assume that the `for` statement executes in three cycles (one to increment `i`, one to compare it to 8, and one to perform the conditional branch); the `while` statement executes in 2 cycles (one to compute `!(UCSR0A & 0x20)` and one to perform the conditional branch); and the assigment to `UDR0` executes in one cycle.

2. Figure 10.9 gives the sketch of a program for an Atmel AVR microcontroller that performs some function repeatedly for three seconds. The function is invoked by calling the procedure foo(). The program begins by setting up a timer interrupt to occur once per second (the code to do this setup is not shown). Each time the interrupt occurs, the specified interrupt service routine is called. That routine decrements a counter until the counter reaches zero. The main() procedure initializes the counter with value 3 and then invokes foo() until the counter reaches zero.

   (a) We wish to assume that the segments of code in the grey boxes, labeled **A**, **B**, and **C**, are atomic. State conditions that make this assumption valid.

   (b) Construct a state machine model for this program, assuming as in part (a) that **A**, **B**, and **C**, are atomic. The transitions in your state machine should be labeled with "guard/action", where the action can be any of **A**, **B**, **C**, or nothing. The actions **A**, **B**, or **C** should correspond to the sections of code in the grey boxes with the corresponding labels. You may assume these actions are atomic.

   (c) Is your state machine deterministic? What does it tell you about how many times foo() may be invoked? Do all the possible behaviors of your model correspond to what the programmer likely intended?

```
#include <avr/interrupt.h>
volatile uint16_t timer_count = 0;

// Interrupt service routine.
SIGNAL(SIG_OUTPUT_COMPARE1A) {
  if(timer_count > 0) {
    timer_count--;                  A
  }
}

// Main program.
int main(void) {
  // Set up interrupts to occur
  // once per second.
  ...

  // Start a 3 second timer.        B
  timer_count = 3;

  // Do something repeatedly
   // for 3 seconds.
  while(timer_count > 0) {
    foo();                          C
  }
}
```

Figure 10.9: Sketch of a C program that performs some function by calling procedure foo() repeatedly for 3 seconds, using a timer interrupt to determine when to stop.

Note that there are many possible answers. Simple models are preferred over elaborate ones, and complete ones (where everything is defined) over incomplete ones. Feel free to give more than one model.

3. In a manner similar to example 10.8, create a C program for the ARM Cortex^{TM} - M3 to use the SysTick timer to invoke a system-clock ISR with a jiffy interval of 10 ms that records the time since system start in a 32-bit int. How long can this program run before your clock overflows?

p.289

4. Consider a dashboard display that displays “normal” when brakes in the car operate normally and “emergency” when there is a failure. The intended behavior is that once “emergency” has been displayed, “normal” will not again be displayed. That is, “emergency” remains on the display until the system is reset.

   In the following code, assume that the variable `display` defines what is displayed. Whatever its value, that is what appears on the dashboard.

```
volatile static uint8_t alerted;
volatile static char* display;
void ISRA() {
    if (alerted == 0) {
        display = "normal";
    }
}
void ISRB() {
    display = "emergency";
    alerted = 1;
}
void main() {
    alerted = 0;
    ...set up interrupts...
    ...enable interrupts...
    ...
}
```

   Assume that `ISRA` is an interrupt service routine that is invoked when the brakes are applied by the driver. Assume that `ISRB` is invoked if a sensor indicates that the brakes are being applied at the same time that the accelerator pedal is depressed. Assume that neither ISR can interrupt itself, but that `ISRB` has higher priority than `ISRA`, and hence `ISRB` can interrupt `ISRA`, but `ISRA` cannot interrupt `ISRB`. Assume further (unrealistically) that each line of code is atomic.

(a) Does this program always exhibit the intended behavior? Explain. In the remaining parts of this problem, you will construct various models that will either demonstrate that the behavior is correct or will illustrate how it can be incorrect.

(b) Construct a determinate extended state machine modeling `ISRA`. Assume that:
- `alerted` is a variable of type $\{0,1\} \subset$ `uint8_t`,
- there is a pure input A that when present indicates an interrupt request for `ISRA`, and
- `display` is an output of type `char*`.

(c) Give the size of the state space for your solution.

(d) Explain your assumptions about when the state machine in (b) reacts. Is this time triggered, event triggered, or neither? p.50

(e) Construct a determinate extended state machine modeling `ISRB`. This one has a pure input B that when present indicates an interrupt request for `ISRB`.

(f) Construct a flat (non-hierarchical) determinate extended state machine describing the joint operation of the these two ISRs. Use your model to argue the correctness of your answer to part (a).

(g) Give an equivalent hierarchical state machine. Use your model to argue the correctness of your answer to part (a).

5. Suppose a processor handles interrupts as specified by the following FSM:

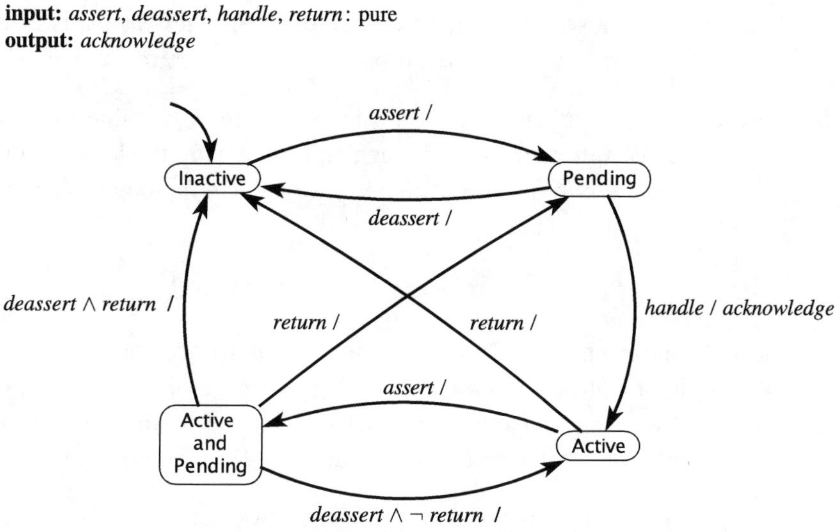

Here, we assume a more complicated interrupt controller than that considered in Example 10.12, where there are several possible interrupts and an arbiter that decides which interrupt to service. The above state machine shows the state of one interrupt. When the interrupt is asserted, the FSM transitions to the Pending state, and remains there until the arbiter provides a *handle* input. At that time, the FSM transitions to the Active state and produces an *acknowledge* output. If another interrupt is asserted while in the Active state, then it transitions to Active and Pending. When the ISR returns, the input *return* causes a transition to either Inactive or Pending, depending on the starting point. The *deassert* input allows external hardware to cancel an interrupt request before it gets serviced.

Answer the following questions.

(a) If the state is Pending and the input is *return*, what is the reaction?

(b) If the state is Active and the input is $assert \wedge deassert$, what is the reaction?

(c) Suppose the state is Inactive and the input sequence in three successive reactions is:

i. *assert* ,

ii. $deassert \wedge handle$,

iii. *return* .

What are all the possible states after reacting to these inputs? Was the interrupt handled or not?

(d) Suppose that an input sequence never includes *deassert*. Is it true that every *assert* input causes an *acknowledge* output? In other words, is every interrupt request serviced? If yes, give a proof. If no, give a counterexample.

6. Suppose you are designing a processor that will support two interrupts whose logic is given by the FSM in Exercise 5. Design an FSM giving the logic of an arbiter that assigns one of these two interrupts higher priority than the other. The inputs should be the following pure signals:

$$assert1, return1, assert2, return2$$

to indicate requests and return from interrupt for interrupts 1 and 2, respectively. The outputs should be pure signals *handle1* and *handle2*. Assuming the *assert* inputs are generated by two state machines like that in Exercise 5, can you be sure that this arbiter will handle every request that is made? Justify your answer.

7. Consider the following program that monitors two sensors. Here `sensor1` and `sensor2` denote the variables storing the readouts from two sensors. The actual

read is performed by the functions `readSensor1()` and `readSensor2()`, respectively, which are called in the interrupt service routine `ISR`.

```
char flag = 0;
volatile char* display;
volatile short sensor1, sensor2;

void ISR() {
  if (flag) {
     sensor1 = readSensor1();
  } else {
     sensor2 = readSensor2();
  }
}

int main() {
  // ... set up interrupts ...
  // ... enable interrupts ...
  while(1) {
     if (flag) {
        if isFaulty2(sensor2) {
           display = "Sensor2 Faulty";
        }
     } else {
        if isFaulty1(sensor1) {
           display = "Sensor1 Faulty";
        }
     }
     flag = !flag;
  }
}
```

Functions `isFaulty1()` and `isFaulty2()` check the sensor readings for any discrepancies, returning 1 if there is a fault and 0 otherwise. Assume that the variable `display` defines what is shown on the monitor to alert a human operator about faults. Also, you may assume that `flag` is modified only in the body of `main`.

Answer the following questions:

(a) Is it possible for the `ISR` to update the value of `sensor1` while the main function is checking whether `sensor1` is faulty? Why or why not?

(b) Suppose a spurious error occurs that causes `sensor1` or `sensor2` to be a faulty value for one measurement. Is it possible for that this code would not report "Sensor1 faulty" or "Sensor2 faulty"?

(c) Assuming the interrupt source for `ISR()` is timer-driven, what conditions would cause this code to never check whether the sensors are faulty?

(d) Suppose that instead being interrupt driven, `ISR` and `main` are executed concurrently, each in its own thread. Assume a microkernel that can interrupt any thread at any time and switch contexts to execute another thread. In this scenario, is it possible for the `ISR` to update the value of `sensor1` while the main function is checking whether `sensor1` is faulty? Why or why not?

11 Multitasking

Contents

In this chapter, we discuss mid-level mechanisms that are used in software to provide concurrent execution of sequential code. There are a number of reasons for executing multiple sequential programs concurrently, but they all involve timing. One reason is to improve responsiveness by avoiding situations where long-running programs can block a program that responds to external stimuli, such as sensor data or a user request. Improved responsiveness reduces **latency**, the time between the occurrence of a stimulus and the response. Another reason is to improve performance by allowing a program to run simultaneously on multiple processors or cores. This is also a timing issue, since it presumes that it is better to complete tasks earlier than later. A third reason is to directly control the timing of external

p.130

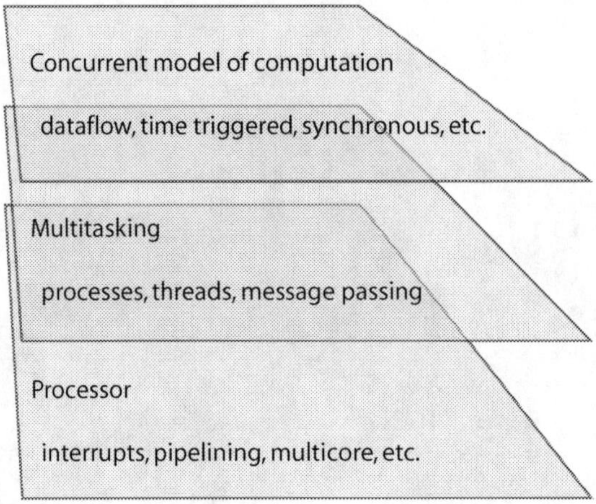

Figure 11.1: Layers of abstraction for concurrency in programs.

interactions. A program may need to perform some action, such as updating a display, at particular times, regardless of what other tasks might be executing at that time.

We have already discussed concurrency in a variety of contexts. Figure 11.1 shows the relationship between the subject of this chapter and those of other chapters. Chapters 8 and 10 cover the lowest layer in Figure 11.1, which represents how hardware provides concurrent mechanisms to the software designer. Chapters 5 and 6 cover the highest layer, which consists of abstract models of concurrency, including synchronous composition, dataflow, and time-triggered models. This chapter bridges these two layers. It describes mechanisms that are implemented using the low-level mechanisms and can provide infrastructure for realizing the high-level mechanisms. Collectively, these mid-level techniques are called **multitasking**, meaning the simultaneous execution of multiple tasks.

Embedded system designers frequently use these mid-level mechanisms directly to build applications, but it is becoming increasingly common for designers to use instead the high-level mechanisms. The designer constructs a model using a software tool that supports a p.130 model of computation (or several models of computation). The model is then automatically or semi-automatically translated into a program that uses the mid-level or low-level mechanisms. This translation process is variously called **code generation** or **autocoding**.

The mechanisms described in this chapter are typically provided by an operating system, a p.293 microkernel, or a library of procedures. They can be rather tricky to implement correctly, and hence the implementation should be done by experts (for some of the pitfalls, see Boehm (2005)). Embedded systems application programmers often find themselves having

```
#include <stdlib.h>
#include <stdio.h>
int x;                                   // Value that gets updated.
typedef void notifyProcedure(int);  // Type of notify proc.
struct element {
  notifyProcedure* listener;       // Pointer to notify procedure.
  struct element* next;            // Pointer to the next item.
};
typedef struct element element_t;   // Type of list elements.
element_t* head = 0;                // Pointer to start of list.
element_t* tail = 0;                // Pointer to end of list.

// Procedure to add a listener.
void addListener(notifyProcedure* listener) {
  if (head == 0) {
    head = malloc(sizeof(element_t));
    head->listener = listener;
    head->next = 0;
    tail = head;
  } else {
    tail->next = malloc(sizeof(element_t));
    tail = tail->next;
    tail->listener = listener;
    tail->next = 0;
  }
}
// Procedure to update x.
void update(int newx) {
  x = newx;
  // Notify listeners.
  element_t* element = head;
  while (element != 0) {
    (*(element->listener))(newx);
    element = element->next;
  }
}
// Example of notify procedure.
void print(int arg) {
  printf("%d ", arg);
}
```

Figure 11.2: A C program used in a series of examples in this chapter.

to implement such mechanisms on **bare iron** (a processor without an operating system). Doing so correctly requires deep understanding of concurrency issues.

This chapter begins with a brief description of models for sequential programs, which enable models of concurrent compositions of such sequential programs. We then progress to discuss threads, processes, and message passing, which are three styles of composition of sequential programs.

11.1 Imperative Programs

A programming language that expresses a computation as a sequence of operations is called p.212 an imperative language. C is an imperative language.

Example 11.1: In this chapter, we illustrate several key points using the example C program shown in Figure 11.2. This program implements a commonly used design pattern called the **observer pattern** (Gamma et al., 1994). In this pattern, an `update` procedure changes the value of a variable `x`. Observers (which are other programs or other parts of the program) will be notified whenever `x` is changed by calling a **callback** procedure. For example, the value of `x` might be displayed by an observer on a screen. Whenever the value changes, the observer needs to be notified so that it can update the display on the screen. The following `main` procedure uses the procedures defined in Figure 11.2:

```
int main(void) {
    addListener(&print);
    addListener(&print);
    update(1);
    addListener(&print);
    update(2);
    return 0;
}
```

This test program registers the `print` procedure as a callback twice, then performs an update (setting `x = 1`), then registers the print procedure again, and finally performs another update (setting `x = 2`). The `print` procedure simply prints the current value, so the output when executing this test program is `1 1 2 2 2`.

A C program specifies a sequence of steps, where each step changes the state of the memory in the machine. In C, the state of the memory in the machine is represented by the values of variables.

Example 11.2: In the program in Figure 11.2, the state of the memory of the machine includes the value of variable `x` (which is a global variable) and a list of elements pointed to by the variable `head` (another global variable). The list itself is represented as a linked list, where each element in the list contains a function pointer referring to a procedure to be called when `x` changes. p.244 p.285

During execution of the C program, the state of the memory of the machine will need to include also the state of the stack, which includes any local variables. p.241

Using extended state machines, we can model the execution of certain simple C programs, assuming the programs have a fixed and bounded number of variables. The variables of the C program will be the variables of the state machine. The states of the state machine will represent positions in the program, and the transitions will represent execution of the program. p.56

Example 11.3: Figure 11.3 shows a model of the `update` procedure in Figure 11.2. The machine transitions from the initial Idle state when the `update` procedure is called. The call is signaled by the input *arg* being present; its value will be the `int` argument to the `update` procedure. When this transition is taken, *newx* (on the stack) will be assigned the value of the argument. In addition, *x* (a global variable) will be updated. p.244

After this first transition, the machine is in state 31, corresponding to the program counter position just prior to the execution of line 31 in Figure 11.2. It then unconditionally transitions to state 32 and sets the value of *element*. From state 32, there are two possibilities; if *element* = 0, then the machine transitions back to Idle and produces the pure output *return*. Otherwise, it transitions to 33.

On the transition from 33 to 34, the action is a procedure call to the listener with the argument being the stack variable *newx*. The transition from 34 back to 32 occurs upon receiving the pure input *returnFromListener*, which indicates that the listener procedure returns. p.51

inputs: *arg*: `int`, *returnFromListener*: pure
outputs: *return*: pure
local variables: *newx*: `int`, *element*: `element_t*`
global variables: *x*: `int`, *head*: `element_t*`

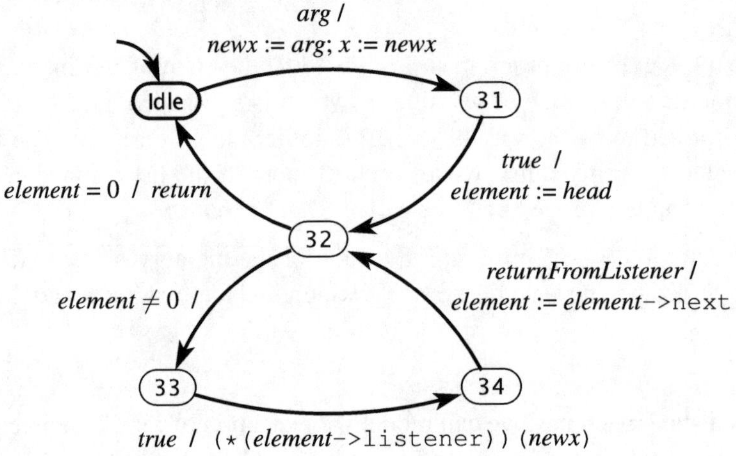

Figure 11.3: Model of the `update` procedure in Figure 11.2.

The model in Figure 11.3 is not the only model we could have constructed of the `update` procedure. In constructing such a model, we need to decide on the level of detail, and we p.264 need to decide which actions can be safely treated as atomic operations. Figure 11.3 uses lines of code as a level of detail, but there is no assurance that a line of C code executes atomically (it usually does not).

In addition, accurate models of C programs are often not finite state systems. Considering only the code in Figure 11.2, a finite-state model is not appropriate because the code supports adding an arbitrary number of listeners to the list. If we combine Figure 11.2 with the `main` procedure in Example 11.1, then the system is finite state because only three listeners are put on the list. An accurate finite-state model, therefore, would need to include the complete program, making modular reasoning about the code very difficult.

The problems get much worse when we add concurrency to the mix. We will show in this chapter that accurate reasoning about C programs with mid-level concurrency mechanisms such as threads is astonishingly difficult and error prone. It is for this reason that designers are tending towards the upper layer in Figure 11.1.

Linked Lists in C

A **linked list** is a data structure for storing a list of elements that varies in length during execution of a program. Each element in the list contains a **payload** (the value of the element) and a pointer to the next element in the list (or a null pointer if the element is the last one). For the program in Figure 11.2, the linked list data structure is defined by:

```
1 typedef void notifyProcedure(int);
2 struct element {
3   notifyProcedure* listener;
4   struct element* next;
5 };
6 typedef struct element element_t;
7 element_t* head = 0;
8 element_t* tail = 0;
```

The first line declares that `notifyProcedure` is a type whose value is a C procedure that takes an `int` and returns nothing. Lines 2–5 declare a **struct**, a composite data type in C. It has two pieces, `listener` (with type `notifyProcedure*`, which is a function pointer, a pointer to a C procedure) and `next` (a pointer to an instance of the same struct). Line 6 declares that `element_t` is a type referring to an instance of the structure `element`. p.262

Line 7 declares `head`, a pointer to a list element. It is initialized to 0, a value that indicates an empty list. The `addListener` procedure in Figure 11.2 creates the first list element using the following code:

```
1 head = malloc(sizeof(element_t));
2 head->listener = listener;
3 head->next = 0;
4 tail = head;
```

Line 1 allocates memory from the heap using `malloc` to store a list element and sets `head` to point to that element. Line 2 sets the payload of the element, and line 3 indicates that this is the last element in the list. Line 4 sets `tail`, a pointer to the last list element. When the list is not empty, the `addListener` procedure will use the `tail` pointer rather than `head` to append an element to the list. p.243

11.2 Threads

Threads are imperative programs that run concurrently and share a memory space. They can access each others' variables. Many practitioners in the field use the term "threads" more narrowly to refer to particular ways of constructing programs that share memory, but here we will use the term broadly to refer to any mechanism where imperative programs run p.261 concurrently and share memory. In this broad sense, threads exist in the form of interrupts p.282 on almost all microprocessors, even without any operating system at all (bare iron).

11.2.1 Creating Threads

Most operating systems provide a higher-level mechanism than interrupts to realize imperative programs that share memory. The mechanism is provided in the form of a collection of procedures that a programmer can use. Such procedures typically conform to a standardized **API** (**application program interface**), which makes it possible to write programs that are portable (they will run on multiple processors and/or multiple operating systems). **Pthreads** (or **POSIX threads**) is such an API; it is integrated into many modern operating systems. Pthreads defines a set of C programming language types, functions and constants. It was standardized by the IEEE in 1988 to unify variants of Unix. In Pthreads, a thread is defined by a C procedure and created by invoking the `pthread_create` procedure.[1]

Example 11.4: A simple multithreaded C program using Pthreads is shown in Figure 11.4. The `printN` procedure (lines 3–9) — the procedure that the thread begins executing — is called the **start routine**; in this case, the start routine prints the argument passed to it 10 times and then exits, which will cause the thread to terminate. The `main` procedure creates two threads, each of which will execute the start routine. The first one, created on line 14, will print the value 1. The second one, created on line 15, will print the value 2. When you run this program, values 1 and 2 will be printed in some interleaved order that depends on the thread scheduler. Typically, repeated runs will yield different interleaved orders of 1's and 2's.

[1] For brevity, in the examples in this text we do not check for failures, as any well-written program using Pthreads should. For example, `pthread_create` will return 0 if it succeeds, and a non-zero error code if it fails. It could fail, for example, due to insufficient system resources to create another thread. Any program that uses `pthread_create` should check for this failure and handle it in some way. Refer to the Pthreads documentation for details.

```
1  #include <pthread.h>
2  #include <stdio.h>
3  void* printN(void* arg) {
4      int i;
5      for (i = 0; i < 10; i++) {
6          printf("My ID: %d\n", *(int*)arg);
7      }
8      return NULL;
9  }
10 int main(void) {
11     pthread_t threadID1, threadID2;
12     void* exitStatus;
13     int x1 = 1, x2 = 2;
14     pthread_create(&threadID1, NULL, printN, &x1);
15     pthread_create(&threadID2, NULL, printN, &x2);
16     printf("Started threads.\n");
17     pthread_join(threadID1, &exitStatus);
18     pthread_join(threadID2, &exitStatus);
19     return 0;
20 }
```

Figure 11.4: Simple multithreaded C program using Pthreads.

The `pthread_create` procedure creates a thread and returns immediately. The start routine may or may not have actually started running when it returns. Lines 17 and 18 use `pthread_join` to ensure that the main program does not terminate before the threads have finished. Without these two lines, running the program may not yield any output at all from the threads.

A start routine may or may not return. In embedded applications, it is quite common to define start routines that never return. For example, the start routine might execute forever and update a display periodically. If the start routine does not return, then any other thread that calls its `pthread_join` will be blocked indefinitely.

As shown in Figure 11.4, the start routine can be provided with an argument and can return a value. The fourth argument to `pthread_create` is the address of the argument to be passed to the start routine. It is important to understand the memory model of C, explained in Section 9.3.5, or some very subtle errors could occur, as illustrated in the next example.

Example 11.5: Suppose we attempt to create a thread inside a procedure like this:

```
pthread_t createThread(int x) {
    pthread_t ID;
    pthread_create(&ID, NULL, printN, &x);
    return ID;
}
```

This code would be incorrect because the argument to the start routine is given by a pointer to a variable on the stack. By the time the thread accesses the specified memory address, the `createThread` procedure will likely have returned and the memory address will have been overwritten by whatever went on the stack next.

11.2.2 Implementing Threads

p.309 The core of an implementation of threads is a scheduler that decides which thread to execute next when a processor is available to execute a thread. The decision may be based on **fairness**, where the principle is to give every active thread an equal opportunity to run, on timing constraints, or on some measure of importance or priority. Scheduling algorithms are discussed in detail in Chapter 12. In this section, we simply describe how a thread scheduler will work without worrying much about how it makes a decision on which thread to execute.

The first key question is how and when the scheduler is invoked. A simple technique called **cooperative multitasking** does not interrupt a thread unless the thread itself calls a certain procedure or one of a certain set of procedures. For example, the scheduler may intervene whenever any operating system service is invoked by the currently executing thread. An operating system service is invoked by making a call to a library procedure. Each thread has p.241 its own stack, and when the procedure call is made, the return address will be pushed onto the stack. If the scheduler determines that the currently executing thread should continue to execute, then the requested service is completed and the procedure returns as normal. If instead the scheduler determines that the thread should be **suspended** and another thread should be selected for execution, then instead of returning, the scheduler makes a record p.241 of the stack pointer of the currently executing thread, and then modifies the stack pointer to point to the stack of the selected thread. It then returns as normal by popping the return address off the stack and resuming execution, but now in a new thread.

The main disadvantage of cooperative multitasking is that a program may execute for a long time without making any operating system service calls, in which case other threads will be **starved**. To correct for this, most operating systems include an interrupt service routine that runs at fixed time intervals. This routine will maintain a **system clock**, which provides application programmers with a way to obtain the current time of day and enables periodic invocation of the scheduler via a timer interrupt. For an operating system with a system clock, a **jiffy** is the time interval at which the system-clock ISR is invoked. p.263

Example 11.6: The jiffy values in Linux versions have typically varied between 1 ms and 10 ms.

The value of a jiffy is determined by balancing performance concerns with required timing precision. A smaller jiffy means that scheduling functions are performed more often, which can degrade overall performance. A larger jiffy means that the precision of the system clock is coarser and that task switching occurs less often, which can cause real-time constraints to be violated. Sometimes, the jiffy interval is dictated by the application.

Example 11.7: Game consoles will typically use a jiffy value synchronized to the frame rate of the targeted television system because the major time-critical task for such systems is to generate graphics at this frame rate. For example, NTSC is the analog television system historically used in most of the Americas, Japan, South Korea, Taiwan, and a few other places. It has a frame rate of 59.94 Hz, so a suitable jiffy would be $1/59.94$ or about 16.68 ms. With the **PAL** (phase alternating line) television standard, used in most of Europe and much of the rest of the world, the frame rate is 50 Hz, yielding a jiffy of 20 ms. p.209

Analog television is steadily being replaced by digital formats such as ATSC. ATSC supports a number of frame rates ranging from just below 24 Hz to 60 Hz and a number of resolutions. Assuming a standard-compliant TV, a game console designer can choose the frame rate and resolution consistent with cost and quality objectives. p.209

In addition to periodic interrupts and operating service calls, the scheduler might be invoked when a thread blocks for some reason. We discuss some of the mechanisms for such blocking next.

11.2.3 Mutual Exclusion

p.264 A thread may be suspended between any two atomic operations to execute another thread and/or an interrupt service routine. This fact can make it extremely difficult to reason about interactions among threads.

Example 11.8: Recall the following procedure from Figure 11.2:

```
14 void addListener(notifyProcedure* listener) {
15   if (head == 0) {
16     head = malloc(sizeof(element_t));
17     head->listener = listener;
18     head->next = 0;
19     tail = head;
20   } else {
21     tail->next = malloc(sizeof(element_t));
22     tail = tail->next;
23     tail->listener = listener;
24     tail->next = 0;
25   }
26 }
```

Suppose that `addListener` is called from more than one thread. Then what could go wrong? First, two threads may be simultaneously modifying the linked list data structure, which can easily result in a corrupted data structure. Suppose for example that a thread is suspended just prior to executing line 23. Suppose that while the thread is suspended, another thread calls `addListener`. When the first thread resumes executing at line 23, the value of `tail` has changed. It is no longer the value that was set in line 22! Careful analysis reveals that this could result in a list where the second to last element of the list points to a random address for the listener (whatever was in the memory allocated by `malloc`), and the second listener that was added to the list is no longer on the list. When `update` is called, it will try to execute a procedure at the random address, which could result in a segmentation fault, or worse, execution of random memory contents as if they were instructions!

p.243

The problem illustrated in the previous example is known as a **race condition**. Two concurrent pieces of code race to access the same resource, and the exact order in which their accesses occurs affects the results of the program. Not all race conditions are as bad as the previous example, where some outcomes of the race cause catastrophic failure. One way to prevent such disasters is by using a **mutual exclusion lock** (or **mutex**), as illustrated in the next example.

Example 11.9: In Pthreads, mutexes are implemented by creating an instance of a structure called a `pthread_mutex_t`. For example, we could modify the `addListener` procedure as follows:

```
pthread_mutex_t lock = PTHREAD_MUTEX_INITIALIZER;

void addListener(notifyProcedure* listener) {
  pthread_mutex_lock(&lock);
  if (head == 0) {
    ...
  } else {
    ...
  }
  pthread_mutex_unlock(&lock);
}
```

The first line creates and initializes a global variable called `lock`. The first line within the `addListener` procedure **acquires** the lock. The principle is that only one thread can **hold** the lock at a time. The `pthread_mutex_lock` procedure will block until the calling thread can acquire the lock. p.244

In the above code, when `addListener` is called by a thread and begins executing, `pthread_mutex_lock` does not return until no other thread holds the lock. Once it returns, this calling thread holds the lock. The `pthread_mutex_unlock` call at the end **releases** the lock. It is a serious error in multithreaded programming to fail to release a lock.

A mutual exclusion lock prevents any two threads from simultaneously accessing or modifying a shared resource. The code between the lock and unlock is a **critical section**. At any one time, only one thread can be executing code in such a critical section. A programmer may need to ensure that all accesses to a shared resource are similarly protected by locks.

Example 11.10: The `update` procedure in Figure 11.2 does not modify the list of listeners, but it does read the list. Suppose that thread *A* calls `addListener` and gets suspended just after line 21, which does this:

```
21      tail->next = malloc(sizeof(element_t));
```

Suppose that while *A* is suspended, another thread *B* calls `update`, which includes the following code:

```
31      element_t* element = head;
32      while (element != 0) {
33        (*(element->listener))(newx);
34        element = element->next;
35      }
```

What will happen on line 33 when `element == tail->next`? At that point, thread *B* will treat whatever random contents were in the memory returned by `malloc` on line 21 as a function pointer and attempt to execute a procedure pointed to by that pointer. Again, this will result in a segmentation fault or worse.

p.243

The mutex added in Example 11.9 is not sufficient to prevent this disaster. The mutex does not prevent thread *A* from being suspended. Thus, we need to protect *all* accesses of the data structure with mutexes, which we can do by modifying `update` as follows

```
void update(int newx) {
  x = newx;
  // Notify listeners.
  pthread_mutex_lock(&lock);
  element_t* element = head;
  while (element != 0) {
    (*(element->listener))(newx);
    element = element->next;
  }
  pthread_mutex_unlock(&lock);
}
```

This will prevent the `update` procedure from reading the list data structure while it is being modified by any other thread.

Operating Systems

The computers in embedded systems often do not interact directly with humans in the same way that desktop or handheld computers do. As a consequence, the collection of services that they need from an **operating system** (**OS**) may be very different. The dominant **general-purpose OS**s for desktops today, Microsoft Windows, Mac OS X, and Linux, provide services that may or may not be required in an embedded processor. For example, many embedded applications do not require a graphical user interface (**GUI**), a file system, font management, or even a network stack.

Several operating systems have been developed specifically for embedded applications, including Windows CE (WinCE) (from Microsoft), VxWorks (from Wind River Systems, acquired by Intel in 2009), QNX (from QNX Software Systems, acquired in 2010 by Research in Motion (RIM)), Embedded Linux (an open source community effort), and FreeRTOS (another open source community effort). These share many features with general-purpose OSs, but typically have specialized the kernel to become a **real-time operating system** (**RTOS**). An RTOS provides bounded latency on interrupt servicing and a scheduler for processes that takes into account real-time constraints. p.309

Mobile operating systems are a third class of OS designed specifically for handheld devices. The smart phone operating systems iOS (from Apple) and Android (from Google) dominate today, but there is a long history of such software for cell phones and PDAs. Examples include Symbian OS (an open-source effort maintained by the Symbian Foundation), BlackBerry OS (from RIM), Palm OS (from Palm, Inc., acquired by Hewlett Packard in 2010), and Windows Mobile (from Microsoft). These OSs have specialized support for wireless connectivity and media formats. p.218

The core of any operating system is the **kernel**, which controls the order in which processes are executed, how memory is used, and how information is communicated to peripheral devices and networks (via device drivers). A **microkernel** is a very small operating system that provides only these services (or even a subset of these services). OSs may provide many other services, however. These could include user interface infrastructure (integral to Mac OS X and Windows), virtual memory, memory allocation and deallocation, memory protection (to isolate applications from the kernel and from each other), a file system, and services for programs to interact such as semaphores, mutexes, and message passing libraries. p.269 p.232 p.243 p.291

11.2.4 Deadlock

As mutex locks proliferate in programs, the risk of **deadlock** increases. A deadlock occurs when some threads become permanently blocked trying to acquire locks. This can occur, for example, if thread *A* holds `lock1` and then blocks trying to acquire `lock2`, which is held by thread *B*, and then thread *B* blocks trying to acquire `lock1`. Such deadly embraces have no clean escape. The program needs to be aborted.

Example 11.11: Suppose that both `addListener` and `update` in Figure 11.2 are protected by a mutex, as in the two previous examples. The `update` procedure includes the line

```
33        (*(element->listener))(newx);
```

which calls a procedure pointed to by the list element. It would not be unreasonable for that procedure to itself need to acquire a mutex lock. Suppose for example that the listener procedure needs to update a display. A display is typically a shared resource, and therefore will likely have to be protected with its own mutex lock. Suppose that thread *A* calls `update`, which reaches line 33 and then blocks because the listener procedure tries to acquire a different lock held by thread *B*. Suppose then that thread *B* calls `addListener`. Deadlock!

Deadlock can be difficult to avoid. In a classic paper, Coffman et al. (1971) give necessary conditions for deadlock to occur, any of which can be removed to avoid deadlock. One simple technique is to use only one lock throughout an entire multithreaded program. This technique does not lead to very modular programming, however. Moreover, it can make it difficult to meet real-time constraints because some shared resources (e.g., displays) may need to be held long enough to cause deadlines to be missed in other threads.

p.261 In a very simple microkernel, we can sometimes use the enabling and disabling of interrupts as a single global mutex. Assume that we have a single processor (not a multicore), and that interrupts are the only mechanism by which a thread may be suspended (i.e., they do not get suspended when calling kernel services or blocking on I/O). With these assumptions, disabling interrupts prevents a thread from being suspended. In most OSs, however, threads can be suspended for many reasons, so this technique won't work.

A third technique is to ensure that when there are multiple mutex locks, every thread acquires the locks in the same order. This can be difficult to guarantee, however, for several reasons (see Exercise 2). First, most programs are written by multiple people, and the locks acquired within a procedure are not part of the signature of the procedure. So this technique relies on very careful and consistent documentation and cooperation across a development team. And any time a lock is added, then all parts of the program that acquire locks may have to be modified.

Second, it can make correct coding extremely difficult. If a programmer wishes to call a procedure that acquires `lock1`, which by convention in the program is always the first lock acquired, then it must first release any locks it holds. As soon as it releases those locks, it may be suspended, and the resource that it held those locks to protect may be modified. Once it has acquired `lock1`, it must then reacquire those locks, but it will then need to assume it no longer knows anything about the state of the resources, and it may have to redo considerable work.

There are many more ways to prevent deadlock. For example, a particularly elegant technique synthesizes constraints on a scheduler to prevent deadlock (Wang et al., 2009). Nevertheless, most available techniques either impose severe constraints on the programmer or require considerable sophistication to apply, which suggests that the problem may be with the concurrent programming model of threads.

11.2.5 Memory Consistency Models

As if race conditions and deadlock were not problematic enough, threads also suffer from potentially subtle problems with the memory model of the programs. Any particular implementation of threads offers some sort of **memory consistency** model, which defines how variables that are read and written by different threads appear to those threads. Intuitively, reading a variable should yield the last value written to the variable, but what does "last" mean? Consider a scenario, for example, where all variables are initialized with value zero, and thread *A* executes the following two statements: p.240

```
x = 1;
w = y;
```

while thread *B* executes the following two statements:

```
y = 1;
z = x;
```

Intuitively, after both threads have executed these statements, we would expect that at least one of the two variables `w` and `z` has value 1. Such a guarantee is referred to as **sequential consistency** (Lamport, 1979). Sequential consistency means that the result of any execution is the same as if the operations of all threads are executed in some sequential order, and the operations of each individual thread appear in this sequence in the order specified by the thread.

However, sequential consistency is not guaranteed by most (or possibly all) implementations of Pthreads. In fact, providing such a guarantee is rather difficult on modern processors using modern compilers. A compiler, for example, is free to re-order the instructions in each of these threads because there is no dependency between them (that is visible to the compiler). Even if the compiler does not reorder them, the hardware might. A good defensive tactic is to very carefully guard such accesses to shared variables using mutual exclusion locks (and to hope that those mutual exclusion locks themselves are implemented correctly).

An authoritative overview of memory consistency issues is provided by Adve and Gharachorloo (1996), who focus on multiprocessors. Boehm (2005) provides an analysis of the memory consistency problems with threads on a single processor.

11.2.6 The Problem with Threads

Multithreaded programs can be very difficult to understand. Moreover, it can be difficult to build confidence in the programs because problems in the code may not show up in testing. A program may have the possibility of deadlock, for example, but nonetheless run correctly for years without the deadlock ever appearing. Programmers have to be very cautious, but reasoning about the programs is sufficiently difficult that programming errors are likely to persist.

In the example of Figure 11.2, we can avoid the potential deadlock of Example 11.11 using a simple trick, but the trick leads to a more **insidious error** (an error that may not occur in testing, and may not be noticed when it occurs, unlike a deadlock, which is almost always noticed when it occurs).

Example 11.12: Suppose we modify the `update` procedure as follows:

```
void update(int newx) {
  x = newx;
  // Copy the list
  pthread_mutex_lock(&lock);
  element_t* headc = NULL;
  element_t* tailc = NULL;
```

```
    element_t* element = head;
    while (element != 0) {
      if (headc == NULL) {
        headc = malloc(sizeof(element_t));
        headc->listener = head->listener;
        headc->next = 0;
        tailc = headc;
      } else {
        tailc->next = malloc(sizeof(element_t));
        tailc = tailc->next;
        tailc->listener = element->listener;
        tailc->next = 0;
      }
      element = element->next;
    }
    pthread_mutex_unlock(&lock);

    // Notify listeners using the copy
    element = headc;
    while (element != 0) {
      (*(element->listener))(newx);
      element = element->next;
    }
  }
```

This implementation does not hold `lock` when it calls the listener procedure. Instead, it holds the lock while it constructs a copy of the list of the listeners, and then it releases the lock. After releasing the lock, it uses the copy of the list of listeners to notify the listeners.

This code, however, has a potentially serious problem that may not be detected in testing. Specifically, suppose that thread *A* calls `update` with argument `newx = 0`, indicating "all systems normal." Suppose that *A* is suspended just after releasing the `lock`, but before performing the notifications. Suppose that while it is suspended, thread *B* calls `update` with argument `newx = 1`, meaning "emergency! the engine is on fire!" Suppose that this call to `update` completes before thread *A* gets a chance to resume. When thread *A* resumes, it will notify all the listeners, but it will notify them of the wrong value! If one of the listeners is updating a pilot display for an aircraft, the display will indicate that all systems are normal, when in fact the engine is on fire.

Many programmers are familiar with threads and appreciate the ease with which they exploit underlying parallel hardware. It is possible, but not easy, to construct reliable and correct multithreaded programs. See for example Lea (1997) for an excellent "how to" guide to using threads in Java. By 2005, standard Java libraries included concurrent data structures and mechanisms based on threads (Lea, 2005). Libraries like OpenMP (Chapman et al., 2007) also provide support for commonly used multithreaded patterns such as parallel loop constructs. However, embedded systems programmers rarely use Java or large sophisticated packages like OpenMP. And even if they did, the same deadlock risks and insidious errors would occur.

Threads have a number of difficulties that make it questionable to expose them to programmers as a way to build concurrent programs (Ousterhout, 1996; Sutter and Larus, 2005; Lee, 2006; Hayes, 2007). In fact, before the 1990s, threads were not used at all by application programmers. It was the emergence of libraries like Pthreads and languages like Java and C# that exposed these mechanisms to application programmers.

Nontrivial multithreaded programs are astonishingly difficult to understand, and can yield p.296 insidious errors, race conditions, and deadlock. Problems can lurk in multithreaded programs through years of even intensive use of the programs. These concerns are particularly important for embedded systems that affect the safety and livelihood of humans. Since virtually every embedded system involves concurrent software, engineers that design embedded systems must confront the pitfalls.

11.3 Processes and Message Passing

Processes are imperative programs with their own memory spaces. These programs cannot refer to each others' variables, and consequently they do not exhibit the same difficulties as threads. Communication between the programs must occur via mechanisms provided by the operating system, microkernel, or a library.

Implementing processes correctly generally requires hardware support in the form of a p.236 memory management unit or MMU. The MMU protects the memory of one process from p.232 accidental reads or writes by another process. It typically also provides address translation, providing for each process the illusion of a fixed memory address space that is the same for all processes. When a process accesses a memory location in that address space, the MMU shifts the address to refer to a location in the portion of physical memory allocated to that process.

To achieve concurrency, processes need to be able to communicate. Operating systems typically provide a variety of mechanisms, often even including the ability to create shared memory spaces, which of course opens the programmer to all the potential difficulties of multithreaded programming.

One such mechanism that has fewer difficulties is a **file system**. A file system is simply a way to create a body of data that is persistent in the sense that it outlives the process that creates it. One process can create data and write it to a file, and another process can read data from the same file. It is up to the implementation of the file system to ensure that the process reading the data does not read it before it is written. This can be done, for example, by allowing no more than one process to operate on a file at a time.

A more flexible mechanism for communicating between processes is **message passing**. Here, one process creates a chunk of data, deposits it in a carefully controlled section of memory that is shared, and then notifies other processes that the message is ready. Those other processes can block waiting for the data to become ready. Message passing requires some memory to be shared, but it is implemented in libraries that are presumably written by experts. An application programmer invokes a library procedure to send a message or to receive a message.

Example 11.13: A simple example of a message passing program is shown in Figure 11.5. This program uses a **producer/consumer pattern**, where one thread produces a sequence of messages (a **stream**), and another thread consumes the messages. This pattern can be used to implement the observer pattern without deadlock risk and without the insidious error discussed in the previous section. The p.296
`update` procedure would always execute in a different thread from the observers, and would produce messages that are consumed by the observers.

In Figure 11.5, the code executed by the producing thread is given by the `producer` procedure, and the code for the consuming thread by the `consumer` procedure. The producer invokes a procedure called `send` (to be defined) on line 4 to send an integer-valued message. The consumer uses `get` (also to be defined) on line 10 to receive the message. The consumer is assured that `get` does not return until it has actually received the message. Notice that in this case, `consumer` never returns, so this program will not terminate on its own.

An implementation of `send` and `get` using Pthreads is shown in Figure 11.6. This implementation uses a linked list similar to that in Figure 11.2, but where p.285
the payload is an `int`. Here, the linked list is implementing an unbounded **first-** p.285

```
void* producer(void* arg) {
    int i;
    for (i = 0; i < 10; i++) {
        send(i);
    }
    return NULL;
}
void* consumer(void* arg) {
    while(1) {
        printf("received %d\n", get());
    }
    return NULL;
}
int main(void) {
    pthread_t threadID1, threadID2;
    void* exitStatus;
    pthread_create(&threadID1, NULL, producer, NULL);
    pthread_create(&threadID2, NULL, consumer, NULL);
    pthread_join(threadID1, &exitStatus);
    pthread_join(threadID2, &exitStatus);
    return 0;
}
```

Figure 11.5: Example of a simple message-passing application.

in, first-out (FIFO) queue, where new elements are inserted at the tail and old elements are removed from the head.

p.291

Consider first the implementation of `send`. It uses a mutex to ensure that `send` and `get` are not simultaneously modifying the linked list, as before. But in addition, it uses a **condition variable** to communicate to the consumer process that the size of the queue has changed. The condition variable called `sent` is declared and initialized on line 7. On line 23, the producer thread calls `pthread_cond_signal`, which will "wake up" another thread that is blocked on the condition variable, if there is such a thread.

To see what it means to "wake up" another thread, look at the `get` procedure. On line 31, if the thread calling `get` has discovered that the current size of the queue is zero, then it calls `pthread_cond_wait`, which will block the thread until some other thread calls `pthread_cond_signal`.

```
#include <pthread.h>
struct element {int payload; struct element* next;};
typedef struct element element_t;
element_t *head = 0, *tail = 0;
int size = 0;
pthread_mutex_t mutex = PTHREAD_MUTEX_INITIALIZER;
pthread_cond_t sent = PTHREAD_COND_INITIALIZER;

void send(int message) {
    pthread_mutex_lock(&mutex);
    if (head == 0) {
        head = malloc(sizeof(element_t));
        head->payload = message;
        head->next = 0;
        tail = head;
    } else {
        tail->next = malloc(sizeof(element_t));
        tail = tail->next;
        tail->payload = message;
        tail->next = 0;
    }
    size++;
    pthread_cond_signal(&sent);
    pthread_mutex_unlock(&mutex);
}
int get() {
    element_t* element;
    int result;
    pthread_mutex_lock(&mutex);
    while (size == 0) {
        pthread_cond_wait(&sent, &mutex);
    }
    result = head->payload;
    element = head;
    head = head->next;
    free(element);
    size--;
    pthread_mutex_unlock(&mutex);
    return result;
}
```

Figure 11.6: Message-passing procedures to send and get messages.

Notice that the `get` procedure acquires the mutex before testing the `size` variable. Notice further on line 31 that `pthread_cond_wait` takes `&mutex` as an argument. In fact, while the thread is blocked on the wait, it releases the `mutex` lock temporarily. If it were not to do this, then the producer thread would be unable to enter its critical section, and therefore would be unable to send a message. The program would deadlock. Before `pthread_cond_wait` returns, it will re-acquire the `mutex` lock.

Programmers have to be very careful when calling `pthread_cond_wait`, because the `mutex` lock is temporarily released during the call. As a consequence, the value of any shared variable after the call to `pthread_cond_wait` may not be the same as it was before the call (see Exercise 3). Hence, the call to `pthread_cond_wait` lies within a while loop (line 30) that repeatedly tests the `size` variable. This accounts for the possibility that there could be multiple threads simultaneously blocked on line 31 (which is possible because of the temporary release of the mutex). When a thread calls `pthread_cond_signal`, all threads that are waiting will be notified. But exactly one will re-acquire the mutex before the others and consume the sent message, causing `size` to be reset to zero. The other notified threads, when they eventually acquire the mutex, will see that `size == 0` and will just resume waiting.

The condition variable used in the previous example is a generalized form of a **semaphore**. Semaphores are named after mechanical signals traditionally used on railroad tracks to signal that a section of track has a train on it. Using such semaphores, it is possible to use a single section of track for trains to travel in both directions (the semaphore implements p.291 mutual exclusion, preventing two trains from simultaneously being on the same section of track).

In the 1960s, Edsger W. Dijkstra, a professor in the Department of Mathematics at the Eindhoven University of Technology, Netherlands, borrowed this idea to show how programs could safely share resources. A counting semaphore (which Dijkstra called a PV semaphore) is a variable whose value is a non-negative integer. A value of zero is treated as distinctly different from a value greater than zero. In fact, the `size` variable in Example 11.13 functions as such a semaphore. It is incremented by sending a message, and a value of zero blocks the consumer until the value is non-zero. Condition variables generalize this idea by supporting arbitrary conditions, rather than just zero or non-zero, as the gating criterion for blocking. Moreover, at least in Pthreads, condition variables also coordinate with mutexes to make patterns like that in Example 11.13 easier to write. Dijkstra received the 1972 Turing Award for his work on concurrent programming. p.291

Using message passing in applications can be easier than directly using threads and shared variables. But even message passing is not without peril. The implementation of the producer/consumer pattern in Example 11.13, in fact, has a fairly serious flaw. Specifically, it imposes no constraints on the size of the message queue. Any time a producer thread calls `send`, memory will be allocated to store the message, and that memory will not be deallocated until the message is consumed. If the producer thread produces messages faster than the consumer consumes them, then the program will eventually exhaust available memory. This can be fixed by limiting the size of the buffer (see Exercise 4), but what size is appropriate? Choosing buffers that are too small can cause a program to deadlock, and choosing buffers that are too large is wasteful of resources. This problem is not trivial to solve (Lee, 2009b).

There are other pitfalls as well. Programmers may inadvertently construct message-passing programs that deadlock, where a set of threads are all waiting for messages from one another. In addition, programmers can inadvertently construct message-passing programs that are nondeterminate, in the sense that the results of the computation depend on the (arbitrary) order in which the thread scheduler happens to schedule the threads. p.56

The simplest solution is for application programmers to use higher-levels of abstraction for concurrency, the top layer in Figure 11.1, as described in Chapter 6. Of course, they can only use that strategy if they have available a reliable implementation of a higher-level concurrent model of computation.

11.4 Summary

This chapter has focused on mid-level abstractions for concurrent programs, above the level of interrupts and parallel hardware, but below the level of concurrent models of computation. Specifically, it has explained threads, which are sequential programs that execute concurrently and share variables. We have explained mutual exclusion and the use of semaphores. We have shown that threads are fraught with peril, and that writing correct multithreaded programs is extremely difficult. Message passing schemes avoid some of the difficulties, but not all, at the expense of being somewhat more constraining by prohibiting direct sharing of data. In the long run, designers will be better off using higher-levels of abstraction, as discussed in Chapter 6.

Exercises

1. Give an extended state-machine model of the `addListener` procedure in Figure 11.2 similar to that in Figure 11.3,

p.244
p.291

2. Suppose that two `int` global variables `a` and `b` are shared among several threads. Suppose that `lock_a` and `lock_b` are two mutex locks that guard access to `a` and `b`. Suppose you cannot assume that reads and writes of `int` global variables are atomic. Consider the following code:

```
int a, b;
pthread_mutex_t lock_a
    = PTHREAD_MUTEX_INITIALIZER;
pthread_mutex_t lock_b
    = PTHREAD_MUTEX_INITIALIZER;

void procedure1(int arg) {
  pthread_mutex_lock(&lock_a);
  if (a == arg) {
     procedure2(arg);
  }
  pthread_mutex_unlock(&lock_a);
}

void procedure2(int arg) {
  pthread_mutex_lock(&lock_b);
  b = arg;
  pthread_mutex_unlock(&lock_b);
}
```

Suppose that to ensure that deadlocks do not occur, the development team has agreed that `lock_b` should always be acquired before `lock_a` by any thread that acquires both locks. Note that the code listed above is not the only code in the program. Moreover, for performance reasons, the team insists that no lock be acquired unnecessarily. Consequently, it would not be acceptable to modify `procedure1` as follows:

```
void procedure1(int arg) {
  pthread_mutex_lock(&lock_b);
  pthread_mutex_lock(&lock_a);
  if (a == arg) {
     procedure2(arg);
  }
  pthread_mutex_unlock(&lock_a);
  pthread_mutex_unlock(&lock_b);
}
```

A thread calling `procedure1` will acquire `lock_b` unnecessarily when `a` is not equal to `arg`. [2] Give a design for `procedure1` that minimizes unnecessary acquisitions of `lock_b`. Does your solution eliminate unnecessary acquisitions of `lock_b`? Is there any solution that does this?

3. The implementation of `get` in Figure 11.6 permits there to be more than one thread calling `get`.

   However, if we change the code on lines 30-32 to:

```
    if (size == 0) {
      pthread_cond_wait(&sent, &mutex);
    }
```

   then this code would only work if two conditions are satisfied:

   - `pthread_cond_wait` returns *only* if there is a matching call to `pthread_cond_signal`, and
   - there is only one consumer thread.

   Explain why the second condition is required.

4. The producer/consumer pattern implementation in Example 11.13 has the drawback that the size of the queue used to buffer messages is unbounded. A program could fail by exhausting all available memory (which will cause `malloc` to fail). Construct a variant of the `send` and `get` procedures of Figure 11.6 that limits the buffer size to 5 messages. p.299

5. An alternative form of message passing called rendezvous is similar to the producer/consumer pattern of Example 11.13, but it synchronizes the producer and consumer more tightly. In particular, in Example 11.13, the `send` procedure returns immediately, regardless of whether there is any consumer thread ready to receive the message. In a rendezvous-style communication, the `send` procedure will not return until a consumer thread has reached a corresponding call to `get`. Consequently, no buffering of the messages is needed. Construct implementations of `send` and `get` that implement such a rendezvous. p.154 p.299

6. Consider the following code.

[2] In some thread libraries, such code is actually incorrect, in that a thread will block trying to acquire a lock it already holds. But we assume for this problem that if a thread attempts to acquire a lock it already holds, then it is immediately granted the lock.

```
int x = 0;
int a;
pthread_mutex_t lock_a = PTHREAD_MUTEX_INITIALIZER;
pthread_cond_t go = PTHREAD_COND_INITIALIZER; // used in part c

void proc1(){
    pthread_mutex_lock(&lock_a);
    a = 1;
    pthread_mutex_unlock(&lock_a);
    <proc3>(); // call to either proc3a or proc3b
               // depending on the question
}

void proc2(){
    pthread_mutex_lock(&lock_a);
    a = 0;
    pthread_mutex_unlock(&lock_a);
    <proc3>();
}

void proc3a(){
    if(a == 0){
        x = x + 1;
    } else {
        x = x - 1;
    }
}

void proc3b(){
    pthread_mutex_lock(&lock_a);
    if(a == 0){
        x = x + 1;
    } else {
        x = x - 1;
    }
    pthread_mutex_unlock(&lock_a);
}
```

Suppose `proc1` and `proc2` run in two separate threads and that each procedure is called in its respective thread exactly once. Variables `x` and `a` are global and shared between threads and `x` is initialized to 0. Further, assume the increment and decrement operations are atomic.

The calls to `proc3` in `proc1` and `proc2` should be replaced with calls to `proc3a` and `proc3b` depending on the part of the question.

(a) If `proc1` and `proc2` call `proc3a` in lines 10 and 18, is the final value of global variable `x` guaranteed to be 0? Justify your answer.

(b) What if `proc1` and `proc2` call `proc3b`? Justify your answer.

(c) With `proc1` and `proc2` still calling `proc3b`, modify `proc1` and `proc2` with condition variable `go` to guarantee the final value of `x` is 2. Specifically, give the lines where `pthread_cond_wait` and `pthread_cond_signal` should be inserted into the code listing. Justify your answer briefly. Make the assumption that `proc1` acquires lock_a before `proc2`.

Also recall that

`pthread_cond_wait(&go, &lock_a);`

will temporarily release `lock_a` and block the calling thread until

`pthread_cond_signal(&go);`

is called in another thread, at which point the waiting thread will be unblocked and reacquire `lock_a`.

(This problem is due to Matt Weber.)

12 Scheduling

Contents

Chapter 11 has explained multitasking, where multiple imperative tasks execute concurrently, either interleaved on a single processor or in parallel on multiple processors. When there are fewer processors than tasks (the usual case), or when tasks must be performed at a particular time, a **scheduler** must intervene. A scheduler makes the decision about what to do next at certain points in time, such as the time when a processor becomes available. p.212

Real-time systems are collections of tasks where in addition to any ordering constraints imposed by precedences between the tasks, there are also timing constraints. These constraints relate the execution of a task to **real time**, which is physical time in the environment of the p.313 computer executing the task. Typically, tasks have deadlines, which are values of physical time by which the task must be completed. More generally, real-time programs can have all manner of **timing constraints**, not just deadlines. For example, a task may be required to be executed no earlier than a particular time; or it may be required to be executed no more than a given amount of time after another task is executed; or it may be required to execute periodically with some specified period. Tasks may be dependent on one another, and may cooperatively form an application. Or they may be unrelated except that they share processor resources. All of these situations require a scheduling strategy.

12.1 Basics of Scheduling

In this section, we discuss the range of possibilities for scheduling, the properties of tasks that a scheduler uses to guide the process, and the implementation of schedulers in an operating system or microkernel.

12.1.1 Scheduling Decisions

A scheduler decides what task to execute next when faced with a choice in the execution of a concurrent program or set of programs. In general, a scheduler may have more than one p.223 processor available to it (for example in a multicore system). A **multiprocessor scheduler** needs to decide not only which task to execute next, but also on which processor to execute it. The choice of processor is called **processor assignment**.

A **scheduling decision** is a decision to execute a task, and it has the following three parts:

- **assignment**: which processor should execute the task;
- **ordering**: in what order each processor should execute its tasks; and
- **timing**: the time at which each task executes.

Each of these three decisions may be made at **design time**, before the program begins executing, or at **run time**, during the execution of the program.

Depending on when the decisions are made, we can distinguish a few different types of schedulers (Lee and Ha, 1989). A **fully-static scheduler** makes all three decisions at design time. The result of scheduling is a precise specification for each processor of what to do when. A fully-static scheduler typically does not need semaphores or locks. p.291 It can use timing instead to enforce mutual exclusion and precedence constraints. However, fully-static schedulers are difficult to realize with most modern microprocessors because the time it takes to execute a task is difficult to predict precisely, and because tasks will typically have data-dependent execution times (see Chapter 16).

A **static order scheduler** performs the task assignment and ordering at design time, but defers until run time the decision of when in physical time to execute a task. That decision may be affected, for example, by whether a mutual exclusion p.291 lock can be acquired, or whether precedence constraints have been satisfied. In static order scheduling, each processor is given its marching orders before the program begins executing, and it simply executes those orders as quickly as it can. It does not, for example, change the order of tasks based on the state of a semaphore or a lock. A task itself, however, may block on a semaphore or lock, in which case it blocks the entire sequence of tasks on that processor. A static order scheduler is often called an **off-line scheduler**.

A **static assignment scheduler** performs the assignment at design time and everything else at run time. Each processor is given a set of tasks to execute, and a **run-time scheduler** decides during execution what task to execute next.

A **fully-dynamic scheduler** performs all decisions at run time. When a processor becomes available (e.g., it finishes executing a task, or a task blocks acquiring a mutex p.291), the scheduler makes a decision at that point about what task to execute next on that processor. Both static assignment and fully-dynamic schedulers are often called **on-line schedulers**.

There are, of course, other scheduler possibilities. For example, the assignment of a task may be done once for a task, at run time just prior to the first execution of the task. For subsequent runs of the same task, the same assignment is used. Some combinations do not make much sense. For example, it does not make sense to determine the time of execution of a task at design time and the order at run time.

A **preemptive** scheduler may make a scheduling decision during the execution of a task, assigning a new task to the same processor. That is, a task may be in the middle of executing when the scheduler decides to stop that execution and begin execution of another task. The interruption of the first task is called **preemption**. A scheduler that always lets tasks run to completion before assigning another task to execute on the same processor is called a **non-preemptive** scheduler.

In preemptive scheduling, a task may be preempted if it attempts to acquire a mutual exclusion lock and the lock is not available. When this occurs, the task is said to be **blocked** on the lock. When another task releases the lock, the blocked task may resume. Moreover, a task may be preempted when it releases a lock. This can occur for example if there is a higher priority task that is blocked on the lock. We will assume in this chapter well-structured programs, where any task that acquires a lock eventually releases it.

12.1.2 Task Models

For a scheduler to make its decisions, it needs some information about the structure of the program. A typical assumption is that the scheduler is given a finite set T of tasks. Each task may be assumed to be finite (it terminates in finite time), or not. A typical operating system scheduler does not assume that tasks terminate, but real-time schedulers often do. A scheduler may make many more assumptions about tasks, a few of which we discuss in this section. The set of assumptions is called the **task model** of the scheduler.

Some schedulers assume that all tasks to be executed are known before scheduling begins, and some support **arrival of tasks**, meaning tasks become known to the scheduler as other tasks are being executed. Some schedulers support scenarios where each task $\tau \in T$ executes repeatedly, possibly forever, and possibly periodically. A task could also be **sporadic**, which means that it repeats, and its timing is irregular, but that there is a lower bound on the time between task executions. In situations where a task $\tau \in T$ executes repeatedly, we need to make a distinction between the task τ and the **task executions** $\tau_1, \tau_2, \cdots$. If each task executes exactly once, then no such distinction is necessary.

Task executions may have **precedence constraints**, a requirement that one execution precedes another. If execution i must precede j, we can write $i < j$. Here, i and j may be distinct executions of the same task, or executions of different tasks.

A task execution i may have some **preconditions** to start or resume execution. These are conditions that must be satisfied before the task can execute. When the preconditions are satisfied, the task execution is said to be **enabled**. Precedences, for example, specify preconditions to start a task execution. Availability of a lock may be a precondition for resumption of a task.

We next define a few terms that are summarized in Figure 12.1.

For a task execution i, we define the **release time** r_i (also called the **arrival time**) to be the earliest time at which a task is enabled. We define the **start time** s_i to be the time at which the execution actually starts. Obviously, we require that

$$s_i \geq r_i \,.$$

We define the **finish time** f_i to be the time at which the task completes execution. Hence,

$$f_i \geq s_i \,.$$

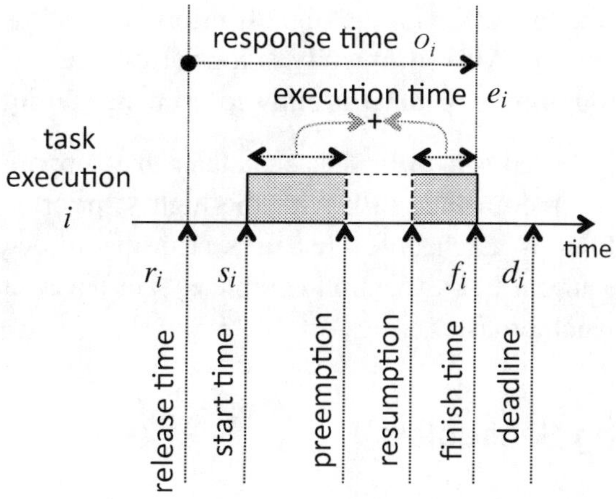

Figure 12.1: Summary of times associated with a task execution.

The **response time** o_i is given by

$$o_i = f_i - r_i \,.$$

The response time, therefore, is the time that elapses between when the task is first enabled and when it completes execution.

The **execution time** e_i of τ_i is defined to be the total time that the task is actually executing. It does not include any time that the task may be blocked or preempted. Many scheduling strategies assume (often unrealistically) that the execution time of a task is known and fixed. If the execution time is variable, it is common to assume (often unrealistically) that the worst-case execution time (WCET) is known. Determining execution times of software can be quite challenging, as discussed in Chapter 16. p.409

The **deadline** d_i is the time by which a task must be completed. Sometimes, a deadline is a real physical constraint imposed by the application, where missing the deadline is considered an error. Such a deadline is called a **hard deadline**. Scheduling with hard deadlines is called **hard real-time scheduling**.

Often, a deadline reflects a design decision that need not be enforced strictly. It is better to meet the deadline, but missing the deadline is not an error. Generally it is better to not miss the deadline by much. This case is called **soft real-time scheduling**.

A scheduler may use **priority** rather than (or in addition to) a deadline. A priority-based scheduler assumes each task is assigned a number called a priority, and the scheduler will always choose to execute the task with the highest priority (which is often represented by the lowest priority number). A **fixed priority** is a priority that remains constant over all executions of a task. A **dynamic priority** is allowed to change during execution.

A **preemptive priority-based scheduler** is a scheduler that supports arrivals of tasks and at all times is executing the enabled task with the highest priority. A **non-preemptive priority-based scheduler** is a scheduler that uses priorities to determine which task to execute next after the current task execution completes, but never interrupts a task during execution to schedule another task.

12.1.3 Comparing Schedulers

The choice of scheduling strategy is governed by considerations that depend on the goals of the application. A rather simple goal is that all task executions meet their deadlines, $f_i \leq d_i$. A schedule that accomplishes this is called a **feasible schedule**. A scheduler that yields p.312 a feasible schedule for any task set (that conforms to its task model) for which there is a feasible schedule is said to be **optimal with respect to feasibility**.

A criterion that might be used to compare scheduling algorithms is the achievable processor **utilization**. The utilization is the percentage of time that the processor spends executing tasks (vs. being idle). This metric is most useful for tasks that execute periodically. A scheduling algorithm that delivers a feasible schedule whenever processor utilization is less than or equal to 100% is obviously optimal with respect to feasibility. It only fails to deliver a feasible schedule in circumstances where *all* scheduling algorithms will fail to deliver a feasible schedule.

Another criterion that might be used to compare schedulers is the maximum **lateness**, defined for a set of task executions T as

$$L_{\max} = \max_{i \in T} (f_i - d_i) \ .$$

For a feasible schedule, this number is zero or negative. But maximum lateness can also be used to compare infeasible schedules. For soft real-time problems, it may be tolerable for this number to be positive, as long as it does not get too large.

A third criterion that might be used for a finite set T of task executions is the **total completion time** or **makespan**, defined by

$$M = \max_{i \in T} f_i - \min_{i \in T} r_i \ .$$

p.213 If the goal of scheduling is to minimize the makespan, this is really more of a performance goal rather than a real-time requirement.

12.1.4 Implementation of a Scheduler

A scheduler may be part of a compiler or code generator (for scheduling decisions made at design time), part of an operating system or microkernel (for scheduling decisions made p.293 at run time), or both (if some scheduling decisions are made at design time and some at run time).

A run-time scheduler will typically implement tasks as threads (or as processes, but the p.286 distinction is not important here). Sometimes, the scheduler assumes these threads complete in finite time, and sometimes it makes no such assumption. In either case, the scheduler is a procedure that gets invoked at certain times. For very simple, non-preemptive schedulers, the scheduling procedure may be invoked each time a task completes. For preemptive schedulers, the scheduling procedure is invoked when any of several things occur:

- A timer interrupt occurs, for example at a jiffy interval. p.289
- An I/O interrupt occurs. p.261
- An operating system service is invoked.
- A task attempts to acquire a mutex. p.291
- A task tests a semaphore. p.302

For interrupts, the scheduling procedure is called by the interrupt service routine (ISR). In p.261 the other cases, the scheduling procedure is called by the operating system procedure that provides the service. In both cases, the stack contains the information required to resume p.241 execution. However, the scheduler may choose not to simply resume execution. I.e., it may choose not to immediately return from the interrupt or service procedure. It may choose instead to preempt whatever task is currently running and begin or resume another task.

To accomplish this preemption, the scheduler needs to record the fact that the task is preempted (and, perhaps, why it is preempted), so that it can later resume this task. It can then adjust the stack pointer to refer to the state of the task to be started or resumed. At p.241 that point, a return is executed, but instead of resuming execution with the task that was preempted, execution will resume for another task.

Implementing a preemptive scheduler can be quite challenging. It requires very careful control of concurrency. For example, interrupts may need to be disabled for significant parts of the process to avoid ending up with a corrupted stack. This is why scheduling is one of the most central functions of an operating system kernel or microkernel. The quality of the implementation strongly affects system reliability and stability.

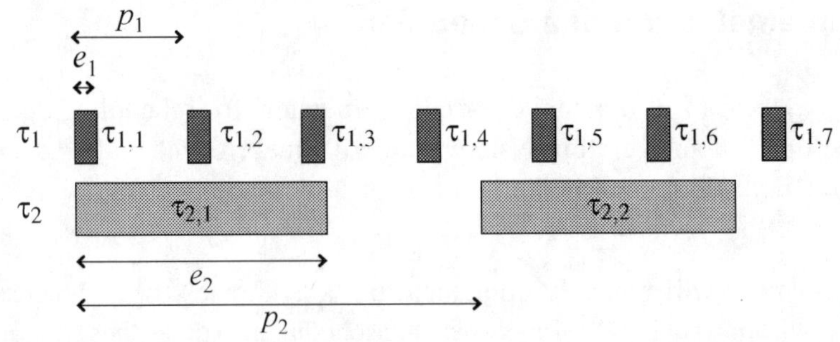

Figure 12.2: Two periodic tasks $T = \{\tau_1, \tau_2\}$ with execution times e_1 and e_2 and periods p_1 and p_2.

12.2 Rate Monotonic Scheduling

Consider a scenario with $T = \{\tau_1, \tau_2, \cdots, \tau_n\}$ of n tasks, where the tasks must execute periodically. Specifically, we assume that each task τ_i must execute to completion exactly once in each time interval p_i. We refer to p_i as the **period** of the task. Thus, the deadline for the j-th execution of τ_i is $r_{i,1} + jp_i$, where $r_{i,1}$ is the release time of the first execution.

Liu and Layland (1973) showed that a simple preemptive scheduling strategy called **rate** p.314 **monotonic** (**RM**) scheduling is optimal with respect to feasibility among fixed priority uniprocessor schedulers for the above task model. This scheduling strategy gives higher priority to a task with a smaller period.

The simplest form of the problem has just two tasks, $T = \{\tau_1, \tau_2\}$ with execution times e_1 and e_2 and periods p_1 and p_2, as depicted in Figure 12.2. In the figure, the execution time e_2 of task τ_2 is longer than the period p_1 of task τ_1. Thus, if these two tasks are to execute on the same processor, then it is clear that a non-preemptive scheduler will not yield p.314 a feasible schedule. If task τ_2 must execute to completion without interruption, then task τ_1 will miss some deadlines.

A preemptive schedule that follows the rate monotonic principle is shown in Figure 12.3. In that figure, task τ_1 is given higher priority, because its period is smaller. So it executes at the beginning of each period interval, regardless of whether τ_2 is executing. If τ_2 is executing, then τ_1 preempts it. The figure assumes that the time it takes to perform the preemption,

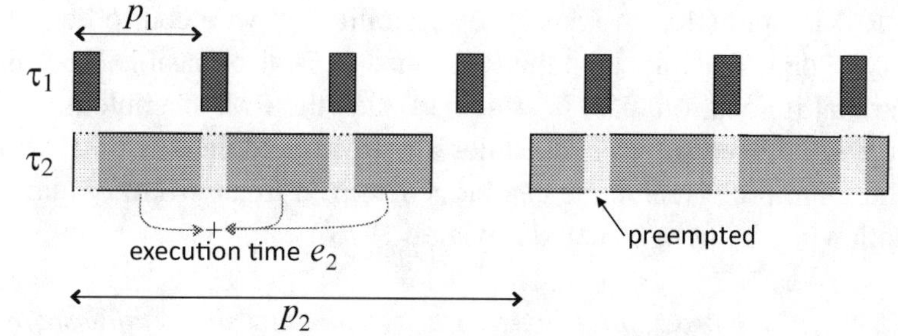

Figure 12.3: Two periodic tasks $T = \{\tau_1, \tau_2\}$ with a preemptive schedule that gives higher priority to τ_1.

called the **context switch time**, is negligible.[1] This schedule is feasible, whereas if τ_2 had been given higher priority, then the schedule would not be feasible.

For the two task case, it is easy to show that among all preemptive fixed priority schedulers, RM is optimal with respect to feasibility, under the assumed task model with negligible context switch time. This is easy to show because there are only two fixed priority schedules for this simple case, the RM schedule, which gives higher priority to task τ_1, and the non-RM schedule, which gives higher priority to task τ_2. To show optimality, we simply need to show that if the non-RM schedule is feasible, then so is the RM schedule.

Before we can do this, we need to consider the possible alignments of task executions that can affect feasibility. As shown in Figure 12.4, the response time of the lower priority task p.313 is worst when its starting phase matches that of higher priority tasks. That is, the worst-case scenario occurs when all tasks start their cycles at the same time. Hence, we only need to consider this scenario.

Under this worst-case scenario, where release times align, the non-RM schedule is feasible p.312 if and only if

$$e_1 + e_2 \leq p_1 \ . \tag{12.1}$$

This scenario is illustrated in Figure 12.5. Since task τ_1 is preempted by τ_2, for τ_1 to not miss its deadline, we require that $e_2 \leq p_1 - e_1$, so that τ_2 leaves enough time for τ_1 to execute before its deadline.

[1] The assumption that context switch time is negligible is problematic in practice. On processors with caches, a context switch often causes substantial cache-related delays. In addition, the operating system overhead for context switching can be substantial.

To show that RM is optimal with respect to feasibility, all we need to do is show that if the non-RM schedule is feasible, then the RM schedule is also feasible. Examining Figure 12.6, it is clear that if equation (12.1) is satisfied, then the RM schedule is feasible. Since these are the only two fixed priority schedules, the RM schedule is optimal with respect to feasibility. The same proof technique can be generalized to an arbitrary number of tasks, yielding the following theorem (Liu and Layland, 1973):

p.311 **Theorem 12.1.** *Given a preemptive, fixed priority scheduler and a finite set of repeating tasks $T = \{\tau_1, \tau_2, \cdots, \tau_n\}$ with associated periods $p_1, p_2, \cdots, p_n$ and no precedence constraints, if any priority assignment yields a feasible schedule, then the rate monotonic priority assignment yields a feasible schedule.* p.314

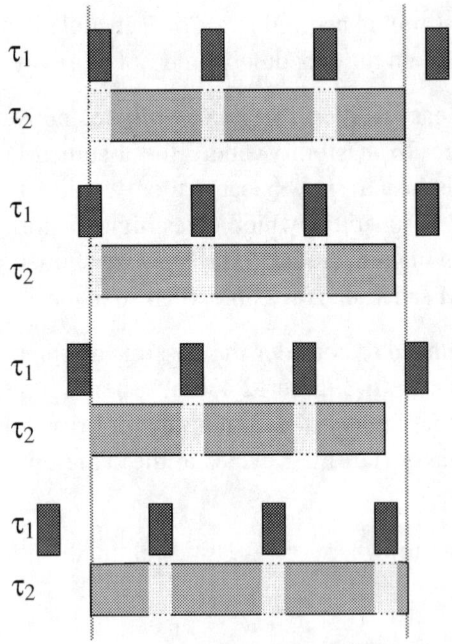

Figure 12.4: Response time o_2 of task τ_2 is worst when its cycle starts at the same time that the cycle of τ_1 starts.

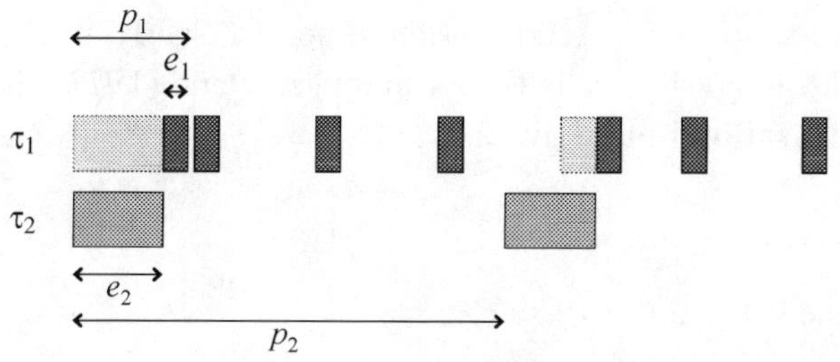

Figure 12.5: The non-RM schedule gives higher priority to τ_2. It is feasible if and only if $e_1 + e_2 \leq p_1$ for this scenario.

RM schedules are easily implemented with a timer interrupt with a time interval equal to the greatest common divisor of the periods of the tasks. They can also be implemented with multiple timer interrupts. p.263

It turns out that RM schedulers cannot always achieve 100% utilization. In particular, RM schedulers are constrained to have fixed priority. This constraint results in situations where a task set that yields a feasible schedule has less than 100% utilization and yet cannot tolerate any increase in execution times or decrease in periods. This means that there are idle processor cycles that cannot be used without causing deadlines to be missed. An example is studied in Exercise 3. p.314

Fortunately, Liu and Layland (1973) show that this effect is bounded. First note that the utilization of n independent tasks with execution times e_i and periods p_i can be written

$$\mu = \sum_{i=1}^{n} \frac{e_i}{p_i}.$$

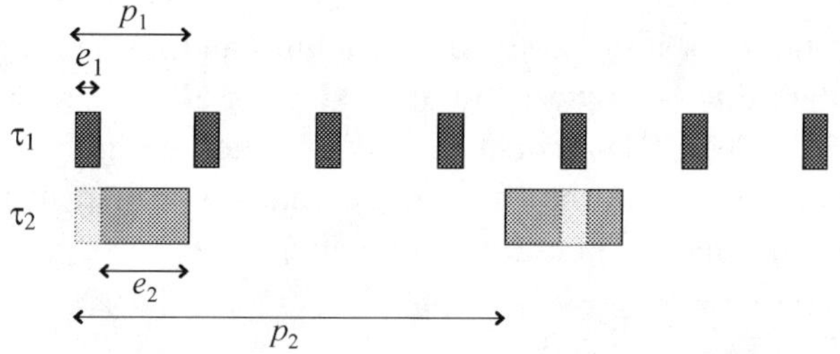

Figure 12.6: The RM schedule gives higher priority to τ_1. For the RM schedule to be feasible, it is sufficient, but not necessary, for $e_1 + e_2 \leq p_1$.

If $\mu = 1$, then the processor is busy 100% of the time. So clearly, if $\mu > 1$ for any task set, then that task set has no feasible schedule. Liu and Layland (1973) show that if μ is less than or equal to a **utilization bound** given by

$$\mu \leq n(2^{1/n} - 1), \tag{12.2}$$

then the RM schedule is feasible.

To understand this (rather remarkable) result, consider a few cases. First, if $n = 1$ (there is only one task), then $n(2^{1/n} - 1) = 1$, so the result tells us that if utilization is 100% or less, then the RM schedule is feasible. This is obvious, of course, because with only one task, $\mu = e_1/p_1$, and clearly the deadline can only be met if $e_1 \leq p_1$.

If $n = 2$, then $n(2^{1/n} - 1) \approx 0.828$. Thus, if a task set with two tasks does not attempt to use more than 82.8% of the available processor time, then the RM schedule will meet all deadlines.

As n gets large, the utilization bound approaches $\ln(2) \approx 0.693$. That is

$$\lim_{n \to \infty} n(2^{1/n} - 1) = \ln(2) \approx 0.693.$$

This means that if a task set with any number of tasks does not attempt to use more than 69.3% of the available processor time, then the RM schedule will meet all deadlines.

In the next section, we relax the fixed-priority constraint and show that dynamic priority schedulers can do better than fixed priority schedulers, in the sense that they can achieve higher utilization. The cost is a somewhat more complicated implementation.

12.3 Earliest Deadline First

Given a finite set of non-repeating tasks with deadlines and no precedence constraints, a simple scheduling algorithm is **earliest due date (EDD)**, also known as **Jackson's algorithm** (Jackson, 1955). The EDD strategy simply executes the tasks in the same order as their deadlines, with the one with the earliest deadline going first. If two tasks have the same deadline, then their relative order does not matter.

Theorem 12.2. *Given a finite set of non-repeating tasks $T = \{\tau_1, \tau_2, \cdots, \tau_n\}$ with associated deadlines $d_1, d_2, \cdots, d_n$ and no precedence constraints, an EDD schedule is optimal in the sense that it minimizes the maximum lateness, compared to all other possible orderings of the tasks.* p.314

Proof. This theorem is easy to prove with a simple **interchange argument**. Consider an arbitrary schedule that is not EDD. In such a schedule, because it is not EDD, there must be two tasks τ_i and τ_j where τ_i immediately precedes τ_j, but $d_j < d_i$. This is depicted here:

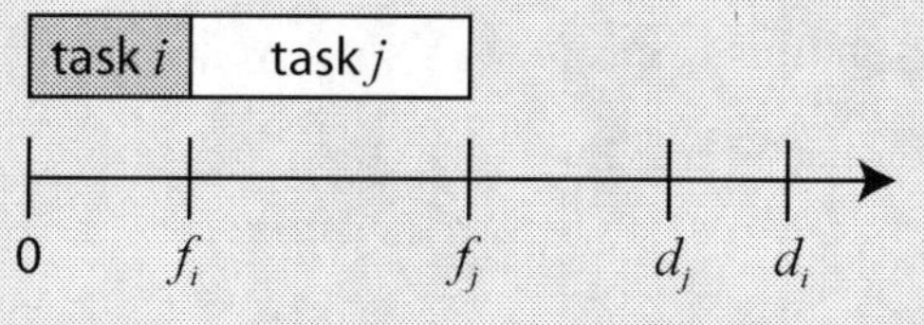

Since the tasks are independent (there are no precedence constraints), reversing the order of these two tasks yields another valid schedule, depicted here:

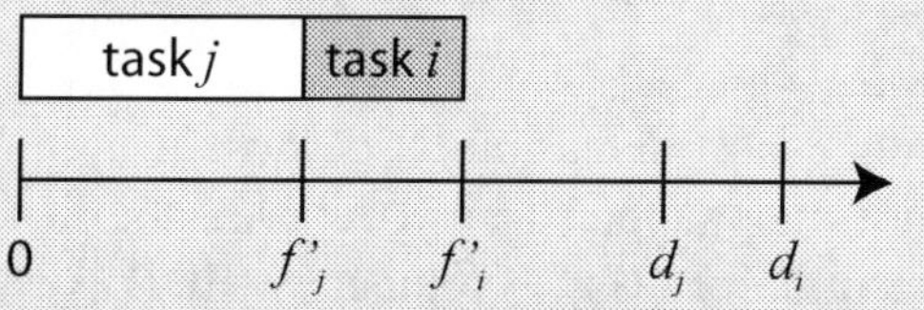

We can show that the new schedule has a maximum lateness no greater than that of the original schedule. If we repeat the above interchange until there are no more tasks eligible for such an interchange, then we have constructed the EDD schedule. Since this schedule has a maximum lateness no greater than that of the original schedule, the EDD schedule has the minimum maximum lateness of all schedules.

To show that the second schedule has a maximum lateness no greater than that of the first schedule, first note that if the maximum lateness is determined by some task other than τ_i or τ_j, then the two schedules have the same maximum lateness, and we are done. Otherwise, it must be that the maximum lateness of the first schedule is

$$L_{\max} = \max(f_i - d_i, f_j - d_j) = f_j - d_j,$$

where the latter equality is obvious from the picture and follows from the facts that $f_i \leq f_j$ and $d_j < d_i$.

The maximum lateness of the second schedule is given by

$$L'_{\max} = \max(f'_i - d_i, f'_j - d_j) .$$

Consider two cases:

Case 1: $L'_{max} = f'_i - d_i$. In this case, since $f'_i = f_j$, we have

$$L'_{max} = f_j - d_i \leq f_j - d_j ,$$

where the latter inequality follows because $d_j < d_i$. Hence, $L'_{max} \leq L_{max}$.

Case 2: $L'_{max} = f'_j - d_j$. In this case, since $f'_j \leq f_j$, we have

$$L'_{max} \leq f_j - d_j ,$$

and again $L'_{max} \leq L_{max}$.

In both cases, the second schedule has a maximum lateness no greater than that of the first schedule. QED.

□

p.314 EDD is also optimal with respect to feasibility, because it minimizes the maximum lateness. p.312 However, EDD does not support arrival of tasks, and hence also does not support periodic or repeated execution of tasks. Fortunately, EDD is easily extended to support these, yielding what is known as **earliest deadline first** (**EDF**) or **Horn's algorithm** (Horn, 1974).

Theorem 12.3. *Given a set of n independent tasks* $T = \{\tau_1, \tau_2, \cdots, \tau_n\}$ *with associated deadlines* $d_1, d_2, \cdots, d_n$ *and arbitrary arrival times, any algorithm that at any instant executes the task with the earliest deadline among all arrived tasks is optimal with respect to minimizing the maximum lateness.*

The proof of this uses a similar interchange argument. Moreover, the result is easily extended to support an unbounded number of arrivals. We leave it as an exercise.

p.314 Note that EDF is a dynamic priority scheduling algorithm. If a task is repeatedly executed, it may be assigned a different priority on each execution. This can make it more complex to implement. Typically, for periodic tasks, the deadline used is the end of the period of the task, though it is certainly possible to use other deadlines for tasks.

Although EDF is more expensive to implement than RM, in practice its performance is generally superior (Buttazzo, 2005b). First, RM is optimal with respect to feasibility only among fixed priority schedulers, whereas EDF is optimal w.r.t. feasibility among dynamic priority schedulers. In addition, EDF also minimizes the maximum lateness. Also, in practice, EDF results in fewer preemptions (see Exercise 2), which means less overhead for

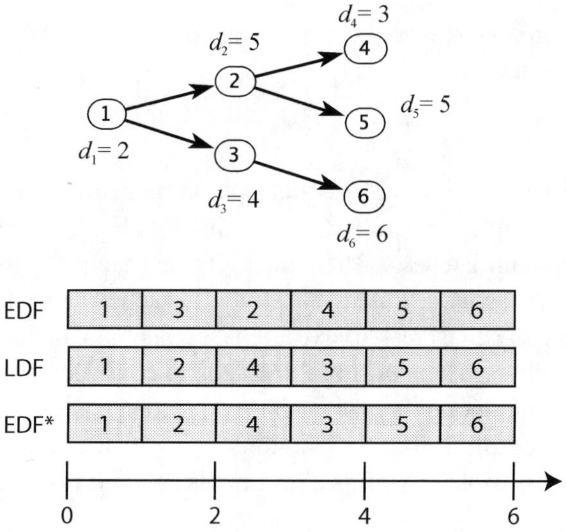

Figure 12.7: An example of a precedence graph for six tasks and the schedule under three scheduling policies. Execution times for all tasks are one time unit.

context switching. This often compensates for the greater complexity in the implementation. In addition, unlike RM, any EDF schedule with less than 100% utilization can tolerate increases in execution times and/or reductions in periods and still be feasible. p.314

12.3.1 EDF with Precedences

Theorem 12.2 shows that EDF is optimal (it minimizes maximum lateness) for a task set without precedences. What if there are precedences? Given a finite set of tasks, precedences between them can be represented by a **precedence graph**.

Example 12.1: Consider six tasks $T = \{1, \cdots, 6\}$, each with execution time $e_i = 1$, with precedences as shown in Figure 12.7. The diagram means that task 1 must execute before either 2 or 3 can execute, that 2 must execute before either 4 or 5, and that 3 must execute before 6. The deadline for each task is shown in the figure. The schedule labeled EDF is the EDF schedule. This schedule is not feasible. Task

4 misses its deadline. However, there is a feasible schedule. The schedule labeled LDF meets all deadlines.

The previous example shows that EDF is not optimal if there are precedences. In 1973, Lawler (1973) gave a simple algorithm that is optimal with precedences, in the sense that it minimizes the maximum lateness. The strategy is very simple. Given a fixed, finite set of tasks with deadlines, Lawler's strategy constructs the schedule backwards, choosing first the *last* task to execute. The last task to execute is the one on which no other task depends that has the latest deadline. The algorithm proceeds to construct the schedule backwards, each time choosing from among the tasks whose dependents have already been scheduled the one with the latest deadline. For the previous example, the resulting schedule, labeled LDF in Figure 12.7, is feasible. Lawler's algorithm is called **latest deadline first** (**LDF**).

LDF is optimal in the sense that it minimizes the maximum lateness, and hence it is also optimal with respect to feasibility. However, it does not support arrival of tasks. Fortunately, there is a simple modification of EDF, proposed by Chetto et al. (1990). **EDF*** (EDF with precedences), supports arrivals and minimizes the maximal lateness. In this modification, we adjust the deadlines of all the tasks. Suppose the set of all tasks is T. For a task execution $i \in T$, let $D(i) \subset T$ be the set of task executions that immediately depend on i in the precedence graph. For all executions $i \in T$, we define a modified deadline

$$d'_i = \min(d_i, \min_{j \in D(i)} (d'_j - e_j)) .$$

EDF* is then just like EDF except that it uses these modified deadlines.

Example 12.2: In Figure 12.7, we see that the EDF* schedule is the same as the LDF schedule. The modified deadlines are as follows:

$$d'_1 = 1, \quad d'_2 = 2, \quad d'_3 = 4, \quad d'_4 = 3, \quad d'_5 = 5, \quad d'_6 = 6 .$$

The key is that the deadline of task 2 has changed from 5 to 2, reflecting the fact that its successors have early deadlines. This causes EDF* to schedule task 2 before task 3, which results in a feasible schedule.

EDF* can be thought of as a technique for rationalizing deadlines. Instead of accepting arbitrary deadlines as given, this algorithm ensures that the deadlines take into account deadlines of successor tasks. In the example, it makes little sense for task 2 to have a later deadline, 5, than its successors. So EDF* corrects this anomaly before applying EDF.

12.4 Scheduling and Mutual Exclusion

Although the algorithms given so far are conceptually simple, the effects they have in practice are far from simple and often surprise system designers. This is particularly true when tasks share resources and use mutual exclusion to guard access to those resources. p.291

12.4.1 Priority Inversion

In principle, a **priority-based preemptive scheduler** is executing at all times the high-priority enabled task. However, when using mutual exclusion, it is possible for a task to become blocked during execution. If the scheduling algorithm does not account for this p.311 possibility, serious problems can occur.

> **Example 12.3:** The Mars Pathfinder, shown in Figure 12.8, landed on Mars on July 4th, 1997. A few days into the mission, the Pathfinder began sporadically missing deadlines, causing total system resets, each with loss of data. Engineers on the ground diagnosed the problem as priority inversion, where a low priority meteorological task was holding a lock and blocking a high-priority task, while medium priority tasks executed. (**Source**: What Really Happened on Mars? Mike Jones, RISKS-19.49 on the comp.programming.threads newsgroup, Dec. 07, 1997, and What Really Happened on Mars? Glenn Reeves, Mars Pathfinder Flight Software Cognizant Engineer, email message, Dec. 15, 1997.)

Priority inversion is a scheduling anomaly where a high-priority task is blocked while unrelated lower-priority tasks are executing. The phenomenon is illustrated in Figure 12.9. In the figure, task 3, a low priority task, acquires a lock at time 1. At time 2, it is preempted by task 1, a high-priority task, which then at time 3 blocks trying to acquire the same lock. Before task 3 reaches the point where it releases the lock, however, it gets preempted by an unrelated task 2, which has medium priority. Task 2 can run for an unbounded amount of time, and effectively prevents the higher-priority task 1 from executing. This is almost certainly not desirable.

Figure 12.8: The Mars Pathfinder and a view of the surface of Mars from the camera of the lander (image from the Wikipedia Commons).

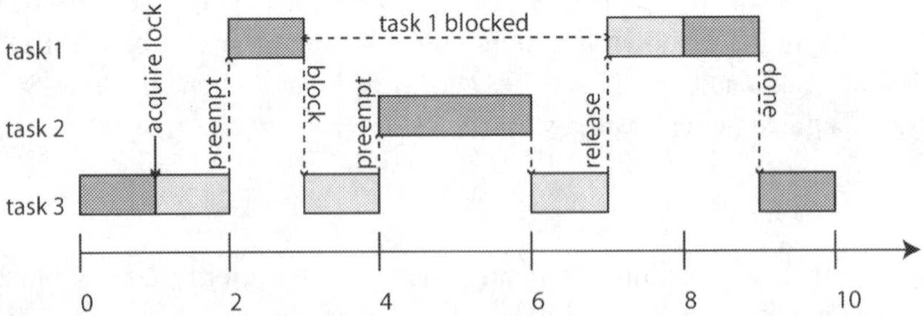

Figure 12.9: Illustration of priority inversion. Task 1 has highest priority, task 3 lowest. Task 3 acquires a lock on a shared object, entering a critical section. It gets preempted by task 1, which then tries to acquire the lock and blocks. Task 2 preempts task 3 at time 4, keeping the higher priority task 1 blocked for an unbounded amount of time. In effect, the priorities of tasks 1 and 2 get inverted, since task 2 can keep task 1 waiting arbitrarily long.

12.4.2 Priority Inheritance Protocol

In 1990, Sha et al. (1990) gave a solution to the priority inversion problem called **priority inheritance**. In their solution, when a task blocks attempting to acquire a lock, then the task that holds the lock inherits the priority of the blocked task. Thus, the task that holds the lock cannot be preempted by a task with lower priority than the one attempting to acquire the lock.

Example 12.4: Figure 12.10 illustrates priority inheritance. In the figure, when task 1 blocks trying to acquire the lock held by task 3, task 3 resumes executing, but now with the higher priority of task 1. Thus, when task 2 becomes enabled at time 4, it does not preempt task 3. Instead, task 3 runs until it releases the lock at time 5. At that time, task 3 reverts to its original (low) priority, and task 1 resumes executing. Only when task 1 completes is task 2 able to execute.

12.4.3 Priority Ceiling Protocol

Priorities can interact with mutual exclusion locks in even more interesting ways. In particular, in 1990, Sha et al. (1990) showed that priorities can be used to prevent certain kinds of deadlocks. p.294

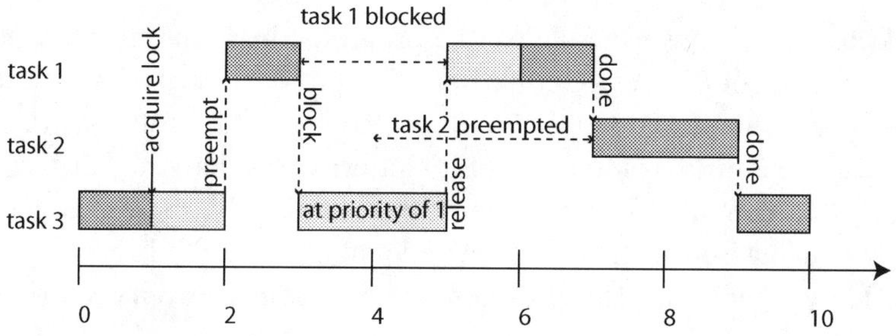

Figure 12.10: Illustration of the priority inheritance protocol. Task 1 has highest priority, task 3 lowest. Task 3 acquires a lock on a shared object, entering a critical section. It gets preempted by task 1, which then tries to acquire the lock and blocks. Task 3 inherits the priority of task 1, preventing preemption by task 2.

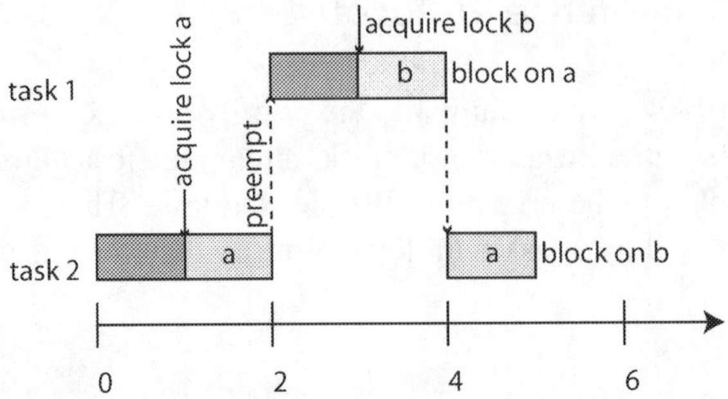

Figure 12.11: Illustration of deadlock. The lower priority task starts first and acquires lock a, then gets preempted by the higher priority task, which acquires lock b and then blocks trying to acquire lock a. The lower priority task then blocks trying to acquire lock b, and no further progress is possible.

Example 12.5: Figure 12.11 illustrates a scenario in which two tasks deadlock. In the figure, task 1 has higher priority. At time 1, task 2 acquires lock *a*. At time 2, task 1 preempts task 2, and at time 3, acquires lock *b*. While holding lock *b*, it attempts to acquire lock *a*. Since *a* is held by task 2, it blocks. At time 4, task 2 resumes executing. At time 5, it attempts to acquire lock *b*, which is held by task 1. Deadlock!

The deadlock in the previous example can be prevented by a clever technique called the p.302 **priority ceiling** protocol (Sha et al., 1990). In this protocol, every lock or semaphore is assigned a priority ceiling equal to the priority of the highest-priority task that can lock it. A task τ can acquire a lock a only if the task's priority is strictly higher than the priority ceilings of all locks currently held by other tasks. Intuitively, if we prevent task τ from acquiring lock a, then we ensure that task τ will not hold lock a while later trying to acquire other locks held by other tasks. This prevents certain deadlocks from occurring.

Example 12.6: The priority ceiling protocol prevents the deadlock of Example 12.5, as shown in Figure 12.12. In the figure, when task 1 attempts to acquire lock *b* at time 3, it is prevented from doing so. At that time, lock *a* is currently held by

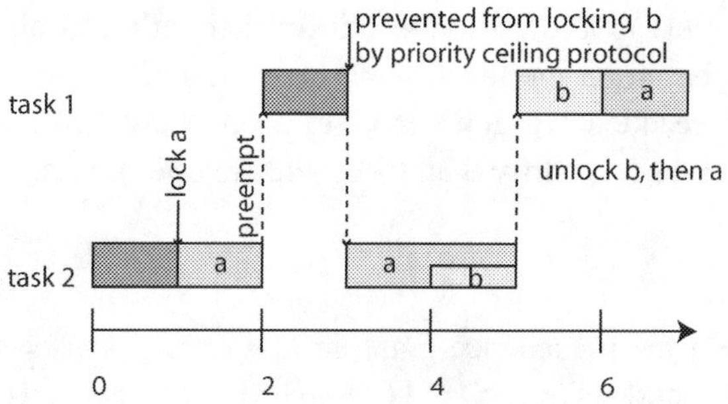

Figure 12.12: Illustration of the priority ceiling protocol. In this version, locks a and b have priority ceilings equal to the priority of task 1. At time 3, task 1 attempts to lock b, but it cannot because task 2 currently holds lock a, which has priority ceiling equal to the priority of task 1.

another task (task 2). The priority ceiling assigned to lock a is equal to the priority of task 1, since task 1 is the highest priority task that can acquire lock a. Since the priority of task 1 is not *strictly higher* than this priority ceiling, task 1 is not permitted to acquire lock b. Instead, task 1 becomes blocked, allowing task 2 to run to completion. At time 4, task 2 acquires lock b unimpeded, and at time 5, it releases both locks. Once it has released both locks, task 1, which has higher priority, is no longer blocked, so it resumes executing, preempting task 2.

Of course, implementing the priority ceiling protocol requires being able to determine in advance which tasks acquire which locks. A simple conservative strategy is to examine the source code for each task and inventory the locks that are acquired in the code. This is conservative because a particular program may or may not execute any particular line of code, so just because a lock is mentioned in the code does not necessarily mean that the task will attempt to acquire the lock.

12.5 Multiprocessor Scheduling

Scheduling tasks on a single processor is hard enough. Scheduling them on multiple processors is even harder. Consider the problem of scheduling a fixed finite set of tasks with precedences on a finite number of processors with the goal of minimizing the makespan. This problem is known to be NP-hard. Nonetheless, effective and efficient scheduling strategies p.492

exist. One of the simplest is known as the **Hu level scheduling** algorithm. It assigns a priority to each task τ based on the **level**, which is the greatest sum of execution times of tasks on a path in the precedence graph from τ to another task with no dependents. Tasks with larger levels have higher priority than tasks with smaller levels.

Example 12.7: For the precedence graph in Figure 12.7, task 1 has level 3, tasks 2 and 3 have level 2, and tasks 4, 5, and 6 have level 1. Hence, a Hu level scheduler will give task 1 highest priority, tasks 2 and 3 medium priority, and tasks 4, 5, and 6 lowest priority.

Hu level scheduling is one of a family of **critical path** methods because it emphasizes the path through the precedence graph with the greatest total execution time. Although it is not optimal, it is known to closely approximate the optimal solution for most graphs (Kohler, 1975; Adam et al., 1974).

Once priorities are assigned to tasks, a **list scheduler** sorts the tasks by priorities and assigns them to processors in the order of the sorted list as processors become available.

Example 12.8: A two-processor schedule constructed with the Hu level scheduling algorithm for the precedence graph shown in Figure 12.7 is given in Figure 12.13. The makespan is 4.

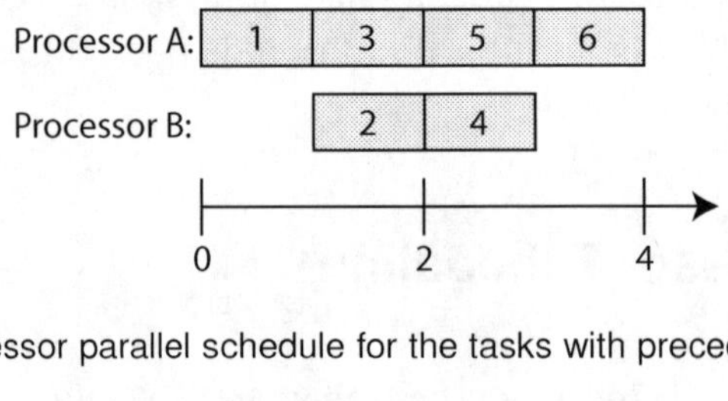

Figure 12.13: A two-processor parallel schedule for the tasks with precedence graph shown in Figure 12.7.

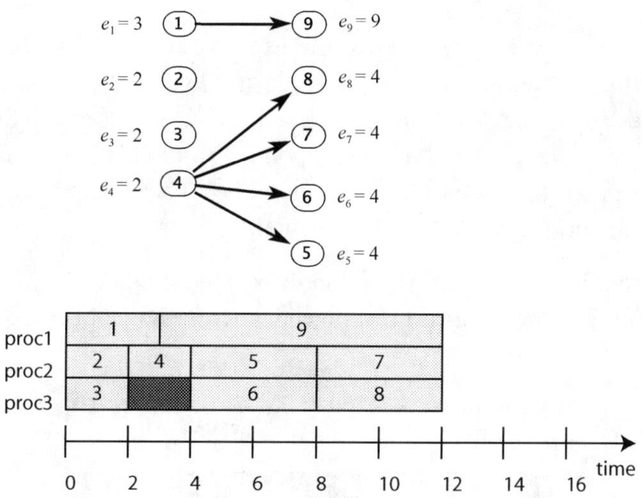

Figure 12.14: A precedence graph with nine tasks, where the lower numbered tasks have higher priority than the higher numbered tasks.

12.5.1 Scheduling Anomalies

Among the worst pitfalls in embedded systems design are **scheduling anomalies**, where unexpected or counterintuitive behaviors emerge due to small changes in the operating conditions of a system. We have already illustrated two such anomalies, priority inversion and deadlock. There are many others. The possible extent of the problems that can arise are well illustrated by the so-called **Richard's anomalies** (Graham, 1969). These show that multiprocessor schedules are **non-montonic**, meaning that improvements in performance at a local level can result in degradations in performance at a global level, and **brittle**, meaning that small changes can have big consequences.

Richard's anomalies are summarized in the following theorem.

Theorem 12.4. *If a task set with fixed priorities, execution times, and precedence constraints is scheduled on a fixed number of processors in accordance with the priorities, then increasing the number of processors, reducing execution times, or weakening precedence constraints can increase the schedule length.*

Proof. The theorem can be proved with the example in Figure 12.14. The example has nine tasks with execution times as shown in the figure. We assume the tasks are assigned priorities so that the lower numbered tasks have higher priority than the higher numbered tasks. Note that this does not correspond to a critical path priority assignment, but it suffices to prove the theorem. The figure shows a three-processor schedule in accordance with the priorities. Notice that the makespan is 12.

First, consider what happens if the execution times are all reduced by one time unit. A schedule conforming to the priorities and precedences is shown below:

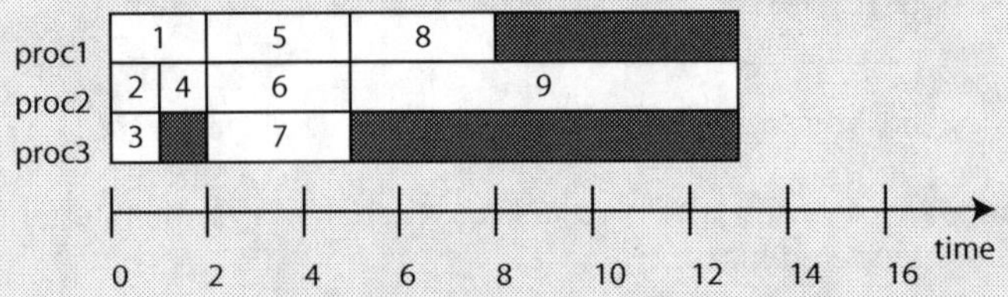

Notice that the makespan has *increased* to 13, even though the total amount of computation has decreased significantly. Since computation times are rarely known exactly, this form of brittleness is particularly troubling.

Consider next what happens if we add a fourth processor and keep everything else the same as in the original problem. A resulting schedule is shown below:

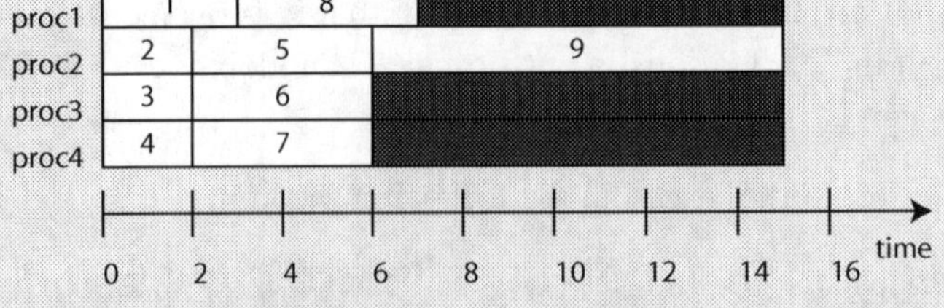

Again, the makespan has increased (to 15 this time) even though we have added 33% more processing power than originally available.

Consider finally what happens if we weaken the precedence constraints by removing the precedences between task 4 and tasks 7 and 8. A resulting schedule is shown below:

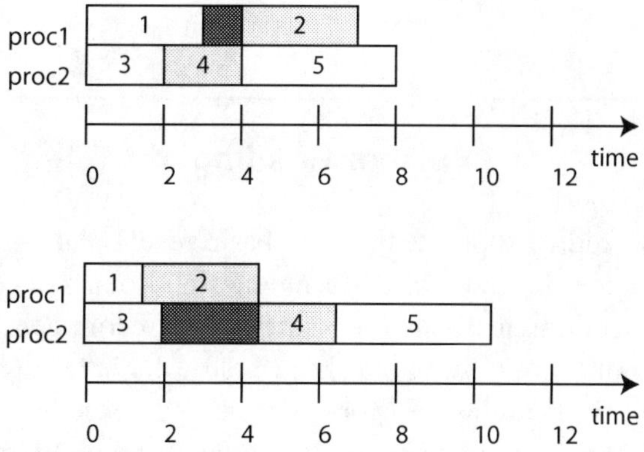

Figure 12.15: Anomaly due to mutual exclusion locks, where a reduction in the execution time of task 1 results in an increased makespan.

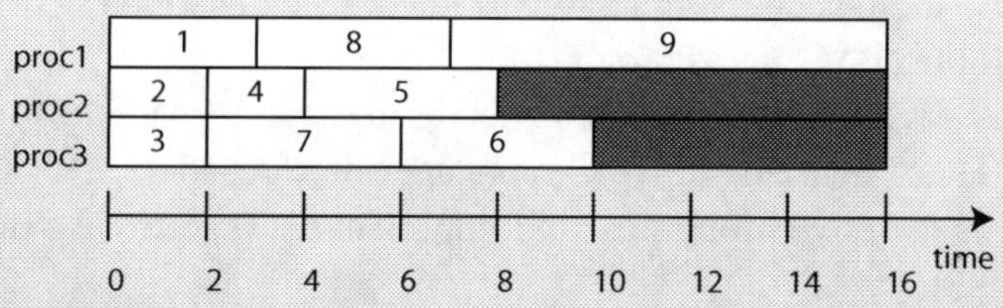

The makespan has now increased to 16, even though weakening precedence constraints increases scheduling flexibility. A simple priority-based scheduling scheme such as this does not take advantage of the weakened constraints.

□

This theorem is particularly troubling when we realize that execution times for software are rarely known exactly (see Chapter 16). Scheduling policies will be based on approximations, and behavior at run time may be quite unexpected.

Another form of anomaly arises when there are mutual exclusion locks. An illustration is p.291
given in Figure 12.15. In this example, five tasks are assigned to two processors using a static assignment scheduler. Tasks 2 and 4 contend for a mutex. If the execution time of p.311
task 1 is reduced, then the order of execution of tasks 2 and 4 reverses, which results in an increased execution time. This kind of anomaly is quite common in practice.

Further Reading

Scheduling is a well-studied topic, with many basic results dating back to the 1950s. This chapter covers only the most basic techniques and omits several important topics. For real-time scheduling textbooks, we particularly recommend Buttazzo (2005a), Stankovic and Ramamritham (1988), and Liu (2000), the latter of which has particularly p.312 good coverage of scheduling of sporadic tasks. An excellent overview article is Sha et al. (2004). A hands-on practical guide can be found in Klein et al. (1993). For an excellent overview of the evolution of fixed-priority scheduling techniques through 2003, see Audsley et al. (2005). For soft real-time scheduling, we recommend studying time utility functions, introduced by Douglas Jensen in 1977 as a way to overcome the limited expressiveness in classic deadline constraints in real-time systems (see, for example, Jensen et al. (1985); Ravindran et al. (2007)).

There are many more scheduling strategies than those described here. For example, p.316 **deadline monotonic (DM)** scheduling modifies rate monotonic to allow periodic tasks to have deadlines less than their periods (Leung and Whitehead, 1982). The **Spring algorithm** is a set of heuristics that support arrivals, precedence relations, resource constraints, non-preemptive properties, and importance levels (Stankovic and Ramamritham, 1987, 1988).

An important topic that we do not cover is **feasibility analysis**, which provides techniques for analyzing programs to determine whether feasible schedules exist. Much of the foundation for work in this area can be found in Harter (1987) and Joseph and Pandya (1986).

Multiprocessor scheduling is also a well-studied topic, with many core results originating in the field of operations research. Classic texts on the subject are Conway et al. (1967) and Coffman (1976). Sriram and Bhattacharyya (2009) focus on embedded multiprocessors and include innovative techniques for reducing synchronization overhead in multiprocessor schedules.

It is also worth noting that a number of projects have introduced programming language constructs that express real-time behaviors of software. Most notable among these is **Ada**, a language developed under contract from the US Department of Defense (DoD) from 1977 to 1983. The goal was to replace the hundreds of programming languages then used in DoD projects with a single, unified language. An excellent discussion of language constructs for real time can be found in Lee and Gehlot (1985) and Wolfe et al. (1993).

12.6 Summary

Embedded software is particularly sensitive to timing effects because it inevitably interacts with external physical systems. A designer, therefore, needs to pay considerable attention to the scheduling of tasks. This chapter has given an overview of some of the basic techniques for scheduling real-time tasks and parallel scheduling. It has explained some of the pitfalls, such as priority inversion and scheduling anomalies. A designer that is aware of the pitfalls is better equipped to guard against them.

Exercises

1. This problem studies fixed-priority scheduling. Consider two tasks to be executed periodically on a single processor, where task 1 has period $p_1 = 4$ and task 2 has period $p_2 = 6$. p.314

   (a) Let the execution time of task 1 be $e_1 = 1$. Find the maximum value for the execution time e_2 of task 2 such that the RM schedule is feasible. p.316

   (b) Again let the execution time of task 1 be $e_1 = 1$. Let non-RMS be a fixed-priority schedule that is not an RM schedule. Find the maximum value for the execution time e_2 of task 2 such that non-RMS is feasible.

   (c) For both your solutions to (a) and (b) above, find the processor utilization. Which is better? p.314

   (d) For RM scheduling, are there any values for e_1 and e_2 that yield 100% utilization? If so, give an example.

2. This problem studies dynamic-priority scheduling. Consider two tasks to be executed periodically on a single processor, where task 1 has period $p_1 = 4$ and task 2 has period $p_2 = 6$. Let the deadlines for each invocation of the tasks be the end of their period. That is, the first invocation of task 1 has deadline 4, the second invocation of task 1 has deadline 8, and so on. p.314

   (a) Let the execution time of task 1 be $e_1 = 1$. Find the maximum value for the execution time e_2 of task 2 such that EDF is feasible. p.322

   (b) For the value of e_2 that you found in part (a), compare the EDF schedule against the RM schedule from Exercise 1 (a). Which schedule has less preemption? Which schedule has better utilization?

3. This problem compares RM and EDF schedules. Consider two tasks with periods $p_1 = 2$ and $p_2 = 3$ and execution times $e_1 = e_2 = 1$. Assume that the deadline for each execution is the end of the period.

   p.316
   p.320

   (a) Give the RM schedule for this task set and find the processor utilization. How does this utilization compare to the Liu and Layland utilization bound of (12.2)?

   (b) Show that any increase in e_1 or e_2 makes the RM schedule infeasible. If you hold $e_1 = e_2 = 1$ and $p_2 = 3$ constant, is it possible to reduce p_1 below 2 and still get a feasible schedule? By how much? If you hold $e_1 = e_2 = 1$ and $p_1 = 2$ constant, is it possible to reduce p_2 below 3 and still get a feasible schedule? By how much?

   (c) Increase the execution time of task 2 to be $e_2 = 1.5$, and give an EDF schedule. Is it feasible? What is the processor utilization?

4. This problem, formulated by Hokeun Kim, also compares RM and EDF schedules. Consider two tasks to be executed periodically on a single processor, where task 1 has period $p_1 = 4$ and task 2 has period $p_2 = 10$. Assume task 1 has execution time $e_1 = 1$, and task 2 has execution time $e_2 = 7$.

   (a) Sketch a rate-monotonic schedule (for 20 time units, the least common multiple of 4 and 10). Is the schedule feasible?

   (b) Now suppose task 1 and 2 contend for a mutex lock, assuming that the lock is acquired at the beginning of each execution and released at the end of each execution. Also, suppose that acquiring or releasing locks takes zero time and the priority inheritance protocol is used. Is the rate-monotonic schedule feasible?

   (c) Assume still that tasks 1 and 2 contend for a mutex lock, as in part (b). Suppose that task 2 is running an **anytime algorithm**, which is an algorithm that can be terminated early and still deliver useful results. For example, it might be an image processing algorithm that will deliver a lower quality image when terminated early. Find the maximum value for the execution time e_2 of task 2 such that the rate-monotonic schedule is feasible. Construct the resulting schedule, with the reduced execution time for task 2, and sketch the schedule for 20 time units. You may assume that execution times are always positive integers.

   (d) For the original problem, where $e_1 = 1$ and $e_2 = 7$, and there is no mutex lock, sketch an EDF schedule for 20 time units. For tie-breaking among task executions with the same deadline, assume the execution of task 1 has higher priority than the execution of task 2. Is the schedule feasible?

(e) Now consider adding a third task, task 3, which has period $p_3 = 5$ and execution time $e_3 = 2$. In addition, assume as in part (c) that we can adjust execution time of task 2.

Find the maximum value for the execution time e_2 of task 2 such that the EDF schedule is feasible and sketch the schedule for 20 time units. Again, you may assume that the execution times are always positive integers. For tie-breaking among task executions with the same deadline, assume task i has higher priority than task j if $i < j$.)

5. This problem compares fixed vs. dynamic priorities, and is based on an example by Burns and Baruah (2008). Consider two periodic tasks, where task τ_1 has period $p_1 = 2$, and task τ_2 has period $p_2 = 3$. Assume that the execution times are $e_1 = 1$ and $e_2 = 1.5$. Suppose that the release time of execution i of task τ_1 is given by p.312

$$r_{1,i} = 0.5 + 2(i-1)$$

for $i = 1, 2, \cdots$. Suppose that the deadline of execution i of task τ_1 is given by p.313

$$d_{1,i} = 2i.$$

Correspondingly, assume that the release times and deadlines for task τ_2 are

$$r_{2,i} = 3(i-1)$$

and

$$d_{2,i} = 3i.$$

(a) Give a feasible fixed-priority schedule. p.314

(b) Show that if the release times of all executions of task τ_1 are reduced by 0.5, then no fixed-priority schedule is feasible.

(c) Give a feasible dynamic-priority schedule with the release times of task τ_1 reduced to p.314

$$r_{1,i} = 2(i-1).$$

6. This problem studies scheduling anomalies. Consider the task precedence graph depicted in Figure 12.16 with eight tasks. In the figure, e_i denotes the execution time of task i. Assume task i has higher priority than task j if $i < j$. There is no preemption. The tasks must be scheduled respecting all precedence constraints and priorities. We assume that all tasks arrive at time $t = 0$.

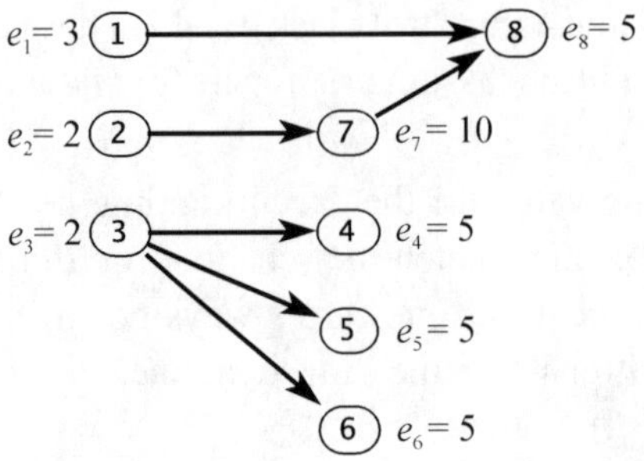

Figure 12.16: Precedence Graph for Exercise 6.

(a) Consider scheduling these tasks on two processors. Draw the schedule for these tasks and report the makespan. p.314

(b) Now consider scheduling these tasks on three processors. Draw the schedule for these tasks and report the makespan. Is the makespan bigger or smaller than that in part (a) above?

(c) Now consider the case when the execution time of each task is reduced by 1 time unit. Consider scheduling these tasks on two processors. Draw the schedule for these tasks and report the makespan. Is the makespan bigger or smaller than that in part (a) above?

7. This problem studies the interaction between real-time scheduling and mutual exclusion, and was formulated by Kevin Weekly.

Consider the following excerpt of code:

```
pthread_mutex_t X; // Resource X: Radio communication
pthread_mutex_t Y; // Resource Y: LCD Screen
pthread_mutex_t Z; // Resource Z: External Memory (slow)

void ISR_A() { // Safety sensor Interrupt Service Routine
  pthread_mutex_lock(&Y);
  pthread_mutex_lock(&X);
  display_alert(); // Uses resource Y
  send_radio_alert(); // Uses resource X
  pthread_mutex_unlock(&X);
  pthread_mutex_unlock(&Y);
}

void taskB() { // Status recorder task
  while (1) {
```

```
      static time_t starttime = time();
      pthread_mutex_lock(&X);
      pthread_mutex_lock(&Z);
      stats_t stat = get_stats();
      radio_report( stat ); // uses resource X
      record_report( stat ); // uses resource Z
      pthread_mutex_unlock(&Z);
      pthread_mutex_unlock(&X);
      sleep(100-(time()-starttime)); // schedule next excecution
    }
  }

  void taskC() { // UI Updater task
    while(1) {
      pthread_mutex_lock(&Z);
      pthread_mutex_lock(&Y);
      read_log_and_display(); // uses resources Y and Z
      pthread_mutex_unlock(&Y);
      pthread_mutex_unlock(&Z);
    }
  }
```

You may assume that the comments fully disclose the resource usage of the procedures. That is, if a comment says "uses resource X", then the relevant procedure uses only resource X. The scheduler running aboard the system is a priority-based preemptive scheduler, where taskB is higher priority than taskC. In this problem, ISR_A can be thought of as an asynchronous task with the highest priority.

The intended behavior is for the system to send out a radio report every 100ms and for the UI to update constantly. Additionally, if there is a safety interrupt, a radio report is sent immediately and the UI alerts the user.

(a) Occasionally, when there is a safety interrupt, the system completely stops working. In a scheduling diagram (like Figure 12.11 in the text), using the tasks {A,B,C}, and resources {X,Y,Z}, explain the cause of this behavior. Execution times do not have to be to scale in your diagram. Label your diagram clearly. You will be graded in part on the clarity of your answer, not just on its correctness.

(b) Using the priority ceiling protocol, show the scheduling diagram for the same sequence of events that you gave in part (a). Be sure to show all resource locks and unlocks until all tasks are finished or reached the end of an iteration. Does execution stop as before?

(c) Without changing the scheduler, how could the code in taskB be reordered to fix the issue? Using an exhaustive search of all task/resource locking scenarios, prove that this system will not encounter deadlock. (Hint: There exists a proof enumerating 6 cases, based on reasoning that the 3 tasks each have 2 possible resources they could block on.)

Part III

Analysis and Verification

This part of this text studies analysis of embedded systems, with emphasis on methods for specifying desired and undesired behaviors and verifying that an implementation conforms to its specification. Chapter 13 covers temporal logic, a formal notation that can express families of input/output behaviors and the evolution of the state of a system over time. This notation can be used to specify unambiguously desired and undesired behaviors. Chapter 14 explains what it means for one specification to be equivalent to another, and what it means for a design to implement a specification. Chapter 15 shows how to check algorithmically whether a design correctly implements a specification. Chapter 16 illustrates how to analyze designs for quantitative properties, with emphasis on execution time analysis for software. Such analysis is essential to achieving real-time behavior in software. Chapter 17 introduces the basics of security and privacy with a focus on concepts relevant to embedded, cyber-physical systems. p.8

13

Invariants and Temporal Logic

Contents

Every embedded system must be designed to meet certain requirements. Such system requirements are also called **properties** or **specifications**. The need for specifications is aptly captured by the following quotation (paraphrased from Young et al. (1985)):

> "A design without specifications cannot be right or wrong, it can only be surprising!"

In present engineering practice, it is common to have system requirements stated in a natural language such as English. As an example, consider the SpaceWire communication protocol that is gaining adoption with several national space agencies (European Cooperation for

Space Standardization, 2002). Here are two properties reproduced from Section 8.5.2.2 of the specification document, stating conditions on the behavior of the system upon reset:

1. "The *ErrorReset* state shall be entered after a system reset, after link operation has been terminated for any reason or if there is an error during link initialization."
2. "Whenever the reset signal is asserted the state machine shall move immediately to the *ErrorReset* state and remain there until the reset signal is de-asserted."

It is important to precisely state requirements to avoid ambiguities inherent in natural languages. For example, consider the first property of the SpaceWire protocol stated above. Observe that there is no mention of *when* the *ErrorReset* state is to be entered. The systems that implement the SpaceWire protocol are synchronous, meaning that transitions of the state machine occur on ticks of a system clock. Given this, must the *ErrorReset* state be entered on the very next tick after one of the three conditions becomes true or on some subsequent tick of the clock? As it turns out, the document intends the system to make the transition to *ErrorReset* on the very next tick, but this is not made precise by the English language description.

This chapter will introduce techniques to specify system properties mathematically and precisely. A mathematical specification of system properties is also known as a **formal specification**. The specific formalism we will use is called **temporal logic**. As the name suggests, temporal logic is a precise mathematical notation with associated rules for representing and reasoning about timing-related properties of systems. While temporal logic has been used by philosophers and logicians since the times of Aristotle, it is only in the last thirty years that it has found application as a mathematical notation for specifying system requirements.

One of the most common kinds of system property is an **invariant**. It is also one of the simplest forms of a temporal logic property. We will first introduce the notion of an invariant and then generalize it to more expressive specifications in temporal logic.

13.1 Invariants

An **invariant** is a property that holds for a system if it remains true at all times during operation of the system. Put another way, an invariant holds for a system if it is true in the initial state of the system, and it remains true as the system evolves, after every reaction, in every state.

In practice, many properties are invariants. Both properties of the SpaceWire protocol stated above are invariants, although this might not be immediately obvious. Both SpaceWire properties specify conditions that must remain true always. Below is an example of an invariant property of a model that we have encountered in Chapter 3.

Example 13.1: Consider the model of a traffic light controller given in Figure 3.10 and its environment as modeled in Figure 3.11. Consider the system formed by the asynchronous composition of these two state machines. An obvious property that the composed system must satisfy is that *there is no pedestrian crossing when the traffic light is green* (when cars are allowed to move). This property must always remain true of this system, and hence is a system invariant.

It is also desirable to specify invariant properties of software and hardware *implementations* of embedded systems. Some of these properties specify correct programming practice on language constructs. For example, the C language property

> "The program never dereferences a null pointer"

is an invariant specifying good programming practice. Typically dereferencing a null pointer in a C program results in a segmentation fault, possibly leading to a system crash. Similarly, several desirable properties of concurrent programs are invariants, as illustrated in the following example. p.243

Example 13.2: Consider the following property regarding an absence of deadlock:

> If a thread A blocks while trying to acquire a mutex lock, then the thread B that holds that lock must not be blocked attempting to acquire a lock held by A. p.291

This property is required to be an invariant on any multithreaded program constructed from threads A and B. The property may or may not hold for a particular program. If it does not hold, there is risk of deadlock.

Many system invariants also impose requirements on program data, as illustrated in the example below.

Example 13.3: Consider the following example of a software task from the open source Paparazzi unmanned aerial vehicle (UAV) project (Nemer et al., 2006):

```
void altitude_control_task(void) {
 if (pprz_mode == PPRZ_MODE_AUTO2
     || pprz_mode == PPRZ_MODE_HOME) {
  if (vertical_mode == VERTICAL_MODE_AUTO_ALT) {
   float err = estimator_z - desired_altitude;
   desired_climb
        = pre_climb + altitude_pgain * err;
   if (desired_climb < -CLIMB_MAX) {
    desired_climb = -CLIMB_MAX;
   }
   if (desired_climb > CLIMB_MAX) {
    desired_climb = CLIMB_MAX;
   }
  }
 }
}
```

For this example, it is required that the value of the `desired_climb` variable at the end of `altitude_control_task` remains within the range [`-CLIMB_MAX`, `CLIMB_MAX`]. This is an example of a special kind of invariant, a **postcondition**, that must be maintained every time `altitude_control_task` returns. Determining whether this is the case requires analyzing the control flow of the program.

13.2 Linear Temporal Logic

We now give a formal description of **temporal logic** and illustrate with examples of how it can be used to specify system behavior. In particular, we study a particular kind of temporal logic known as **linear temporal logic**, or **LTL**. There are other forms of temporal logic, some of which are briefly surveyed in sidebars.

Using LTL, one can express a property over a *single, but arbitrary execution* of a system. For instance, one can express the following kinds of properties in LTL:

- *Occurrence of an event and its properties.* For example, one can express the property that an event A must occur at least once in every trace of the system, or that it must occur infinitely many times.
- *Causal dependency between events.* An example of this is the property that if an event A occurs in a trace, then event B must also occur.
- *Ordering of events.* An example of this kind of property is one specifying that every occurrence of event A is preceded by a matching occurrence of B.

We now formalize the above intuition about the kinds of properties expressible in linear temporal logic. In order to perform this formalization, it is helpful to fix a particular formal model of computation. We will use the theory of finite-state machines, introduced in Chapter 3. p.47

Recall from Section 3.6 that an execution trace of a finite-state machine is a sequence of the form p.67

$$q_0,\ q_1,\ q_2,\ q_3,\ \ldots,$$

where $q_j = (x_j, s_j, y_j)$, s_j is the state, x_j is the input valuation, and y_j is the output valuation at reaction j. p.45

13.2.1 Propositional Logic Formulas

First, we need to be able to talk about conditions at each reaction, such as whether an input or output is present, what the value of an input or output is, or what the state is. Let an **atomic proposition** be such a statement about the inputs, outputs, or states. It is a predicate (an expression that evaluates to true or false). Examples of atomic propositions that are relevant for the state machines in Figure 13.1 are:

true	Always true.
false	Always false.
x	True if input x is *present*.
$x = present$	True if input x is *present*.
$y = absent$	True if y is *absent*.
b	True if the FSM is in state b

In each case, the expression is true or false at a reaction q_i. The proposition b is true at a reaction q_i if $q_i = (x, \mathsf{b}, y)$ for any valuations x and y, which means that the machine is in state b at the *start* of the reaction. I.e., it refers to the current state, not the next state.

A **propositional logic formula** or (more simply) **proposition** is a predicate that combines p.49 atomic propositions using **logical connectives**: conjunction (logical AND, denoted $\wedge$), disjunction (logical OR, denoted $\vee$), negation (logical NOT, denoted $\neg$), and **implies** (logical implication, denoted $\implies$). Propositions for the state machines in Figure 13.1 include any of the above atomic proposition and expressions using the logical connectives together with atomic propositions. Here are some examples:

$x \wedge y$	True if x and y are both *present*.
$x \vee y$	True if either x or y is *present*.
$x = present \wedge y = absent$	True if x is *present* and y is *absent*.
$\neg y$	True if y is *absent*.
$\mathsf{a} \implies y$	True if whenever the FSM is in state a, the output y will be made present by the reaction

Note that if p_1 and p_2 are propositions, the proposition $p_1 \implies p_2$ is true if and only if $\neg p_2 \implies \neg p_1$. In other words, if we wish to establish that $p_1 \implies p_2$ is true, it is equally valid to establish that $\neg p_2 \implies \neg p_1$ is true. In logic, the latter expression is called the **contrapositive** of the former.

Note further that $p_1 \implies p_2$ is true if p_1 is false. This is easy to see by considering the contrapositive. The proposition $\neg p_2 \implies \neg p_1$ is true regardless of p_2 if $\neg p_1$ is true. Thus, another proposition that is equivalent to $p_1 \implies p_2$ is

$$\neg p_1 \vee p_2 \,.$$

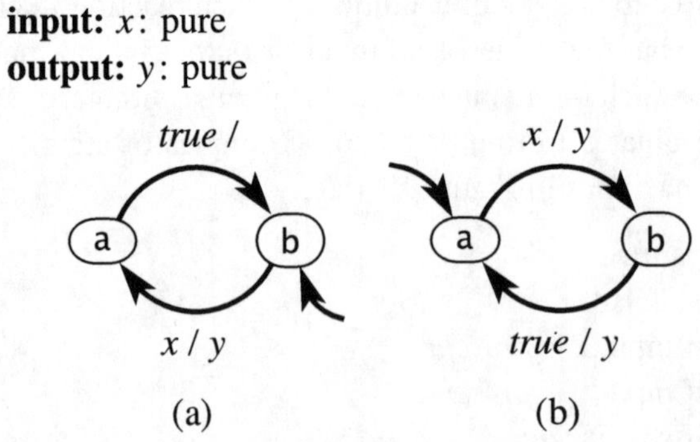

Figure 13.1: Two finite-state machines used to illustrate LTL formulas.

13.2.2 LTL Formulas

An **LTL formula**, unlike the above propositions, applies to an entire trace

$$q_0, q_1, q_2, \ldots,$$

rather than to just one reaction q_i. The simplest LTL formulas look just like the propositions above, but they apply to an entire trace rather than just a single element of the trace. If p is a proposition, then by definition, we say that LTL formula $\phi = p$ **holds** for the trace $q_0, q_1, q_2, \ldots$ if and only if p is true for q_0. It may seem odd to say that the formula holds for the entire trace even though the proposition only holds for the first element of the trace, but we will see that LTL provides ways to reason about the entire trace.

By convention, we will denote LTL formulas by ϕ, ϕ_1, ϕ_2, etc. and propositions by p, p_1, p_2, etc.

Given a state machine M and an LTL formula ϕ, we say that ϕ holds for M if ϕ holds for all possible traces of M. This typically requires considering all possible inputs.

Example 13.4: The LTL formula a holds for Figure 13.1(b), because all traces begin in state a. It does not hold for Figure 13.1(a).

The LTL formula $x \implies y$ holds for both machines. In both cases, in the first reaction, if x is *present*, then y will be *present*.

To demonstrate that an LTL formula is false for an FSM, it is sufficient to give one trace for which it is false. Such a trace is called a **counterexample**. To show that an LTL formula is true for an FSM, you must demonstrate that it is true for all traces, which is often much harder (although not so much harder when the LTL formula is a simple propositional logic formula, because in that case we only have to consider the first element of the trace).

Example 13.5: The LTL formula y is false for both FSMs in Figure 13.1. In both cases, a counterexample is a trace where x is absent in the first reaction.

In addition to propositions, LTL formulas can also have one or more special **temporal operators**. These make LTL much more interesting, because they enable reasoning about entire traces instead of just making assertions about the first element of a trace. There are four main temporal operators, which we describe next.

G Operator

The property $\mathbf{G}\phi$ (which is read as "**globally** ϕ") holds for a trace if ϕ holds for *every* suffix of that trace. (A **suffix** is a tail of a trace beginning with some reaction and including all subsequent reactions.)

In mathematical notation, $\mathbf{G}\phi$ holds for the trace if and only if, for *all* $j \geq 0$, formula ϕ holds in the suffix $q_j, q_{j+1}, q_{j+2}, \cdots$.

Example 13.6: In Figure 13.1(b), $\mathbf{G}(x \implies y)$ is true for all traces of the machine, and hence holds for the machine. $\mathbf{G}(x \wedge y)$ does not hold for the machine, because it is false for any trace where x is absent in any reaction. Such a trace provides a counterexample.

If ϕ is a propositional logic formula, then $\mathbf{G}\phi$ simply means that ϕ holds in every reaction. We will see, however, that when we combine the **G** operator with other temporal logic operators, we can make much more interesting statements about traces and about state machines.

F Operator

The property $\mathbf{F}\phi$ (which is read as "**eventually** ϕ" or "**finally** ϕ") holds for a trace if ϕ holds for *some* suffix of the trace.

Formally, $\mathbf{F}\phi$ holds for the trace if and only if, for *some* $j \geq 0$, formula ϕ holds in the suffix $q_j, q_{j+1}, q_{j+2}, \cdots$.

Example 13.7: In Figure 13.1(a), **F**b is trivially true because the machine starts in state b, hence, for all traces, the proposition b holds for the trace itself (the very first suffix).

More interestingly, $\mathbf{G}(x \implies \mathbf{F}\mathsf{b})$ holds for Figure 13.1(a). This is because if x is *present* in any reaction, then the machine will eventually be in state b. This is true even in suffixes that start in state a.

Notice that parentheses can be important in interpreting an LTL formula. For example, $(\mathbf{G}x) \implies (\mathbf{F}\mathsf{b})$ is trivially true because $\mathbf{F}\mathsf{b}$ is true for all traces (since the initial state is b).

Notice that $\mathbf{F}\neg\phi$ holds if and only if $\neg\mathbf{G}\phi$. That is, stating that ϕ is eventually false is the same as stating that ϕ is not always true.

X Operator

The property $\mathbf{X}\phi$ (which is read as "**next state** ϕ") holds for a trace $q_0, q_1, q_2, \ldots$ if and only if ϕ holds for the trace $q_1, q_2, q_3, \ldots$.

Example 13.8: In Figure 13.1(a), $x \implies \mathbf{X}\mathsf{a}$ holds for the state machine, because if x is *present* in the first reaction, then the next state will be a. $\mathbf{G}(x \implies \mathbf{X}\mathsf{a})$ does not hold for the state machine because it does not hold for any suffix that begins in state a. In Figure 13.1(b), $\mathbf{G}(\mathsf{b} \implies \mathbf{X}\mathsf{a})$ holds for the state machine.

U Operator

The property $\phi_1\mathbf{U}\phi_2$ (which is read as "ϕ_1 **until** ϕ_2") holds for a trace if ϕ_2 holds for some suffix of that trace, and ϕ_1 holds until ϕ_2 becomes *true*.

Formally, $\phi_1\mathbf{U}\phi_2$ holds for the trace if and only if there exists $j \geq 0$ such that ϕ_2 holds in the suffix $q_j, q_{j+1}, q_{j+2}, \ldots$ and ϕ_1 holds in suffixes $q_i, q_{i+1}, q_{i+2}, \ldots$, for all i s.t. $0 \leq i < j$. ϕ_1 may or may not hold for $q_j, q_{j+1}, q_{j+2}, \ldots$.

Example 13.9: In Figure 13.1(b), $\mathsf{a}\mathbf{U}x$ is true for any trace for which $\mathbf{F}x$ holds. Since this does not include all traces, $\mathsf{a}\mathbf{U}x$ does not hold for the state machine.

Some authors define a weaker form of the **U** operator that does not require ϕ_2 to hold. Using our definition, this can be written

$$(\mathbf{G}\phi_1) \vee (\phi_1 \mathbf{U} \phi_2) \; .$$

Probing Further: Alternative Temporal Logics

Amir Pnueli (1977) was the first to formalize temporal logic as a way of specifying program properties. For this he won the 1996 ACM Turing Award, the highest honor in Computer Science. Since his seminal paper, temporal logic has become widespread as a way of specifying properties for a range of systems, including hardware, software, and cyber-physical systems.

p.346 In this chapter, we have focused on LTL, but there are several alternatives. LTL formulas apply to individual traces of an FSM, and in this chapter, by convention, we assert than an LTL formula holds for an FSM if it holds for all possible traces of the FSM. A more general logic called **computation tree logic** (**CTL***) explicitly provides quantifiers over possible traces of an FSM (Emerson and Clarke (1980); Ben-Ari et al. (1981)). For example, we can write a CTL* expression that holds for an FSM if there exists *any* trace that satisfies some property, rather than insisting that the property must hold *for all* traces. CTL* is called a **branching-time logic** because whenever a reaction of the FSM has a nondeterministic choice, it will simultaneously consider all options. LTL, by contrast, considers only one trace at a time, and hence it is called a **linear-time logic**. Our convention of asserting that an LTL formula holds for an FSM if it holds for all traces cannot be expressed directly in LTL, because LTL does not include quantifiers like "for all traces." We have to step outside the logic to apply this convention. With CTL*, this convention is expressible directly in the logic.

Several other temporal logic variants have found practical use. For instance, **real-time temporal logics** (e.g., **timed computation tree logic** or **TCTL**), is used for reasoning about real-time systems (Alur et al., 1991; Alur and Henzinger, 1993) where p.310 the passage of time is not in discrete steps, but is continuous. Similarly, **probabilistic temporal logics** are useful for reasoning about probabilistic models such as Markov chains or Markov decision processes (see, for example, Hansson and Jonsson (1994)), and **signal temporal logic** has proved effective for reasoning about real-time behavior of hybrid systems (Maler and Nickovic, 2004).

Techniques for inferring temporal logic properties from traces, also known as **specification mining**, have also proved useful in industrial practice (see Jin et al. (2015)).

This holds if either ϕ_1 always holds (for any suffix) or, if ϕ_2 holds for some suffix, then ϕ_1 holds for all previous suffixes. This can equivalently be written

$$(\mathbf{F}\neg\phi_1) \implies (\phi_1 \mathbf{U} \phi_2) .$$

Example 13.10: In Figure 13.1(b), $(\mathbf{G}\neg x) \vee (a\mathbf{U}x)$ holds for the state machine.

13.2.3 Using LTL Formulas

Consider the following English descriptions of properties and their corresponding LTL formalizations:

Example 13.11: *"Whenever the robot is facing an obstacle, eventually it moves at least 5 cm away from the obstacle."*

Let p denote the condition that the robot is facing an obstacle, and q denote the condition where the robot is at least 5 cm away from the obstacle. Then, this property can be formalized in LTL as

$$\mathbf{G}(p \implies \mathbf{F}q) .$$

Example 13.12: Consider the SpaceWire property:
"Whenever the reset signal is asserted the state machine shall move immediately to the ErrorReset *state and remain there until the reset signal is de-asserted."*

Let p be *true* when the reset signal is asserted, and q be true when the state of the FSM is *ErrorReset*. Then, the above English property is formalized in LTL as:

$$\mathbf{G}(p \implies \mathbf{X}(q\mathbf{U}\neg p)) .$$

In the above formalization, we have interpreted "immediately" to mean that the state changes to *ErrorReset* in the very next time step. Moreover, the above LTL formula will fail to hold for any execution where the reset signal is asserted and not eventually de-asserted. It was probably the intent of the standard that the reset signal should be eventually de-asserted, but the English language statement does not make this clear.

Example 13.13: Consider the traffic light controller in Figure 3.10. A property of this controller is that the outputs always cycle through *sigG*, *sigY* and *sigR*. We can express this in LTL as follows:

$$\begin{aligned} \mathbf{G} \quad \{ \quad & (sigG \implies \mathbf{X}((\neg sigR \wedge \neg sigG)\,\mathbf{U}\, sigY)) \\ \wedge \quad & (sigY \implies \mathbf{X}((\neg sigG \wedge \neg sigY)\,\mathbf{U}\, sigR)) \\ \wedge \quad & (sigR \implies \mathbf{X}((\neg sigY \wedge \neg sigR)\,\mathbf{U}\, sigG)) \ \} \,. \end{aligned}$$

The following LTL formulas express commonly useful properties.

(a) *Infinitely many occurrences:* This property is of the form $\mathbf{G}\mathbf{F}p$, meaning that it is always the case that p is *true* eventually. Put another way, this means that p is true **infinitely often**.

(b) *Steady-state property:* This property is of the form $\mathbf{F}\mathbf{G}p$, read as "from some point in the future, p holds at all times." This represents a steady-state property, indicating that after some point in time, the system reaches a **steady state** in which p is always *true*.

(c) *Request-response property:* The formula $\mathbf{G}(p \implies \mathbf{F}q)$ can be interpreted to mean that a request p will eventually produce a response q.

13.3 Summary

Dependability and correctness are central concerns in embedded systems design. Formal specifications, in turn, are central to achieving these goals. In this chapter, we have studied temporal logic, one of the main approaches for writing formal specifications. This chapter has provided techniques for precisely stating properties that must hold over time for a system. It has specifically focused on linear temporal logic, which is able to express many safety and liveness properties of systems.

Safety and Liveness Properties

System properties may be **safety** or **liveness** properties. Informally, a safety property is one specifying that "nothing bad happens" during execution. Similarly, a liveness property specifies that "something good will happen" during execution.

More formally, a property p is a **safety property** if a system execution does not satisfy p if and only if there exists a finite-length prefix of the execution that cannot be extended to an infinite execution satisfying p. We say p is a **liveness property** if every finite-length execution trace can be extended to an infinite execution that satisfies p. See Lamport (1977) and Alpern and Schneider (1987) for a theoretical treatment of safety and liveness.

The properties we have seen in Section 13.1 are all examples of safety properties. Liveness properties, on the other hand, specify performance or progress requirements on a system. For a state machine, a property of the form $\mathbf{F}\phi$ is a liveness property. No finite execution can establish that this property is not satisfied.

The following is a slightly more elaborate example of a liveness property:

> "Whenever an interrupt is asserted, the corresponding interrupt service routine (ISR) is eventually executed."

In temporal logic, if p_1 is the property that an interrupt is asserted, and p_2 is the property that the interrupt service routine is executed, then this property can be written

$$\mathbf{G}(p_1 \implies \mathbf{F}p_2) \ .$$

Note that both safety and liveness properties can constitute system invariants. For example, the above liveness property on interrupts is also an invariant; $p_1 \implies \mathbf{F}p_2$ must hold in *every state*.

Liveness properties can be either *bounded* or *unbounded*. A **bounded liveness** property specifies a time bound on something desirable happening (which makes it a safety property). In the above example, if the ISR must be executed within 100 clock cycles of the interrupt being asserted, the property is a bounded liveness property; otherwise, if there is no such time bound on the occurrence of the ISR, it is an **unbounded liveness** property. LTL can express a limited form of bounded liveness properties using the **X** operator, but it does not provide any mechanism for quantifying time directly.

Exercises

1. For each of the following questions, give a short answer and justification.

    (a) TRUE or FALSE: If $\mathbf{GF}p$ holds for a state machine A, then so does $\mathbf{FG}p$.

    (b) TRUE or FALSE: $\mathbf{G}(\mathbf{G}p)$ holds for a trace if and only if $\mathbf{G}p$ holds.

2. Consider the following state machine:

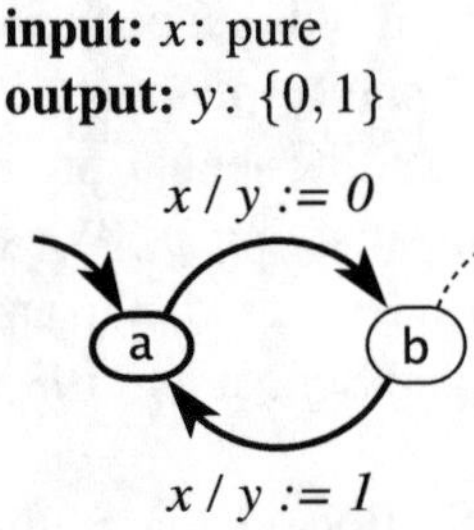

p.53 (Recall that the dashed line represents a default transition.) For each of the following LTL formulas, determine whether it is true or false, and if it is false, give a counterexample:

    (a) $x \implies \mathbf{F}\mathsf{b}$

    (b) $\mathbf{G}(x \implies \mathbf{F}(y = 1))$

    (c) $(\mathbf{G}x) \implies \mathbf{F}(y = 1)$

    (d) $(\mathbf{G}x) \implies \mathbf{GF}(y = 1)$

    (e) $\mathbf{G}((\mathsf{b} \land \neg x) \implies \mathbf{FG}\mathsf{c})$

    (f) $\mathbf{G}((\mathsf{b} \land \neg x) \implies \mathbf{G}\mathsf{c})$

    (g) $(\mathbf{GF}\neg x) \implies \mathbf{FG}\mathsf{c}$

3. Consider the synchronous feedback composition studied in Exercise 6 of Chapter 6. Determine whether the following statement is true or false:

    The following temporal logic formula is satisfied by the sequence w for every possible behavior of the composition and is not satisfied by any sequence that is not a behavior of the composition:

$$(\mathbf{G}w) \lor (w\mathbf{U}(\mathbf{G}\neg w)) .$$

    Justify your answer. If you decide it is false, then provide a temporal logic formula for which the assertion is true.

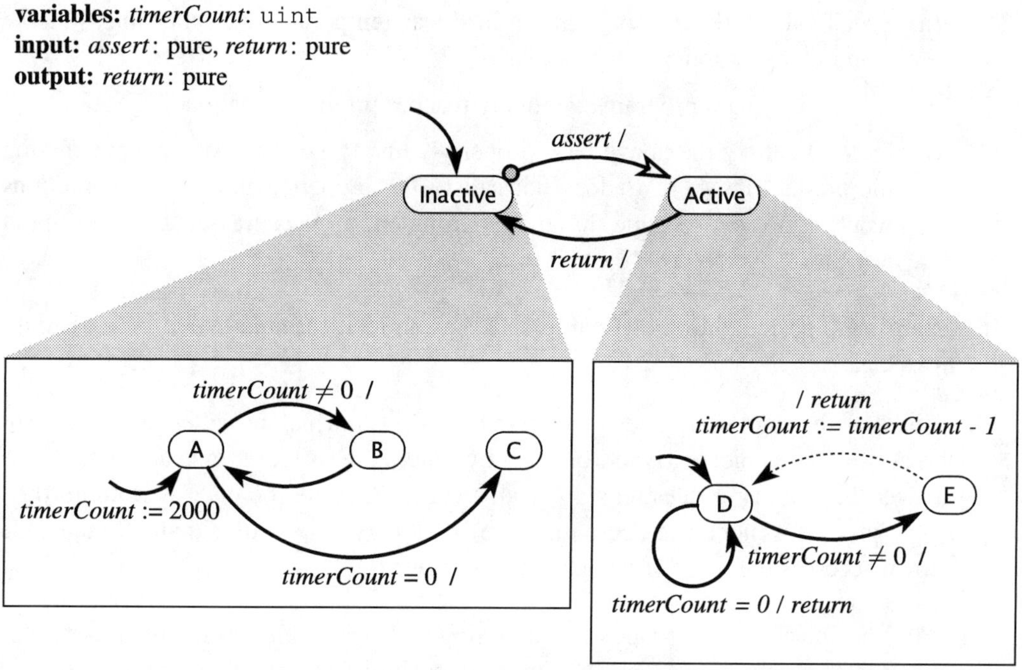

Figure 13.2: Hierarchical state machine modeling a program and its interrupt service routine.

4. This problem is concerned with specifying in linear temporal logic tasks to be performed by a robot. Suppose the robot must visit a set of n locations $l_1, l_2, \ldots, l_n$. Let p_i be an atomic formula that is *true* if and only if the robot visits location l_i.

   Give LTL formulas specifying the following tasks:

   (a) The robot must eventually visit at least one of the n locations.

   (b) The robot must eventually visit all n locations, but in any order.

   (c) The robot must eventually visit all n locations, in the order $l_1, l_2, \ldots, l_n$.

5. Consider a system M modeled by the hierarchical state machine of Figure 13.2, which models an interrupt-driven program. M has two modes: Inactive, in which the main program executes, and Active, in which the interrupt service routine (ISR) executes. The main program and ISR read and update a common variable *timerCount*.

   Answer the following questions:

(a) Specify the following property ϕ in linear temporal logic, choosing suitable atomic propositions:

ϕ: The main program eventually reaches program location C.

(b) Does M satisfy the above LTL property? Justify your answer by constructing the product FSM. If M does not satisfy the property, under what conditions would it do so? Assume that the environment of M can assert the interrupt at any time.

p.346

6. Express the postcondition of Example 13.3 as an LTL formula. State your assumptions clearly.

7. Consider the program fragment shown in Figure 11.6, which provides procedures for threads to communicate asynchronously by sending messages to one another. Please answer the following questions about this code. Assume the code is running on a single processor (not a multicore machine). You may also assume that only the code shown accesses the static variables that are shown.

   (a) Let s be an atomic proposition asserting that `send` releases the mutex (i.e. executes line 24). Let g be an atomic proposition asserting that `get` releases the mutex (i.e. executes line 38). Write an LTL formula asserting that g cannot occur before s in an execution of the program. Does this formula hold for the first execution of any program that uses these procedures?

   (b) Suppose that a program that uses the `send` and `get` procedures in Figure 11.6 is aborted at an arbitrary point in its execution and then restarted at the beginning. In the new execution, it is possible for a call to `get` to return before any call to `send` has been made. Describe how this could come about. What value will `get` return?

   (c) Suppose again that a program that uses the `send` and `get` procedures above is aborted at an arbitrary point in its execution and then restarted at the beginning. In the new execution, is it possible for deadlock to occur, where neither a call to `get` nor a call to `send` can return? If so, describe how this could come about and suggest a fix. If not, give an argument.

14

Equivalence and Refinement

Contents

This chapter discusses some fundamental ways to compare state machines and other modal models, such as trace equivalence, trace containment, simulation, and bisimulation. These mechanisms can be used to check conformance of a state machine against a specification.

14.1 Models as Specifications

The previous chapter provided techniques for unambiguously stating properties that a system must have to be functioning properly and safely. These properties were expressed using p.346 linear temporal logic, which can concisely describe requirements that the trace of a finite-state machine must satisfy. An alternative way to give requirements is to provide a model, p.343 a specification, that exhibits expected behavior of the system. Typically, the specification is quite abstract, and it may exhibit more behaviors than a useful implementation of the system would. But the key to being a useful specification is that it explicitly excludes undesired or dangerous behaviors.

Example 14.1: A simple specification for a traffic light might state: "The lights should always be lighted in the order green, yellow, red. It should never, for example, go directly from green to red, or from yellow to green." This requirement can be given as a temporal logic formula (as is done in Example 13.13) or as an abstract model (as is done in Figure 3.12).

The topic of this chapter is on the use of abstract models as specifications, and on how such models relate to an implementation of a system and to temporal logic formulas.

Example 14.2: We will show how to demonstrate that the traffic light model shown in Figure 3.10 is a valid implementation of the specification in Figure 3.12. Moreover, all traces of the model in Figure 3.10 satisfy the temporal logic formula in Example 13.13, but not all traces of the specification in Figure 3.12 do. Hence, these two specifications are not the same.

This chapter is about comparing models, and about being able to say with confidence that one model can be used in place of another. This enables an engineering design process where we start with abstract descriptions of desired and undesired behaviors, and successively refine our models until we have something that is detailed enough to provide a complete implementation. It also tells when it is safe to change an implementation, replacing it with another that might, for example, reduce the implementation cost.

14.2 Type Equivalence and Refinement

We begin with a simple relationship between two models that compares only the data types p.45 of their communication with their environment. Specifically, the goal is to ensure that a model B can be used in any environment where a model A can be used without causing any conflicts about data types. We will require that B can accept any inputs that A can accept

Abstraction and Refinement

This chapter focuses on relationships between models known as **abstraction** and **refinement**. These terms are symmetric in that the statement "model A is an abstraction of model B" means the same thing as "model B is a refinement of model A." As a general rule, the refinement model B has more detail than the abstraction A, and the abstraction is simpler, smaller, or easier to understand.

An abstraction is **sound** (with respect to some formal system of properties) if properties that are true of the abstraction are also true of the refinement. The formal system of properties could be, for example, a type system, linear temporal logic, or the languages p.346 of state machines. If the formal system is LTL, then if every LTL formula that holds for A also holds for B, then A is a sound abstraction of B. This is useful when it is easier to prove that a formula holds for A than to prove that it holds for B, for example because the state space of B may be much larger than the state space of A.

An abstraction is **complete** (with respect to some formal system of properties) if properties that are true of the refinement are also true of the abstraction. For example, if the formal system of properties is LTL, then A is a complete abstraction of B if every LTL formula that holds for B also holds for A. Useful abstractions are usually sound but not complete, because it is hard to make a complete abstraction that is significantly simpler or smaller.

Consider for example a program B in an imperative language such as C that has multi- p.212 ple threads. We might construct an abstraction A that ignores the values of variables and p.286 replaces all branches and control structures with nondeterministic choices. The abstrac- p.63 tion clearly has less information than the program, but it may be sufficient for proving some properties about the program, for example a mutual exclusion property. p.291

from the environment, and that any environment that can accept any output A can produce can also accept any output that B can produce.

p.26 To make the problem concrete, assume an actor model for A and B, as shown in Figure p.25 14.1. In that figure, A has three ports, two of which are input ports represented by the set $P_A = \{x, w\}$, and one of which is an output port represented by the set $Q_A = \{y\}$. These ports represent communication between A and its environment. The inputs have type V_x and p.44 V_w, which means that at a reaction of the actor, the values of the inputs will be members of the sets V_x or V_w.

If we want to replace A by B in some environment, the ports and their types impose four constraints:

1. The first constraint is that B does not require some input signal that the environment does not provide. If the input ports of B are given by the set P_B, then this is guaranteed by
$$P_B \subseteq P_A. \tag{14.1}$$

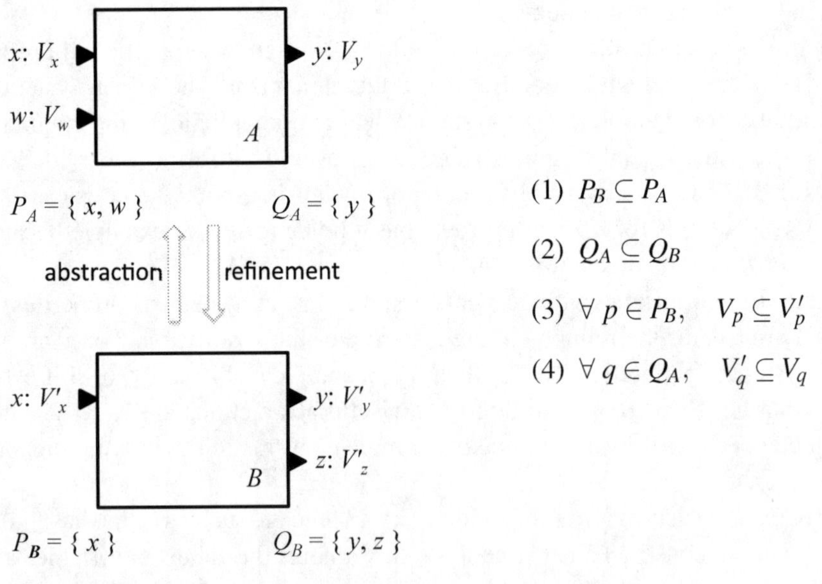

Figure 14.1: Summary of type refinement. If the four constraints on the right are satisfied, then B is a type refinement of A.

The ports of B are a subset of the ports of A. It is harmless for A to have more input ports than B, because if B replaces A in some environment, it can simply ignore any input signals that it does not need.

2. The second constraint is that B produces all the output signals that the environment may require. This is ensured by the constraint

$$Q_A \subseteq Q_B, \tag{14.2}$$

where Q_A is the set of output ports of A, and Q_B is the set of output ports of B. It is harmless for B to have additional output ports because an environment capable of working with A does not expect such outputs and hence can ignore them.

The remaining two constraints deal with the types of the ports. Let the type of an input port $p \in P_A$ be given by V_p. This means that an acceptable input value v on p satisfies $v \in V_p$. Let V_p' denote the type of an input port $p \in P_B$.

3. The third constraint is that if the environment provides a value $v \in V_p$ on an input port p that is acceptable to A, then if p is also an input port of B, then the value is also acceptable to B; i.e., $v \in V_p'$. This constraint can be written compactly as follows,

$$\forall\, p \in P_B, \quad V_p \subseteq V_p'. \tag{14.3}$$

Let the type of an output port $q \in Q_A$ be V_q, and the type of the corresponding output port $q \in Q_B$ be V_q'.

4. The fourth constraint is that if B produces a value $v \in V_q'$ on an output port q, then if q is also an output port of A, then the value must be acceptable to any environment in which A can operate. In other words,

$$\forall\, q \in Q_A, \quad V_q' \subseteq V_q. \tag{14.4}$$

The four constraints of equations (14.1) through (14.4) are summarized in Figure 14.1. When these four constraints are satisfied, we say that B is a **type refinement** of A. If B is a type refinement of A, then replacing A by B in any environment will not cause type system problems. It could, of course, cause other problems, since the behavior of B may not be acceptable to the environment, but that problem will be dealt with in subsequent sections.

If B is a type refinement of A, and A is a type refinement of B, then we say that A and B are **type equivalent**. They have the same input and output ports, and the types of the ports are the same.

Example 14.3: Let A represent the nondeterministic traffic light model in Figure 3.12 and B represent the more detailed deterministic model in Figure 3.10. The ports and their types are identical for both machines, so they are type equivalent. Hence, replacing A with B or vice versa in any environment will not cause type system problems.

Notice that since Figure 3.12 ignores the *pedestrian* input, it might seem reasonable to omit that port. Let A' represent a variant of Figure 3.12 without the *pedestrian* input. It is not safe to replace A' with B in all environments, because B requires an input *pedestrian* signal, but A' can be used in an environment that provides no such input.

14.3 Language Equivalence and Containment

To replace a machine A with a machine B, looking at the data types of the inputs and outputs alone is usually not enough. If A is a specification and B is an implementation, then normally A imposes more constraints than just data types. If B is an optimization of A (e.g., a lower cost implementation or a refinement that adds functionality or leverages new technology), then B normally needs to conform in some way with the functionality of A.

In this section, we consider a stronger form of equivalence and refinement. Specifically,
p.45 equivalence will mean that given a particular sequence of input valuations, the two machines produce the same output valuations.

Example 14.4: The garage counter of Figure 3.4, discussed in Example 3.4, is type
p.56 equivalent to the extended state machine version in Figure 3.8. The actor model is shown below:

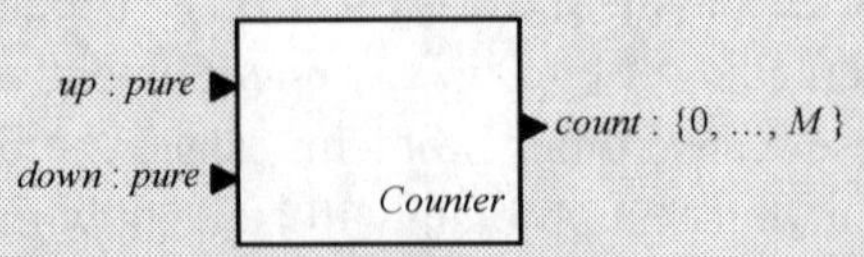

However, these two machines are equivalent in a much stronger sense than simply type equivalence. These two machines behave in exactly the same way, as viewed from the outside. Given the same input sequence, the two machines will produce the same output sequence.

Consider a port p of a state machine with type V_p. This port will have a sequence of values from the set $V_p \cup \{absent\}$, one value at each reaction. We can represent this sequence as a function of the form

$$s_p\colon \mathbb{N} \to V_p \cup \{absent\}.$$

This is the signal received on that port (if it is an input) or produced on that port (if it is an output). Recall that a behavior of a state machine is an assignment of such a signal to each port of such a machine. Recall further that the language $L(M)$ of a state machine M is the set of all behaviors for that state machine. Two machines are said to be **language equivalent** if they have the same language. p.66 p.67

Example 14.5: A behavior of the garage counter is a sequence of *present* and *absent* valuations for the two inputs, *up* and *down*, paired with the corresponding output sequence at the output port, *count*. A specific example is given in Example 3.16. This is a behavior of both Figures 3.4 and 3.8. All behaviors of Figure 3.4 are also behaviors of 3.8 and vice versa. These two machines are language equivalent.

In the case of a nondeterministic machine M, two distinct behaviors may share the same input signals. That is, given an input signal, there is more than one possible output sequence. The language $L(M)$ includes all possible behaviors. Just like deterministic machines, two nondeterministic machines are language equivalent if they have the same language. p.63

Suppose that for two state machines A and B, $L(A) \subset L(B)$. That is, B has behaviors that A does not have. This is called **language containment**. A is said to be a **language refinement** of B. Just as with type refinement, language refinement makes an assertion about the suitability of A as a replacement for B. If every behavior of B is acceptable to an environment, then every behavior of A will also be acceptable to that environment. A can substitute for B. p.363

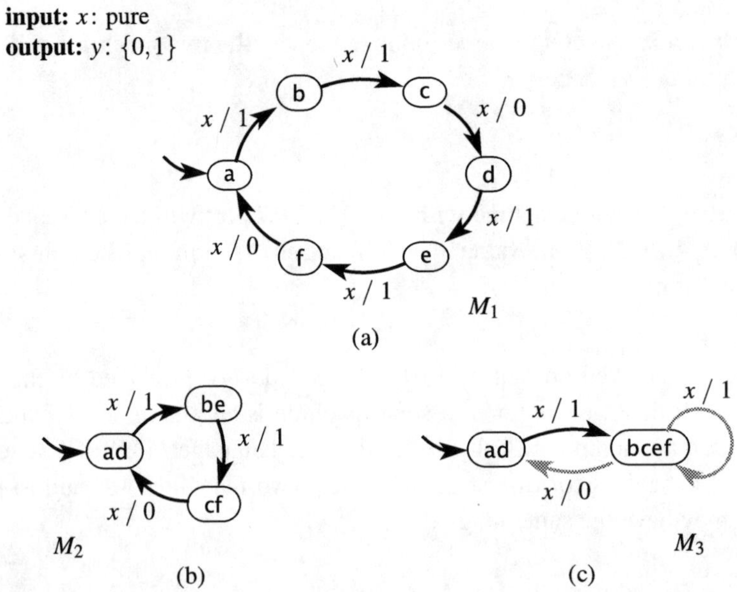

Figure 14.2: Three state machines where (a) and (b) have the same language, and that language is contained by that of (c).

Example 14.6: Machines M_1 and M_2 in Figure 14.2 are language equivalent. Both machines produce output $1,1,0,1,1,0,\cdots$, possibly interspersed with *absent* if the input is absent in some reactions.

Machine M_3, however, has more behaviors. It can produce any output sequence that M_1 and M_2 can produce, but it can also produce other outputs given the same inputs. Thus, M_1 and M_2 are both language refinements of M_3.

p.361 Language containment assures that an abstraction is sound with respect to LTL formulas about input and output sequences. That is, if A is a language refinement of B, then any LTL formula about inputs and outputs that holds for B also holds for A.

Example 14.7: Consider again the machines in Figure 14.2. M_3 might be a
p.343 specification. For example, if we require that any two output values 0 have at

Finite Sequences and Accepting States

A complete execution of the FSMs considered in this text is infinite. Suppose that we are interested in only the finite executions. To do this, we introduce the notion of an **accepting state**, indicated with a double outline as in state b in the example below:

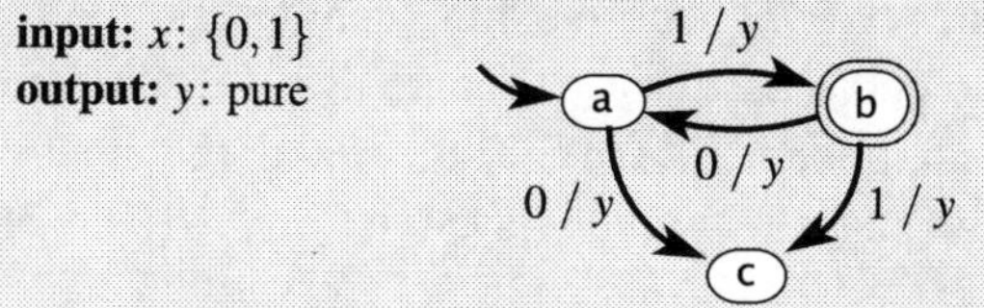

Let $L_a(M)$ denote the subset of the language $L(M)$ that results from executions that terminate in an accepting state. Equivalently, $L_a(M)$ includes only those behaviors in $L(M)$ with an infinite tail of stuttering reactions that remain in an accepting state. All such executions are effectively finite, since after a finite number of reactions, the inputs and outputs will henceforth be *absent*, or in LTL, $\mathbf{FG}\neg p$ for every port p. p.346

We call $L_a(M)$ the **language accepted by an FSM** M. A behavior in $L_a(M)$ specifies for each port p a finite **string**, or a finite sequence of values from the type V_p. For the above example, the input strings (1), $(1,0,1)$, $(1,0,1,0,1)$, etc., are all in $L_a(M)$. So are versions of these with an arbitrary finite number of *absent* values between any two present values. When there is no ambiguity, we can write these strings 1, 101, 10101, etc.

In the above example, in all behaviors in $L_a(M)$, the output is present a finite number of times, in the same reactions when the input is present.

The state machines in this text are receptive, meaning that at each reaction, each input port p can have any value in its type V_p or be *absent*. Hence, the language $L(M)$ of the machine above includes all possible sequences of input valuations. $L_a(M)$ excludes any of these that do not leave the machine in an accepting state. For example, any input sequence with two 1's in a row and the infinite sequence $(1,0,1,0,\cdots)$ are in $L(M)$ but not in $L_a(M)$. p.56

Note that it is sometimes useful to consider language containment when referring to the language *accepted* by the state machine, rather than the language that gives all behaviors of the state machine. p.365

Accepting states are also called **final states**, since for any behavior in $L_a(M)$, it is the last state of the machine. Accepting states are further explored in Exercise 2.

Regular Languages and Regular Expressions

A **language** is a set of sequences of values from some set called its **alphabet**. A
p.367 language accepted by an FSM is called a **regular language**. A classic example of a language that is not regular has sequences of the form $0^n 1^n$, a sequence of n zeros followed by n ones. It is easy to see that no *finite* state machine can accept this language because the machine would have to count the zeros to ensure that the number of ones matches. And the number of zeros is not bounded. On the other hand, the input sequences accepted by the FSM in the box on page 367, which have the form $10101\cdots01$, are regular.

A **regular expression** is a notation for describing regular languages. A central feature of regular expressions is the **Kleene star** (or **Kleene closure**), named after the American mathematician Stephen Kleene (who pronounced his name KLAY-nee). The notation $V*$, where V is a set, means the set of all finite sequences of elements from V. For example, if $V = \{0, 1\}$, then $V*$ is a set that includes the **empty sequence** (often written λ), and every finite sequence of zeros and ones.

The Kleene star may be applied to sets of sequences. For example, if $A = \{00, 11\}$, then $A*$ is the set of all finite sequences where zeros and ones always appear in pairs. In the notation of regular expressions, this is written `(00|11)*`, where the vertical bar means "or." What is inside the parentheses defines the set A.

Regular expressions are sequences of symbols from an alphabet and sets of sequences. Suppose our alphabet is $A = \{\texttt{a}, \texttt{b}, \cdots, \texttt{z}\}$, the set of lower-case characters. Then `grey` is a regular expression denoting a single sequence of four characters. The expression `grey|gray` denotes a set of two sequences. Parentheses can be used to group sequences or sets of sequences. For example, `(grey)|(gray)` and `gr(e|a)y` mean the same thing.

Regular expressions also provide convenience notations to make them more compact and readable. For example, the + operator means "one or more," in contrast to the Kleene star, which means "zero or more." For example, `a+` specifies the sequences `a`, `aa`, `aaa`, etc.; it is the same as `a(a*)`. The `?` operator species "zero or one." For example, `colou?r` specifies a set with two sequences, `color` and `colour`; it is the same as `colo(λ|u)r`, where λ denotes the empty sequence.

Regular expressions are commonly used in software systems for pattern matching. A typical implementation provides many more convenience notations than the ones illustrated here.

least one intervening 1, then M_3 is a suitable specification of this requirement. This requirement can be written as an LTL formula as follows:

$$\mathbf{G}((y=0) \Rightarrow \mathbf{X}((y \neq 0)\mathbf{U}(y=1))).$$

If we prove that this property holds for M_3, then we have implicitly proved that it also holds for M_1 and M_2.

We will see in the next section that language containment is *not sound* with respect to LTL formulas that refer to states of the state machines. In fact, language containment does not require the state machines to have the same states, so an LTL formula that refers to the states of one machine may not even apply to the other machine. A sound abstraction that references states will require simulation. p.372

Language containment is sometimes called **trace containment**, but here the term "trace" refers only to the observable trace, not to the execution trace. As we will see next, things get much more subtle when considering execution traces. p.67

Probing Further: Omega Regular Languages

The regular languages discussed in the boxes on pages 367 and 368 contain only finite sequences. But embedded systems most commonly have infinite executions. To extend the idea of regular languages to infinite runs, we can use a **Büchi automaton**, named after Julius Richard Büchi, a Swiss logician and mathematician. A Büchi automaton is a possibly nondeterministic FSM that has one or more accepting states. The language accepted by the FSM is defined to be the set of behaviors that visit one or more of the accepting states infinitely often; in other words, these behaviors satisfy the LTL formula $\mathbf{GF}(s_1 \vee \cdots \vee s_n)$, where $s_1, \cdots, s_n$ are the accepting states. Such a language is called an **omega-regular language** or **ω-regular language**, a generalization of regular languages. The reason for using ω in the name is because ω is used to construct infinite sequences, as explained in the box on page 477. p.367 p.354

As we will see in Chapter 15, many model checking questions can be expressed by giving a Büchi automaton and then checking to see whether the ω-regular language it defines contains any sequences. p.385

14.4 Simulation

p.64 Two nondeterministic FSMs may be language equivalent but still have observable differences in behavior in some environments. Language equivalence merely states that given p.45 the same sequences of input valuations, the two machines are *capable* of producing the same sequences of output valuations. However, as they execute, they make choices allowed by the nondeterminism. Without being able to see into the future, these choices could result in one of the machines getting into a state where it can no longer match the outputs of the other.

When faced with a nondeterministic choice, each machine is free to use any policy to make that choice. Assume that the machine cannot see into the future; that is, it cannot anticipate future inputs, and it cannot anticipate future choices that any other machine will make. For two machines to be equivalent, we will require that each machine be able to make choices that allow it to match the reaction of the other machine (producing the same outputs), and further allow it to continue to do such matching in the future. It turns out that language equivalence is not strong enough to ensure that this is possible.

Example 14.8: Consider the two state machines in Figure 14.3. Suppose that M_2 is acceptable in some environment (every behavior it can exhibit in that environment is consistent with some specification or design intent). Is it safe for M_1 to replace M_2? The two machines are language equivalent. In all behaviors, the output is one of two finite strings, 01 or 00, for both machines. So it would seem that M_1 can replace M_2. But this is not necessarily the case.

Suppose we compose each of the two machines with its own copy of the environment that finds M_2 acceptable. In the first reaction where x is *present*, M_1 has no choice but to take the transition to state b and produce the output $y = 0$. However, M_2 must choose between f and h. Whichever choice it makes, M_2 matches the output $y = 0$ of M_1 but enters a state where it is no longer able to always match the outputs of M_1. If M_1 can observe the state of M_2 when making its choice, then in the second reaction where x is *present*, it can choose a transition that M_2 can *never* match. Such a policy for M_1 ensures that the behavior of M_1, given the same inputs, is never the same as the behavior of M_2. Hence, it is not safe to replace M_2 with M_1.

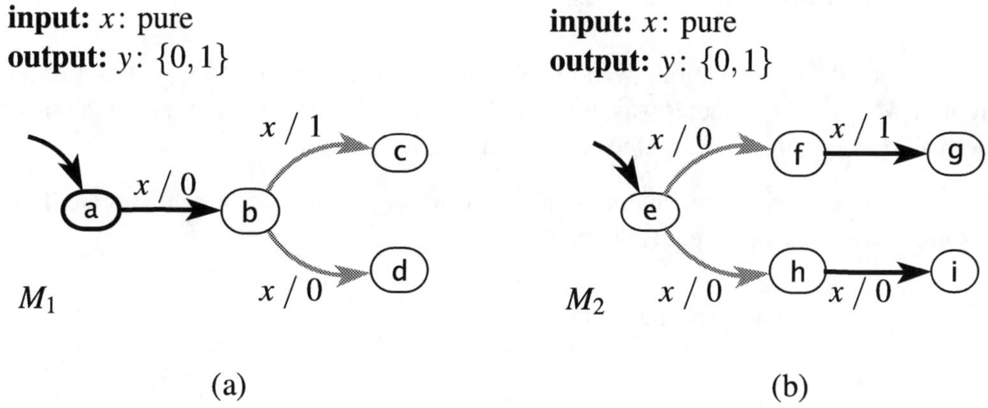

Figure 14.3: Two state machines that are language equivalent but where M_2 does not simulate M_1 (M_1 does simulate M_2).

On the other hand, if M_1 is acceptable in some environment, is it safe for M_2 to replace M_1? What it means for M_1 to be acceptable in the environment is that whatever decisions it makes are acceptable. Thus, in the second reaction where x is *present*, both outputs $y = 1$ and $y = 0$ are acceptable. In this second reaction, M_2 has no choice but to produce one or the other these outputs, and it will inevitably transition to a state where it continues to match the outputs of M_1 (henceforth forever *absent*). Hence it is safe for M_2 to replace M_1.

In the above example, we can think of the machines as maliciously trying to make M_1 look different from M_2. Since they are free to use any policy to make choices, they are free to use policies that are contrary to our goal to replace M_2 with M_1. Note that the machines do not need to know the future; it is sufficient to simply have good visibility of the present. The question that we address in this section is: under what circumstances can we assure that there is no policy for making nondeterministic choices that can make machine M_1 observably different from M_2? The answer is a stronger form of equivalence called bisimulation and a refinement relation called simulation. We begin with the simulation relation.

14.4.1 Simulation Relations

First, notice that the situation given in Example 14.8 is not symmetric. It is safe for M_2 to p.361 replace M_1, but not the other way around. Hence, M_2 is a refinement of M_1, in a sense that we will now establish. M_1, on the other hand, is not a refinement of M_2.

The particular kind of refinement we now consider is a **simulation refinement**. The following statements are all equivalent:

- M_2 is a simulation refinement of M_1.
- M_1 simulates M_2.
- M_1 is a simulation abstraction of M_2.

Simulation is defined by a **matching game**. To determine whether M_1 simulates M_2, we play a game where M_2 gets to move first in each round. The game starts with both machines in their initial states. M_2 moves first by reacting to an input valuation. If this involves a nondeterministic choice, then it is allowed to make any choice. Whatever it choses, an output valuation results and M_2's turn is over.

It is now M_1's turn to move. It must react to the same input valuation that M_2 reacted to. If this involves a nondeterministic choice, then it must make a choice that matches the output valuation of M_2. If there are multiple such choices, it must select one without knowledge of the future inputs or future moves of M_2. Its strategy should be to choose one that enables it to continue to match M_2, regardless of what future inputs arrive or future decisions M_2 makes.

Machine M_1 "wins" this matching game (M_1 simulates M_2) if it can always match the output symbol of machine M_2 for all possible input sequences. If in any reaction M_2 can produce an output symbol that M_1 cannot match, then M_1 does not simulate M_2.

Example 14.9: In Figure 14.3, M_1 simulates M_2 but not vice versa. To see this, first play the game with M_2 moving first in each round. M_1 will always be able to match M_2. Then play the game with M_1 moving first in each round. M_2 will not always be able to match M_1. This is true even though the two machines are language equivalent.

Interestingly, if M_1 simulates M_2, it is possible to compactly record all possible games over all possible inputs. Let S_1 be the states of M_1 and S_2 be the states of M_2. Then a simulation relation $S \subseteq S_2 \times S_1$ is a set of pairs of states occupied by the two machines in each round of the game for all possible inputs. This set summarizes all possible plays of the game.

Example 14.10: In Figure 14.3,

$$S_1 = \{\mathsf{a}, \mathsf{b}, \mathsf{c}, \mathsf{d}\}$$

and

$$S_2 = \{\mathsf{e}, \mathsf{f}, \mathsf{g}, \mathsf{h}, \mathsf{i}\}.$$

The simulation relation showing that M_1 simulates M_2 is

$$S = \{(\mathsf{e}, \mathsf{a}), (\mathsf{f}, \mathsf{b}), (\mathsf{h}, \mathsf{b}), (\mathsf{g}, \mathsf{c}), (\mathsf{i}, \mathsf{d})\}$$

First notice that the pair (e, a) of initial states is in the relation, so the relation includes the state of the two machines in the first round. In the second round, M_2 may be in either f or h, and M_1 will be in b. These two possibilities are also accounted for. In the third round and beyond, M_2 will be in either g or i, and M_1 will be in c or d.

There is no simulation relation showing that M_2 simulates M_1, because it does not.

A simulation relation is complete if it includes all possible plays of the game. It must therefore account for all reachable states of M_2, the machine that moves first, because M_2's moves are unconstrained. Since M_1's moves are constrained by the need to match M_2, it is not necessary to account for all of its reachable states. p.61

14.4.2 Formal Model

Using the formal model of nondeterministic FSMs given in Section 3.5.1, we can formally define a simulation relation. Let p.64

$$M_1 = (States_1, Inputs, Outputs, possibleUpdates_1, initialState_1),$$

and

$$M_2 = (States_2, Inputs, Outputs, possibleUpdates_2, initialState_2).$$

p.363 Assume the two machines are type equivalent. If either machine is deterministic, then its *possibleUpdates* function always returns a set with only one element in it. If M_1 simulates M_2, the simulation relation is given as a subset of $States_2 \times States_1$. Note the ordering here; the machine that moves first in the game, M_2, the one being simulated, is first in $States_2 \times States_1$.

To consider the reverse scenario, if M_2 simulates M_1, then the relation is given as a subset of $States_1 \times States_2$. In this version of the game M_1 must move first.

We can state the "winning" strategy mathematically. We say that M_1 **simulates** M_2 if there is a subset $S \subseteq States_2 \times States_1$ such that

1. $(initialState_2, initialState_1) \in S$, and
2. If $(s_2, s_1) \in S$, then $\forall\, x \in Inputs$, and $\forall\, (s_2', y_2) \in possibleUpdates_2(s_2, x)$, there is a $(s_1', y_1) \in possibleUpdates_1(s_1, x)$ such that:
   (a) $(s_2', s_1') \in S$, and
   (b) $y_2 = y_1$.

This set S, if it exists, is called the **simulation relation**. It establishes a correspondence between states in the two machines. If it does not exist, then M_1 does not simulate M_2.

14.4.3 Transitivity

Simulation is **transitive**, meaning that if M_1 simulates M_2 and M_2 simulates M_3, then M_1 simulates M_3. In particular, if we are given simulation relations $S_{2,1} \subseteq States_2 \times States_1$ (M_1 simulates M_2) and $S_{3,2} \subseteq States_3 \times States_2$ (M_2 simulates M_3), then

$$\begin{aligned} S_{3,1} = \;& \{(s_3, s_1) \in States_3 \times States_1 \mid \text{there exists } s_2 \in States_2 \text{ where} \\ & (s_3, s_2) \in S_{3,2} \text{ and } (s_2, s_1) \in S_{2,1}\} \end{aligned}$$

Example 14.11: For the machines in Figure 14.2, it is easy to show that (c) simulates (b) and that (b) simulates (a). Specifically, the simulation relations are

$$S_{a,b} = \{(\mathsf{a}, \mathsf{ad}), (\mathsf{b}, \mathsf{be}), (\mathsf{c}, \mathsf{cf}), (\mathsf{d}, \mathsf{ad}), (\mathsf{e}, \mathsf{be}), (\mathsf{f}, \mathsf{cf})\}.$$

and

$$S_{b,c} = \{(\mathsf{ad}, \mathsf{ad}), (\mathsf{be}, \mathsf{bcef}), (\mathsf{cf}, \mathsf{bcef})\}.$$

By transitivity, we can conclude that (c) simulates (a), and that the simulation relation is

$$S_{a,c} = \{(\mathsf{a}, \mathsf{ad}), (\mathsf{b}, \mathsf{bcef}), (\mathsf{c}, \mathsf{bcef}), (\mathsf{d}, \mathsf{ad}), (\mathsf{e}, \mathsf{bcef}), (\mathsf{f}, \mathsf{bcef})\},$$

which further supports the suggestive choices of state names.

14.4.4 Non-Uniqueness of Simulation Relations

When a machine M_1 simulates another machine M_2, there may be more than one simulation relation.

Example 14.12: In Figure 14.4, it is easy to check that M_1 simulates M_2. Note that M_1 is nondeterministic, and in two of its states it has two distinct ways of matching the moves of M_2. It can arbitrarily choose from among these possibilities to match the moves. If from state b it always chooses to return to state a, then the simulation relation is

$$S_{2,1} = \{(\mathsf{ac}, \mathsf{a}), (\mathsf{bd}, \mathsf{b})\}.$$

Otherwise, if from state c it always chooses to return to state b, then the simulation relation is

$$S_{2,1} = \{(\mathsf{ac}, \mathsf{a}), (\mathsf{bd}, \mathsf{b}), (\mathsf{ac}, \mathsf{c})\}.$$

Otherwise, the simulation relation is

$$S_{2,1} = \{(\mathsf{ac}, \mathsf{a}), (\mathsf{bd}, \mathsf{b}), (\mathsf{ac}, \mathsf{c}), (\mathsf{bd}, \mathsf{d})\}.$$

All three are valid simulation relations, so the simulation relation is not unique.

input: x: pure
output: y: $\{0,1\}$

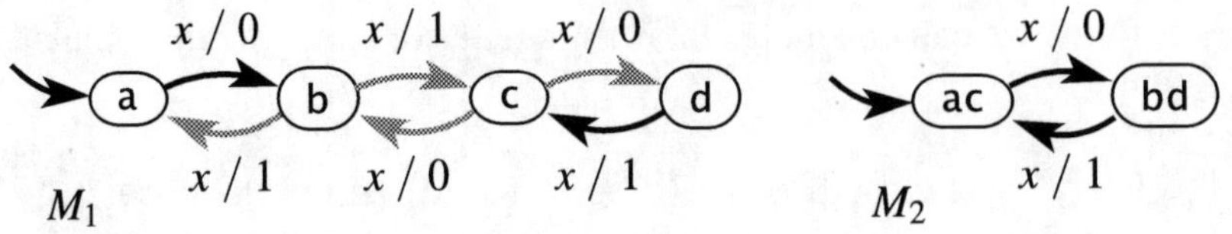

(a) (b)

Figure 14.4: Two state machines that simulate each other, where there is more than one simulation relation.

14.4.5 Simulation vs. Language Containment

As with all abstraction-refinement relations, simulation is typically used to relate a simpler specification M_1 to a more complicated realization M_2. When M_1 simulates M_2, then the language of M_1 contains the language of M_2, but the guarantee is stronger than language containment. This fact is summarized in the following theorem.

Theorem 14.1. *Let M_1 simulate M_2. Then*

$$L(M_2) \subseteq L(M_1).$$

p.66 **Proof.** This theorem is easy to prove. Consider a behavior $(x,y) \in L(M_2)$. We need to show that $(x,y) \in L(M_1)$.

p.67 Let the simulation relation be S. Find all possible execution traces for M_2

$$((x_0,s_0,y_0),(x_1,s_1,y_1),(x_2,s_2,y_2),\cdots),$$

that result in behavior (x,y). (If M_2 is deterministic, then there will be only one execution trace.) The simulation relation assures us that we can find an execution trace for M_1

$$((x_0,s'_0,y_0),(x_1,s'_1,y_1),(x_2,s'_2,y_2),\cdots),$$

where $(s_i,s'_i) \in S$, such that given input valuation x_i, M_1 produces y_i. Thus, $(x,y) \in L(M_1)$.

□

One use of this theorem is to show that M_1 does not simulate M_2 by showing that M_2 has behaviors that M_1 does not have.

Example 14.13: For the examples in Figure 14.2, M_2 does not simulate M_3. To see this, just note that the language of M_2 is a strict subset of the language of M_3,

$$L(M_2) \subset L(M_3),$$

meaning that M_3 has behaviors that M_2 does not have.

It is important to understand what the theorem says, and what it does not say. It does not say, for example, that if $L(M_2) \subseteq L(M_1)$ then M_1 simulates M_2. In fact, this statement is not true, as we have already shown with the examples in Figure 14.3. These two machines have the same language. The two machines are observably different despite the fact that their input/output behaviors are the same.

Of course, if M_1 and M_2 are deterministic and M_1 simulates M_2, then their languages are identical and M_2 simulates M_1. Thus, the simulation relation differs from language containment only for nondeterministic FSMs.

14.5 Bisimulation

It is possible to have two machines M_1 and M_2 where M_1 simulates M_2 and M_2 simulates M_1, and yet the machines are observably different. Note that by the theorem in the previous section, the languages of these two machines must be identical.

Example 14.14: Consider the two machines in Figure 14.5. These two machines simulate each other, with simulation relations as follows:

$$S_{2,1} = \{(\mathsf{e},\mathsf{a}),(\mathsf{f},\mathsf{b}),(\mathsf{h},\mathsf{b}),(\mathsf{j},\mathsf{b}),(\mathsf{g},\mathsf{c}),(\mathsf{i},\mathsf{d}),(\mathsf{k},\mathsf{c}),(\mathsf{m},\mathsf{d})\}$$

(M_1 simulates M_2), and

$$S_{1,2} = \{(\mathsf{a},\mathsf{e}),(\mathsf{b},\mathsf{j}),(\mathsf{c},\mathsf{k}),(\mathsf{d},\mathsf{m})\}$$

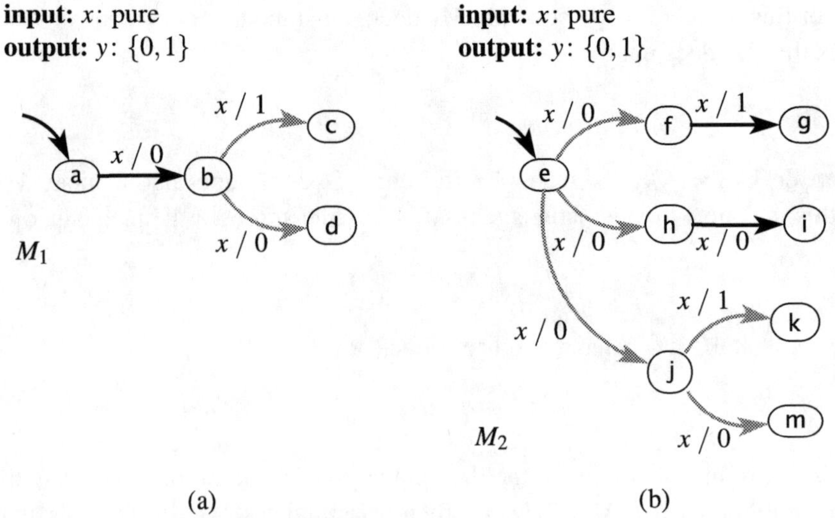

Figure 14.5: An example of two machines where M_1 simulates M_2, and M_2 simulates M_1, but they are not bisimilar.

(M_2 simulates M_1). However, there is a situation in which the two machines will be observably different. In particular, suppose that the policies for making the nondeterministic choices for the two machines work as follows. In each reaction, they flip a coin to see which machine gets to move first. Given an input valuation, that machine makes a choice of move. The machine that moves second must be able to match all of its possible choices. In this case, the machines can end up in a state where one machine can no longer match all the possible moves of the other.

Specifically, suppose that in the first move M_2 gets to move first. It has three possible moves, and M_1 will have to match all three. Suppose it chooses to move to f or h. In the next round, if M_1 gets to move first, then M_2 can no longer match all of its possible moves.

Notice that this argument does not undermine the observation that these machines simulate each other. If in each round, M_2 always moves first, then M_1 will always be able to match its every move. Similarly, if in each round M_1 moves first, then M_2 can always match its every move (by always choosing to move to j in the first round). The observable difference arises from the ability to alternate which machines moves first.

To ensure that two machines are observably identical in all environments, we need a stronger equivalence relation called **bisimulation**. We say that M_1 is **bisimilar** to M_2 (or M_1 **bisimulates** M_2) if we can play the matching game modified so that in each round either machine can move first. p.372

As in Section 14.4.2, we can use the formal model of nondeterministic FSMs to define a bisimulation relation. Let p.64

$$\begin{aligned} M_1 &= (States_1, Inputs, Outputs, possibleUpdates_1, initialState_1), \text{ and} \\ M_2 &= (States_2, Inputs, Outputs, possibleUpdates_2, initialState_2). \end{aligned}$$

Assume the two machines are type equivalent. If either machine is deterministic, then its *possibleUpdates* function always returns a set with only one element in it. If M_1 bisimulates M_2, the simulation relation is given as a subset of $States_2 \times States_1$. The ordering here is not important because if M_1 bisimulates M_2, then M_2 bisimulates M_1. p.363

We say that M_1 **bisimulates** M_2 if there is a subset $S \subseteq States_2 \times States_1$ such that

1. $(initialState_2, initialState_1) \in S$, and
2. If $(s_2, s_1) \in S$, then $\forall\, x \in Inputs$, and
   $\forall\, (s_2', y_2) \in possibleUpdates_2(s_2, x)$,
   there is a $(s_1', y_1) \in possibleUpdates_1(s_1, x)$ such that:
   (a) $(s_2', s_1') \in S$, and
   (b) $y_2 = y_1$, and
3. If $(s_2, s_1) \in S$, then $\forall\, x \in Inputs$, and
   $\forall\, (s_1', y_1) \in possibleUpdates_1(s_1, x)$,
   there is a $(s_2', y_2) \in possibleUpdates_2(s_2, x)$ such that:
   (a) $(s_2', s_1') \in S$, and
   (b) $y_2 = y_1$.

This set S, if it exists, is called the **bisimulation relation**. It establishes a correspondence between states in the two machines. If it does not exist, then M_1 does not bisimulate M_2.

14.6 Summary

In this chapter, we have considered three increasingly strong abstraction-refinement relations for FSMs. These relations enable designers to determine when one design can safely replace another, or when one design correctly implements a specification. The first relation is type refinement, which considers only the existence of input and output ports and their data types. The second relation is language refinement, which considers the sequences of valuations of inputs and outputs. The third relation is simulation, which considers the state trajectories of the machines. In all three cases, we have provided both a refinement relation and an equivalence relation. The strongest equivalence relation is bisimulation, which ensures that two nondeterministic FSMs are indistinguishable from each other.

Exercises

1. In Figure 14.6 are four pairs of actors. For each pair, determine whether
   - A and B are type equivalent,
   - A is a type refinement of B,
   - B is a type refinement of A, or
   - none of the above.

2. In the box on page 367, a state machine M is given that accepts finite inputs x of the form (1), $(1,0,1)$, $(1,0,1,0,1)$, etc.

   p.368 (a) Write a regular expression that describes these inputs. You may ignore stuttering reactions.

   (b) Describe the output sequences in $L_a(M)$ in words, and give a regular expression for those output sequences. You may again ignore stuttering reactions.

   (c) Create a state machine that accepts *output* sequences of the form (1), $(1,0,1)$, $(1,0,1,0,1)$, etc. (see box on page 367). Assume the input x is pure and that p.56 whenever the input is present, a present output is produced. Give a deterministic solution if there is one, or explain why there is no deterministic solution. What *input* sequences does your machine accept?

3. The state machine in Figure 14.7 has the property that it outputs at least one 1 between any two 0's. Construct a two-state nondeterministic state machine that simulates this one and preserves that property. Give the simulation relation. Are the machines bisimilar?

4. Consider the FSM in Figure 14.8, which recognizes an input code. The state machine in Figure 14.9 also recognizes the same code, but has more states than the one in Figure 14.8. Show that it is equivalent by giving a bisimulation relation with the machine in Figure 14.8.

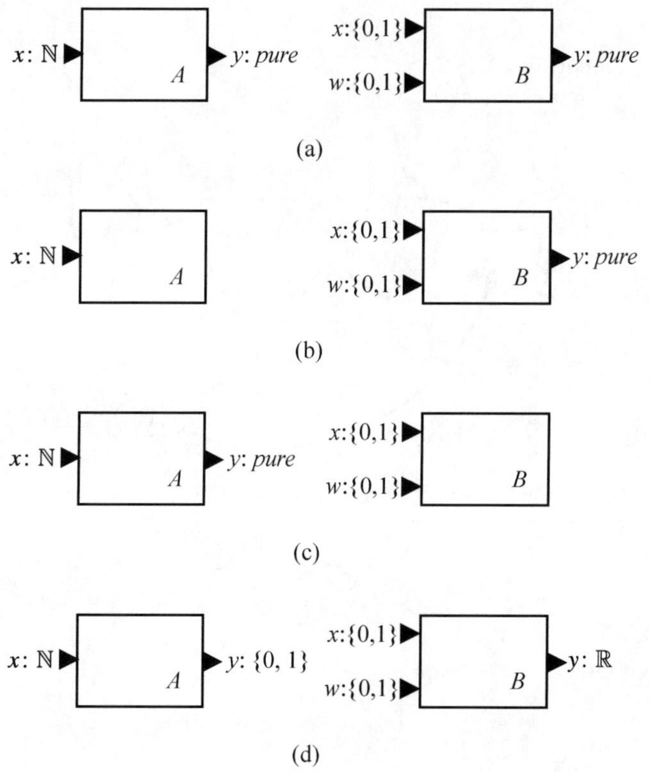

Figure 14.6: Four pairs of actors whose type refinement relationships are explored in Exercise 1.

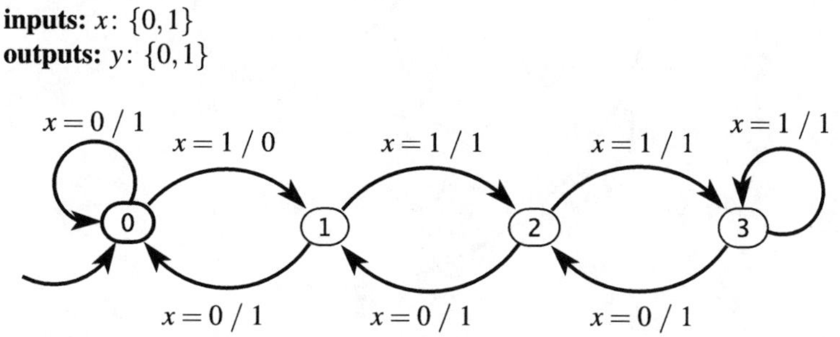

Figure 14.7: Machine that outputs at least one 1 between any two 0's.

5. Consider the state machine in Figure 14.10. Find a bisimilar state machine with only two states, and give the bisimulation relation.

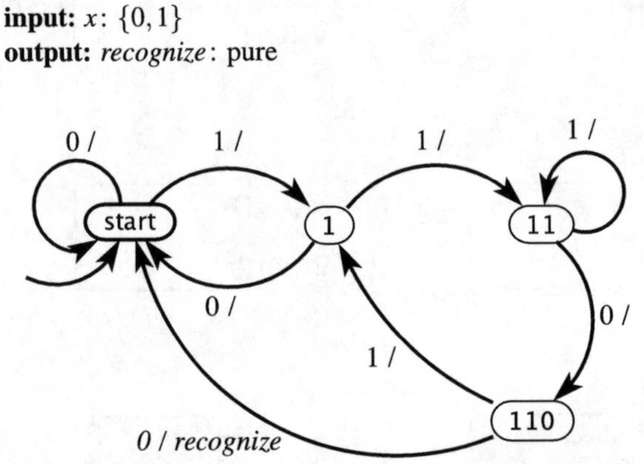

Figure 14.8: A machine that implements a code recognizer. It outputs *recognize* at the end of every input subsequence 1100; otherwise it outputs *absent*.

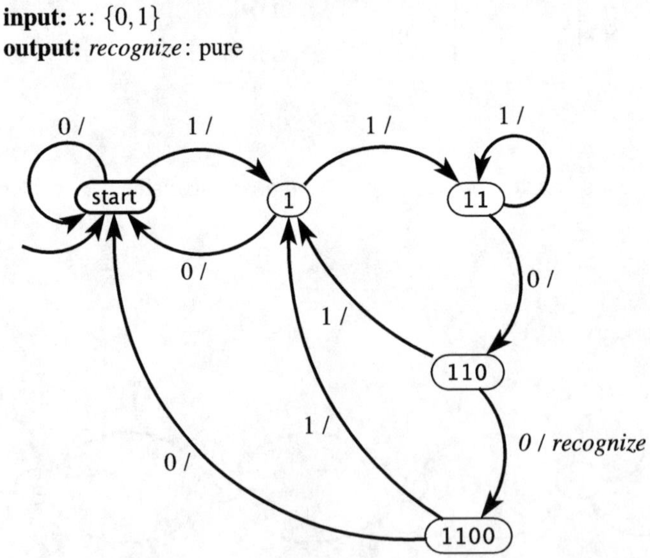

Figure 14.9: A machine that implements a recognizer for the same code as in Figure 14.8, but has more states.

6. You are told that state machine A has one input x, and one output y, both with type $\{1,2\}$, and that it has states $\{a,b,c,d\}$. You are told nothing further. Do you have enough information to construct a state machine B that simulates A? If so, give such a state machine, and the simulation relation.

7. Consider a state machine with a pure input x, and output y of type $\{0,1\}$. Assume the states are

$$States = \{a,b,c,d,e,f\},$$

and the initial state is a. The *update* function is given by the following table (ignoring stuttering):

$(currentState, input)$	$(nextState, output)$
(a,x)	$(b,1)$
(b,x)	$(c,0)$
(c,x)	$(d,0)$
(d,x)	$(e,1)$
(e,x)	$(f,0)$
(f,x)	$(a,0)$

(a) Draw the state transition diagram for this machine.

(b) Ignoring stuttering, give all possible behaviors for this machine. p.66

(c) Find a state machine with three states that is bisimilar to this one. Draw that state machine, and give the bisimulation relation.

8. For each of the following questions, give a short answer and justification.

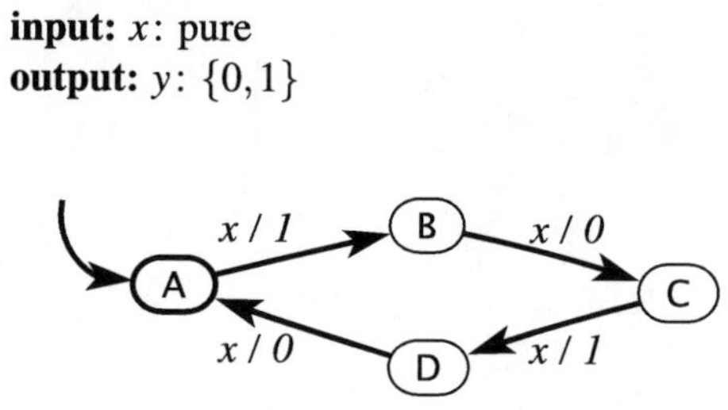

Figure 14.10: A machine that has more states than it needs.

p.45

(a) TRUE or FALSE: Consider a state machine A that has one input x, and one output y, both with type $\{1,2\}$ and a single state s, with two self loops labeled $true/1$ and $true/2$. Then for any state machine B which has exactly the same inputs and outputs (along with types), A simulates B.

p.346

(b) TRUE or FALSE: Suppose that f is an arbitrary LTL formula that holds for state machine A, and that A simulates another state machine B. Then we can safely assert that f holds for B.

(c) TRUE or FALSE: Suppose that A are B are two type-equivalent state machines, and that f is an LTL formula where the atomic propositions refer only to the inputs and outputs of A and B, not to their states. If the LTL formula f holds for state machine A, and A simulates state machine B, then f holds for B.

15

Reachability Analysis and Model Checking

Contents

Chapters 13 and 14 have introduced techniques for formally specifying properties and models of systems, and for comparing such models. In this chapter, we will study algorithmic techniques for **formal verification** — the problem of checking whether a system satisfies its formal specification in its specified operating environment. In particular, we study a technique called **model checking**. Model checking is an algorithmic method for determining whether a system satisfies a formal specification expressed as a temporal logic formula.

It was introduced by Clarke and Emerson (1981) and Queille and Sifakis (1981), which earned the creators the 2007 ACM Turing Award, the highest honor in Computer Science.

p.61 Central to model checking is the notion of the set of reachable states of a system. **Reachability analysis** is the process of computing the set of reachable states of a system. This chapter presents basic algorithms and ideas in reachability analysis and model checking. These algorithms are illustrated using examples drawn from embedded systems design, including verification of high-level models, sequential and concurrent software, as well as control and robot path planning. Model checking is a large and active area of research, and a detailed treatment of the subject is out of the scope of this chapter; we refer the interested reader to Clarke et al. (1999) and Holzmann (2004) for an in-depth introduction to this field.

15.1 Open and Closed Systems

A **closed system** is one with no inputs. An **open system**, in contrast, is one that maintains an ongoing interaction with its environment by receiving inputs and (possibly) generating output to the environment. Figure 15.1 illustrates these concepts.

Techniques for formal verification are typically applied to a model of the closed system M obtained by composing the model of the system S that is to be verified with a model of its environment E. S and E are typically open systems, where all inputs to S are generated by E and vice-versa. Thus, as shown in Figure 15.2, there are three inputs to the verification process:

- A model of the system to be verified, S;
- A model of the environment, E, and
- The property to be verified Φ.

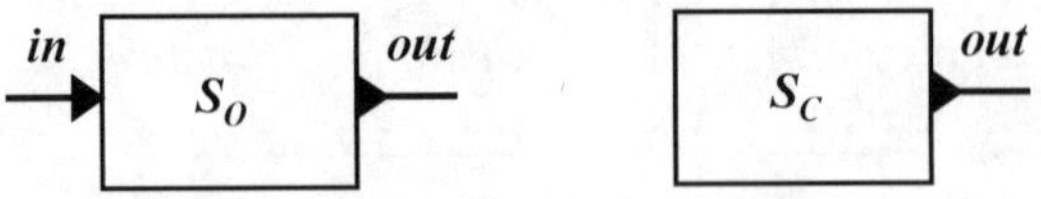

(a) Open system (b) Closed system

Figure 15.1: Open and closed systems.

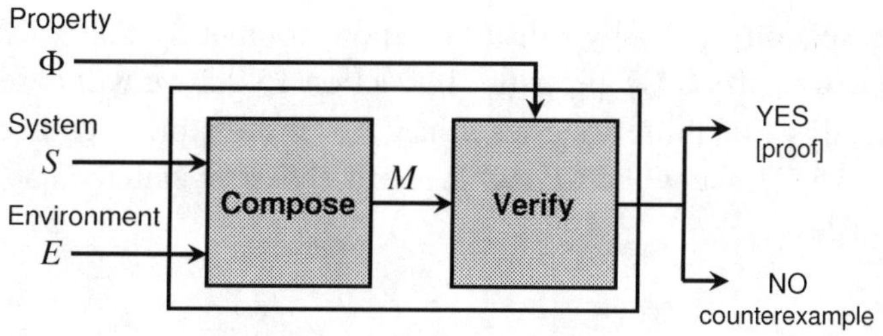

Figure 15.2: Formal verification procedure.

The verifier generates as output a YES/NO answer, indicating whether or not S satisfies the property Φ in environment E. Typically, a NO output is accompanied by a counterexample, p.349 also called an **error trace**, which is a trace of the system that indicates how Φ is violated. p.67 Counterexamples are very useful aids in the debugging process. Some formal verification tools also include a proof or certificate of correctness with a YES answer; such an output can be useful for **certification** of system correctness.

The form of composition used to combine system model S with environment model E depends on the form of the interaction between system and environment. Chapters 5 and 6 describe several ways to compose state machine models. All of these forms of composition can be used in generating a verification model M from S and E. Note that M can be nondeterministic.

For simplicity, in this chapter we will assume that system composition has already been performed using one of the techniques presented in Chapters 5 and 6. All algorithms discussed in the following sections will operate on the combined verification model M, and will be concerned with answering the question of whether M satisfies property Φ. Additionally, we will assume that Φ is specified as a property in linear temporal logic. p.346

15.2 Reachability Analysis

We consider first a special case of the model checking problem which is useful in practice. Specifically, we assume that M is a finite-state machine and Φ is an LTL formula of the p.47 form $\mathbf{G}p$, where p is a proposition. Recall from Chapter 13 that $\mathbf{G}p$ is the temporal logic p.348 formula that holds in a trace when the proposition p holds in every state of that trace. As we have seen in Chapter 13, several system properties are expressible as $\mathbf{G}p$ properties.

We will begin in Section 15.2.1 by illustrating how computing the reachable states of a system enables one to verify a **G***p* property. In Section 15.2.2 we will describe a technique for reachability analysis of finite-state machines based on explicit enumeration of states. Finally, in Section 15.2.3, we will describe an alternative approach to analyze systems with very large state spaces.

15.2.1 Verifying Gp

In order for a system M to satisfy **G**p, where p is a proposition, every trace exhibitable by M must satisfy **G**p. This property can be verified by enumerating all states of M and checking that every state satisfies p.

When M is finite-state, in theory, such enumeration is always possible. As shown in Chapter 3, the state space of M can be viewed as a directed graph where the nodes of the graph correspond to states of M and the edges correspond to transitions of M. This graph is called p.47 the **state graph** of M, and the set of all states is called its state space. With this graph-theoretic viewpoint, one can see that checking **G**p for a finite-state system M corresponds to traversing the state graph for M, starting from the initial state and checking that every state reached in this traversal satisfies p. Since M has a finite number of states, this traversal must terminate.

Example 15.1: Let the system S be the traffic light controller of Figure 3.10 and its environment E be the pedestrian model shown in Figure 3.11. Let M be the synchronous composition of S and E as shown in Figure 15.3. Observe that M is a closed system. Suppose that we wish to verify that M satisfies the property

$$\mathbf{G}\,\neg(\mathsf{green} \wedge \mathsf{crossing})$$

In other words, we want to verify that it is never the case that the traffic light is green while pedestrians are crossing.

The composed system M is shown in Figure 15.4 as an extended FSM. Note that M has no inputs or outputs. M is finite-state, with a total of 188 states (using a similar calculation to that in Example 3.12). The graph in Figure 15.4 is not the full state graph of M, because each node represents a set of states, one for each different value of *count* in that node. However, through visual inspection of this graph we can check for ourselves that no state satisfies the proposition $(\mathsf{green} \wedge \mathsf{crossing})$, and hence every trace satisfies the LTL property $\mathbf{G}\,\neg(\mathsf{green} \wedge \mathsf{crossing})$.

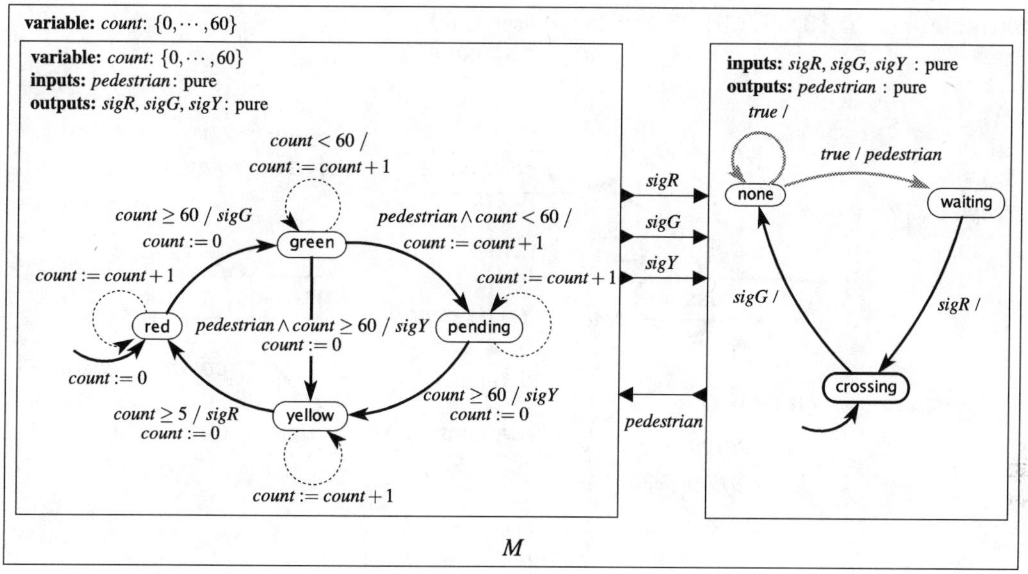

Figure 15.3: Composition of traffic light controller (Figure 3.10) and pedestrian model (Figure 3.11).

In practice, the seemingly simple task of verifying whether a finite-state system M satisfies a $\mathbf{G}p$ property is not as straightforward as in the previous example for the following reasons:

- Typically, one starts only with the initial state and transition function, and the state graph must be constructed on the fly. p.55
- The system might have a huge number of states, possibly exponential in the size of the syntactic description of M. As a consequence, the state graph cannot be represented using traditional data structures such as an adjacency or incidence matrix.

The next two sections describe how these challenges can be handled.

15.2.2 Explicit-State Model Checking

In this section, we discuss how to compute the reachable state set by generating and traversing the state graph on the fly.

First, recall that the system of interest M is closed, finite-state, and can be nondeterministic. Since M has no inputs, its set of possible next states is a function of its current state alone.

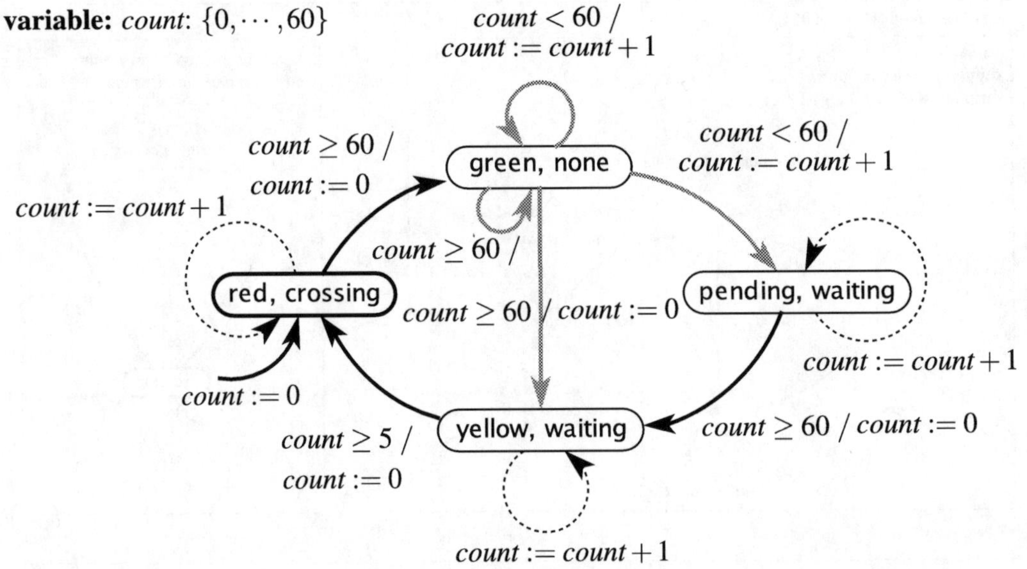

Figure 15.4: Extended state machine obtained from synchronous-reactive composition of traffic light controller and pedestrian models. Note that this is nondeterministic.

p.64 We denote this transition relation of M by δ, which is only a function of the current state of M, in contrast to the *possibleUpdates* function introduced in Chapter 3 which is also a function of the current input. Thus, $\delta(s)$ is the set of possible next states from state s of M.

Algorithm 15.1 computes the set of reachable states of M, given its initial state s_0 and transition relation δ. Procedure **DFS_Search** performs a depth-first traversal of the state graph of M, starting with state s_0. The graph is generated on-the-fly by repeatedly applying δ to states visited during the traversal.

p.241 The main data structures required by the algorithm are Σ, the stack storing the current path in the state graph being explored from s_0, and R, the current set of states reached during traversal. Since M is finite-state, at some point all states reachable from s_0 will be in R, which implies that no new states will be pushed onto Σ and thus Σ will become empty. Hence, procedure **DFS_Search** terminates and the value of R at the end of the procedure is the set of all reachable states of M.

The space and time requirements for this algorithm are linear in the size of the state graph (see Appendix B for an introduction to such complexity notions). However, the number of nodes and edges in the state graph of M can be exponential in the size of the descriptions

Input : Initial state s_0 and transition relation δ for closed finite-state system M
Output: Set R of reachable states of M

```
Initialize: Stack Σ to contain a single state s0; Current set of reached
states R := {s0}.
DFS_Search() {
while Stack Σ is not empty do
    Pop the state s at the top of Σ
    Compute δ(s), the set of all states reachable from s in one transition
    for each s' ∈ δ(s) do
        if s' ∉ R then
            R := R ∪ {s'}
            Push s' onto Σ
        end
    end
end
}
```

Algorithm 15.1: Computing the reachable state set by depth-first explicit-state search.

of S and E. For example, if S and E together have 100 Boolean state variables (a small number in practice!), the state graph of M can have a total of 2^{100} states, far more than what contemporary computers can store in main memory. Therefore, explicit-state search algorithms such as **DFS_Search** must be augmented with **state compression** techniques. Some of these techniques are reviewed in the sidebar on page 404.

A challenge for model checking concurrent systems is the **state-explosion problem**. Recall that the state space of a composition of k finite-state systems $M_1, M_2, \ldots, M_k$ (say, using synchronous composition), is the cartesian product of the state spaces of $M_1, M_2, \ldots, M_k$. In p.107 other words, if $M_1, M_2, \ldots, M_k$ have $n_1, n_2, \ldots, n_k$ states respectively, their composition can have $\Pi_{i=1}^{k} n_i$ states. It is easy to see that the number of states of a concurrent composition of k components grows exponentially with k. Explicitly representing the state space of the composite system does not scale. In the next section, we will introduce techniques that can mitigate this problem in some cases.

15.2.3 Symbolic Model Checking

The key idea in **symbolic model checking** is to represent a set of states *symbolically* as a propositional logic formula, rather than explicitly as a collection of individual states. Spe- p.348

cialized data structures are often used to efficiently represent and manipulate such formulas. Thus, in contrast to explicit-state model checking, in which individual states are manipulated, symbolic model checking operates on sets of states.

Algorithm 15.2 (**Symbolic_Search**) is a symbolic algorithm for computing the set of reachable states of a closed, finite-state system M. This algorithm has the same input-output specification as the previous explicit-state algorithm **DFS_Search**; however, all operations in **Symbolic_Search** are set operations.

Input : Initial state s_0 and transition relation δ for closed finite-state system M, represented symbolically
Output: Set R of reachable states of M, represented symbolically

1 **Initialize:** *Current set of reached states* $R = \{s_0\}$
2 **Symbolic_Search**() {
3 $R_{\text{new}} = R$
4 **while** $R_{\text{new}} \neq \emptyset$ **do**
5 $\quad R_{\text{new}} := \{s' \mid \exists s \in R \text{ s.t. } s' \in \delta(s) \wedge s' \notin R\}$
6 $\quad R := R \cup R_{\text{new}}$
7 **end**
8 }

Algorithm 15.2: Computing the reachable state set by symbolic search.

In algorithm **Symbolic_Search**, R represents the entire set of states reached at any point in the search, and R_{new} represents the *new* states generated at that point. When no more new states are generated, the algorithm terminates, with R storing all states reachable from s_0. The key step of the algorithm is line 5, in which R_{new} is computed as the set of all states s' reachable from any state s in R in one step of the transition relation δ. This operation p.474 is called **image computation**, since it involves computing the image of the function δ. Efficient implementations of image computation that directly operate on propositional logic formulas are central to symbolic reachability algorithms. Apart from image computation, the key set operations in **Symbolic_Search** include set union and emptiness checking.

Example 15.2: We illustrate symbolic reachability analysis using the finite-state system in Figure 15.4.

To begin with, we need to introduce some notation. Let v_l be a variable denoting p.44 the state of the traffic light controller FSM S at the start of each reaction; i.e., $v_l \in \{\mathsf{green}, \mathsf{yellow}, \mathsf{red}, \mathsf{pending}\}$. Similarly, let v_p denote the state of the pedestrian FSM E, where $v_p \in \{\mathsf{crossing}, \mathsf{none}, \mathsf{waiting}\}$.

Given this notation, the initial state set $\{s_0\}$ of the composite system M is represented as the following propositional logical formula:

$$v_l = \text{red} \wedge v_p = \text{crossing} \wedge count = 0$$

From s_0, the only enabled outgoing transition is the self-loop on the initial state of the extended FSM in Figure 15.4. Thus, after one step of reachability computation, the set of reached states R is represented by the following formula:

$$v_l = \text{red} \wedge v_p = \text{crossing} \wedge 0 \leq count \leq 1$$

After two steps, R is given by

$$v_l = \text{red} \wedge v_p = \text{crossing} \wedge 0 \leq count \leq 2$$

and after k steps, $k \leq 60$, R is represented by the formula

$$v_l = \text{red} \wedge v_p = \text{crossing} \wedge 0 \leq count \leq k$$

On the 61st step, we exit the state (red, crossing), and compute R as

$$\begin{aligned} & v_l = \text{red} \wedge v_p = \text{crossing} \wedge 0 \leq count \leq 60 \\ \vee\ & v_l = \text{green} \wedge v_p = \text{none} \wedge count = 0 \end{aligned}$$

Proceeding similarly, the set of reachable states R is grown until there is no further change. The final reachable set is represented as:

$$\begin{aligned} & v_l = \text{red} \wedge v_p = \text{crossing} \wedge 0 \leq count \leq 60 \\ \vee\ & v_l = \text{green} \wedge v_p = \text{none} \wedge 0 \leq count \leq 60 \\ \vee\ & v_l = \text{pending} \wedge v_p = \text{waiting} \wedge 0 < count \leq 60 \\ \vee\ & v_l = \text{yellow} \wedge v_p = \text{waiting} \wedge 0 \leq count \leq 5 \end{aligned}$$

In practice, the symbolic representation is much more compact than the explicit one. The previous example illustrates this nicely because a large number of states are compactly represented by inequalities like $0 < count \leq 60$. Computer programs can be designed to operate directly on the symbolic representation. Some examples of such programs are given in the box on page 404.

Symbolic model checking has been used successfully to address the state-explosion problem for many classes of systems, most notably for hardware models. However, in the worst case, even symbolic set representations can be exponential in the number of system variables.

15.3 Abstraction in Model Checking

p.361 A challenge in model checking is to work with the simplest abstraction of a system that will provide the required proofs of safety. Simpler abstractions have smaller state spaces and can be checked more efficiently. The challenge, of course, is to know what details to omit from the abstraction.

The part of the system to be abstracted away depends on the property to be verified. The following example illustrates this point.

Example 15.3: Consider the traffic light system M in Figure 15.4. Suppose that, as in Example 15.1 we wish to verify that M satisfies the property

$$\mathbf{G}\,\neg(\mathsf{green} \wedge \mathsf{crossing})$$

Suppose we abstract the variable *count* away from M by hiding all references to *count* from the model, including all guards mentioning it and all updates to it. This generates the abstract model M_{abs} shown in Figure 15.5.

We observe that this abstract M_{abs} exhibits more behaviors than M. For instance, from the state (yellow, waiting) we can take the self-loop transition forever, staying in that state perennially, even though in the actual system M this state must be exited within five clock ticks. Moreover, every behavior of M can be exhibited by M_{abs}.

The interesting point is that, even with this approximation, we can prove that M_{abs} satisfies $\mathbf{G}\,\neg(\mathsf{green} \wedge \mathsf{crossing})$. The value of *count* is irrelevant for this property.

Notice that while M has 188 states, M_{abs} has only 4 states. Reachability analysis on M_{abs} is far easier than for M as we have far fewer states to explore.

There are several ways to compute an abstraction. One of the simple and extremely useful approaches is called **localization reduction** or **localization abstraction** (Kurshan (1994)).

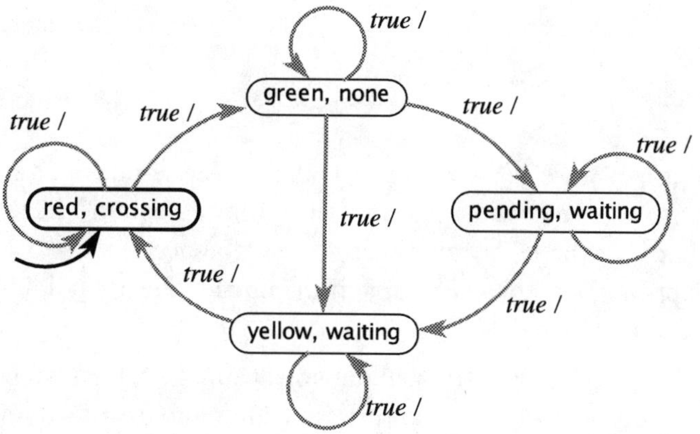

Figure 15.5: Abstraction of the traffic light system in Figure 15.4.

In localization reduction, parts of the design model that are irrelevant to the property being checked are abstracted away by hiding a subset of state variables. Hiding a variable corresponds to freeing that variable to evolve arbitrarily. It is the form of abstraction used in Example 15.3 above, where *count* is allowed to change arbitrarily, and all transitions are made independent of the value of *count*.

Example 15.4: Consider the multithreaded program given below (adapted from Ball et al. (2001)). The procedure `lock_unlock` executes a loop within which it acquires a lock, then calls the function `randomCall`, based on whose result it either releases the lock and executes another loop iteration, or it quits the loop (and then releases the lock). The execution of another loop iteration is ensured by incrementing `new`, so that the condition `old != new` evaluates to true.

```
pthread_mutex_t lock = PTHREAD_MUTEX_INITIALIZER;
unsigned int old, new;

void lock_unlock() {
  do {
    pthread_mutex_lock(&lock);
    old = new;
    if (randomCall()) {
      pthread_mutex_unlock(&lock);
      new++;
    }
```

```
    } while (old != new)
    pthread_mutex_unlock(&lock);
}
```

Suppose the property we want to verify is that the code does not attempt to call `pthread_mutex_lock` twice in a row. Recall from Section 11.2.4 how the system can deadlock if a thread becomes permanently blocked trying to acquire a lock. This could happen in the above example if the thread, already holding lock `lock`, attempts to acquire it again.

If we model this program exactly, without any abstraction, then we need to reason about all possible values of `old` and `new`, in addition to the remaining state of the program. Assuming a word size of 32 in this system, the size of the state space is roughly $2^{32} \times 2^{32} \times n$, where 2^{32} is the number of values of `old` and `new`, and n denotes the size of the remainder of the state space.

However, it is not necessary to reason about the precise values of `old` and `new` to prove that this program is correct. Assume, for this example, that our programming language is equipped with a `boolean` type. Assume further that the program can perform nondeterministic assignments. Then, we can generate the following abstraction of the original program, written in C-like syntax, where the Boolean variable `b` represents the predicate `old == new`.

```
pthread_mutex_t lock = PTHREAD_MUTEX_INITIALIZER;
boolean b; // b represents the predicate (old == new)
void lock_unlock() {
  do {
       pthread_mutex_lock(&lock);
       b = true;
       if (randomCall()) {
         pthread_mutex_unlock(&lock);
         b = false;
       }
  } while (!b)
  pthread_mutex_unlock(&lock);
}
```

It is easy to see that this abstraction retains just enough information to show that the program satisfies the desired property. Specifically, the lock will not be acquired twice because the loop is only iterated if `b` is set to false, which implies that the lock was released before the next attempt to acquire.

Moreover, observe that size of the state space to be explored has reduced to simply $2n$. This is the power of using the "right" abstraction.

A major challenge for formal verification is to *automatically* compute simple abstractions. An effective and widely-used technique is **counterexample-guided abstraction refinement (CEGAR)**, first introduced by Clarke et al. (2000). The basic idea (when using localization reduction) is to start by hiding almost all state variables except those referenced by the temporal logic property. The resulting abstract system will have more behaviors than the original system. Therefore, if this abstract system satisfies an LTL formula Φ (i.e., each of its behaviors satisfies Φ), then so does the original. However, if the abstract system does not satisfy Φ, the model checker generates a counterexample. If this counterexample is p.349 a counterexample for the original system, the process terminates, having found a genuine counterexample. Otherwise, the CEGAR approach analyzes this counterexample to infer which hidden variables must be made visible, and with these additional variables, recomputes an abstraction. The process continues, terminating either with some abstract system being proven correct, or generating a valid counterexample for the original system.

The CEGAR approach and several follow-up ideas have been instrumental in driving progress in the area of software model checking. We review some of the key ideas in the sidebar on page 404.

15.4 Model Checking Liveness Properties

So far, we have restricted ourselves to verifying properties of the form $\mathbf{G}p$, where p is an atomic proposition. An assertion that $\mathbf{G}p$ holds for all traces is a very restricted kind of safety property. However, as we have seen in Chapter 13, several useful system properties p.355 are not safety properties. For instance, the property stating that "the robot must visit location A" is a liveness property: if visiting location A is represented by proposition q, then this p.355 property is an assertion that $\mathbf{F}q$ must hold for all traces. In fact, several problems, including path planning problems for robotics and progress properties of distributed and concurrent systems can be stated as liveness properties. It is therefore useful to extend model checking to handle this class of properties.

Properties of the form $\mathbf{F}p$, though liveness properties, can be partially checked using the techniques introduced earlier in this chapter. Recall from Chapter 13 that $\mathbf{F}p$ holds for a trace if and only if $\neg\mathbf{G}\neg p$ holds for the same trace. In words, "p is true some time in the future" iff "$\neg p$ is always false." Therefore, we can attempt to verify that the system satisfies $\mathbf{G}\neg p$. If the verifier asserts that $\mathbf{G}\neg p$ holds for all traces, then we know that $\mathbf{F}p$ does not hold for any trace. On the other hand, if the verifier outputs "NO", then the accompanying counterexample provides a witness exhibiting how p may become true eventually. This witness provides one trace for which $\mathbf{F}p$ holds, but it does not prove that $\mathbf{F}p$ holds for all traces (unless the machine is deterministic).

More complete checks and more complicated liveness properties require a more sophisticated approach. Briefly, one approach used in explicit-state model checking of LTL properties is as follows:

1. Represent the negation of the property Φ as an automaton B, where certain states are labeled as accepting states. p.367
2. p.107 Construct the synchronous composition of the property automaton B and the system automaton M. The accepting states of the property automaton induce accepting states of the product automaton M_B.
3. If the product automaton M_B can visit an accepting state infinitely often, then it indicates that M does not satisfy Φ; otherwise, M satisfies Φ.

The above approach is known as the **automata-theoretic approach to verification**. We give a brief introduction to this subject in the rest of this section. Further details may be found in the seminal papers on this topic (Wolper et al. (1983); Vardi and Wolper (1986)) and the book on the SPIN model checker (Holzmann (2004)).

15.4.1 Properties as Automata

Consider the first step of viewing properties as automata. Recall the material on omega-regular languages introduced in the box on page 369. The theory of Büchi automata and omega-regular languages, briefly introduced there, is relevant for model checking liveness p.474 properties. Roughly speaking, an LTL property Φ has a one-to-one correspondence with a p.67 set of behaviors that satisfy Φ. This set of behaviors constitutes the language of the Büchi automaton corresponding to Φ.

For the LTL model checking approach we describe here, if Φ is the property that the system must satisfy, then we represent its negation $\neg\Phi$ as a Büchi automaton. We present some illustrative examples below.

Example 15.5: Suppose that an FSM M_1 models a system that executes forever and produces a pure output h (for heartbeat), and that it is required to produce this output at least once every three reactions. That is, if in two successive reactions it fails to produce the output h, then in the third it must.

We can formulate this property in LTL as the property Φ_1 below:

$$\mathbf{G}(h \vee \mathbf{X}h \vee \mathbf{X}^2 h)$$

and the negation of this property is

$$\mathbf{F}(\neg h \wedge \mathbf{X}\neg h \wedge \mathbf{X}^2 \neg h)$$

The Büchi automaton B_1 corresponding to the negation of the desired property is given below:

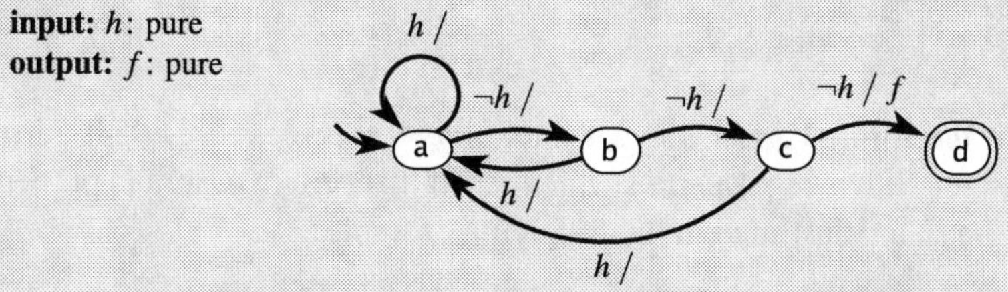

Let us examine this automaton. The language accepted by this automaton includes all behaviors that enter and stay in state d. Equivalently, the language includes all behaviors that produce a *present* output on f in some reaction. When we compose the above machine with M_1, if the resulting composite machine can never produce $f = present$, then the language accepted by the composite machine is empty. If we can prove that the language is empty, then we have proved that M produces the heartbeat h at least once every three reactions.

Observe that the property Φ_1 in the above example is in fact a safety property. We give an example of a liveness property below. p.355 p.355

Example 15.6: Suppose that the FSM M_2 models a controller for a robot that must locate a room and stay there forever. Let p be the proposition that becomes true when the robot is in the target room. Then, the desired property Φ_2 can be expressed in LTL as $\mathbf{FG}p$.

The negation of this property is $\mathbf{GF}\neg p$. The Büchi automaton B_2 corresponding to this negated property is given below:

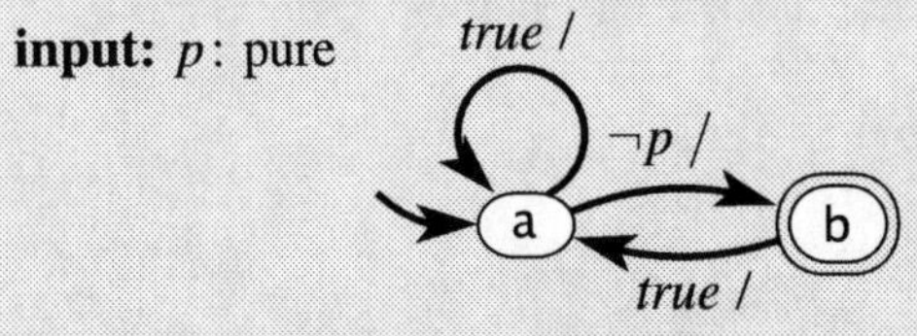

Notice that all accepting behaviors of B_2 correspond to those where $\neg p$ holds infinitely often. These behaviors correspond to a cycle in the state graph for the product automaton where state b of B_2 is visited repeatedly. This cycle is known as an **acceptance cycle**.

p.354

Liveness properties of the form $\mathbf{GF}p$ also occur naturally as specifications. This form of property is useful in stating **fairness** properties which assert that certain desirable properties hold infinitely many times, as illustrated in the following example.

Example 15.7: Consider a traffic light system such as that in Example 3.10. We may wish to assert that the traffic light becomes green infinitely many times in any execution. In other words, the state green is visited infinitely often, which can be expressed as $\Phi_3 = \mathbf{GF}$ green.

The automaton corresponding to Φ_3 is identical to that for the negation of Φ_2 in Example 15.6 above with $\neg p$ replaced by green. However, in this case the accepting behaviors of this automaton are the desired behaviors.

Thus, from these examples we see that the problem of detecting whether a certain accepting state s in an FSM can be visited infinitely often is the workhorse of explicit-state model checking of LTL properties. We next present an algorithm for this problem.

15.4.2 Finding Acceptance Cycles

We consider the following problem:

> Given a finite-state system M, can an accepting state s_a of M be visited infinitely often?

Put another way, we seek an algorithm to check whether (i) state s_a is reachable from the initial state s_0 of M, and (ii) s_a is reachable from itself. Note that asking whether a state *can* be visited infinitely often is not the same as asking whether it *must* be visited infinitely often.

The graph-theoretic viewpoint is useful for this problem, as it was in the case of $\mathbf{G}p$ discussed in Section 15.2.1. Assume for the sake of argument that we have the entire state graph constructed a priori. Then, the problem of checking whether state s_a is reachable from s_0 is simply a graph traversal problem, solvable for example by depth-first search (DFS). Further, the problem of detecting whether s_a is reachable from itself amounts to checking whether there is a cycle in the state graph containing that state.

The main challenges for solving this problem are similar to those discussed in Section 15.2.1: we must perform this search on-the-fly, and we must deal with large state spaces.

The **nested depth-first search** (nested DFS) algorithm, which is implemented in the SPIN model checker (Holzmann (2004)), solves this problem and is shown as Algorithm 15.3. The algorithm begins by calling the procedure called Nested_DFS_Search with argument 1, as shown in the Main function at the bottom. M_B is obtained by composing the original closed system M with the automaton B representing the negation of LTL formula Φ.

As the name suggests, the idea is to perform two depth-first searches, one nested inside the other. The first DFS identifies a path from initial state s_0 to the target accepting state s_a. Then, from s_a we start another DFS to see if we can reach s_a again. The variable `mode` is either 1 or 2 depending on whether we are performing the first DFS or the second. Stacks Σ_1 and Σ_2 are used in the searches performed in modes 1 and 2 respectively. If s_a is encountered in the second DFS, the algorithm generates as output the path leading from s_0 to s_a with a loop on s_a. The path from s_0 to s_a is obtained simply by reading off the contents of stack Σ_1. Likewise, the cycle from s_a to itself is obtained from stack Σ_2. Otherwise, the algorithm reports failure.

Search optimization and state compression techniques that are used in explicit-state reachability analysis can be used with nested DFS also. Further details are available in Holzmann (2004).

Input : Initial state s_0 and transition relation δ for automaton M_B; Target accepting state s_a of M_B

Output: Acceptance cycle containing s_a, if one exists

```
Initialize: (i) Stack Σ_1 to contain a single state s_0, and stack Σ_2 to be empty;
(ii) Two sets of reached states R_1 := R_2 := {s_0}; (iii) Flag found := false.
Nested_DFS_Search(Mode mode) {
while Stack Σ_mode is not empty do
    Pop the state s at the top of Σ_mode
    if (s = s_a and mode = 1) then
        Push s onto Σ_2
        Nested_DFS_Search(2)
        return
    end
    Compute δ(s), the set of all states reachable from s in one transition
    for each s' ∈ δ(s) do
        if (s' = s_a and mode = 2) then
            Output path to s_a with acceptance cycle using contents of stacks
            Σ_1 and Σ_2
            found := true
            return
        end
        if s' ∉ R_mode then
            R_mode := R_mode ∪ {s'}
            Push s' onto Σ_mode
            Nested_DFS_Search(mode)
        end
    end
end
}
Main() {
    Nested_DFS_Search(1)
    if (found = false) then Output "no acceptance cycle with s_a" end }
```

Algorithm 15.3: Nested depth-first search algorithm.

15.5 Summary

This chapter gives some basic algorithms for formal verification, including model checking, a technique for verifying if a finite-state system satisfies a property specified in temporal logic. Verification operates on closed systems, which are obtained by composing a system with its operating environment. The first key concept is that of reachability analysis, which verifies properties of the form $\mathbf{G}p$. The concept of abstraction, central to the scalability of model checking, is also discussed in this chapter. This chapter also shows how explicit-state model checking algorithms can handle liveness properties, where a crucial concept is the correspondence between properties and automata.

Probing Further: Model Checking in Practice

Several tools are available for computing the set of reachable states of a finite-state system and checking that they satisfy specifications in temporal logic. One such tool is **SMV** (symbolic model verifier), which was first developed at Carnegie Mellon University by Kenneth McMillan. SMV was the first model checking tool to use binary decision diagrams (**BDD**s), a compact data structure introduced by Bryant (1986) for representing a Boolean function. The use of BDDs has proved instrumental in enabling analysis of more complex systems. Current symbolic model checkers also rely heavily on Boolean satisfiability (SAT) solvers (see Malik and Zhang (2009)), which are programs for deciding whether a propositional logic formula can evaluate to true. One of the first uses of SAT solvers in model checking was for **bounded model checking** (see Biere et al. (1999)), where the transition relation of the system is unrolled only a bounded number of times. A few different versions of SMV are available online (see for example http://nusmv.fbk.eu/).

p.492 p.348

The SPIN model checker (Holzmann, 2004) developed in the 1980's and 1990's at Bell Labs by Gerard Holzmann and others, is another leading tool for model checking (see http://www.spinroot.com/). Rather than directly representing models as communicating FSMs, it uses a specification language (called Promela, for process meta language) that enables specifications that closely resemble multithreaded programs. SPIN incorporates state-compression techniques such as **hash compaction** (the use of hashing to reduce the size of the stored state set) and **partial-order reduction** (a technique to reduce the number of reachable states to be explored by considering only a subset of the possible process interleavings).

Automatic abstraction has played a big role in applying model checking directly to software. An example of abstraction-based software model checking is the **SLAM** system developed at Microsoft Research (Ball and Rajamani, 2001; Ball et al., 2011). SLAM combines CEGAR with a particular form of abstraction called predicate abstraction, in which predicates in a program are abstracted to Boolean variables. A key step in these techniques is checking whether a counterexample generated on the abstract model is in fact a true counterexample. This check is performed using satisfiability solvers for logics richer than propositional logic. These solvers are called **SAT-based decision procedures** or **satisfiability modulo theories (SMT)** solvers (for more details, see Barrett et al. (2009)).

p.397 p.48

More recently, techniques based on **inductive learning**, that is, generalization from sample data, have started playing an important role in formal verification (see Seshia (2015) for an exposition of this topic).

Exercises

1. Consider the system M modeled by the hierarchical state machine of Figure 13.2, which models an interrupt-driven program.

   Model M in the modeling language of a verification tool (such as SPIN). You will have to construct an environment model that asserts the interrupt. Use the verification tool to check whether M satisfies ϕ, the property stated in Exercise 5:

   ϕ: The main program eventually reaches program location C.

   Explain the output you obtain from the verification tool.

2. Figure 15.3 shows the synchronous-reactive composition of the traffic light controller of Figure 3.10 and the pedestrian model of Figure 3.11.

   Consider replacing the pedestrian model in Figure 15.3 with the alternative model given below where the initial state is nondeterministically chosen to be one of none or crossing:

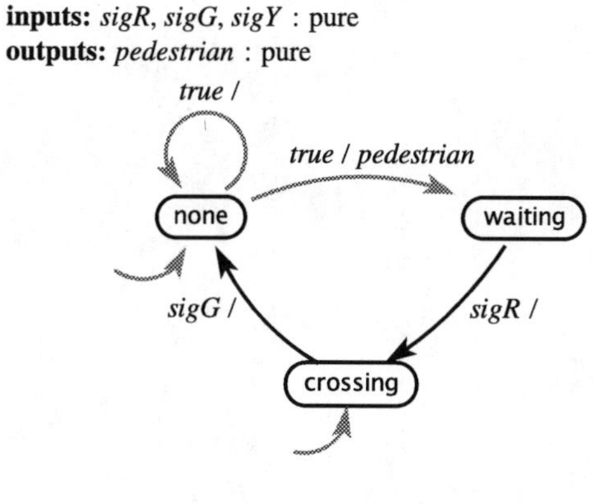

   (a) Model the composite system in the modeling language of a verification tool (such as SPIN). How many reachable states does the combined system have? How many of these are initial states?
   (b) Formulate an LTL property stating that every time a pedestrian arrives, eventually the pedestrian is allowed to cross (i.e., the traffic light enters state red).
   (c) Use the verification tool to check whether the model constructed in part (a) satisfies the LTL property specified in part (b). Explain the output of the tool.

3. The notion of reachability has a nice symmetry. Instead of describing all states that are reachable from some initial state, it is just as easy to describe all states from which

some state can be reached. Given a finite-state system M, the **backward reachable states** of a set F of states is the set B of all states from which some state in F can be reached. The following algorithm computes the set of backward reachable states for a given set of states F:

Input : A set F of states and transition relation δ for closed finite-state system M
Output: Set B of backward reachable states from F in M

Initialize: $B := F$

$B_{\text{new}} := B$
while $B_{\text{new}} \neq \emptyset$ **do**
  $B_{\text{new}} := \{s \mid \exists s' \in B \text{ s.t. } s' \in \delta(s) \wedge s \notin B\}$
  $B := B \cup B_{\text{new}}$
end

Explain how this algorithm can check the property $\mathbf{G}p$ on M, where p is some property that is easily checked for each state s in M. You may assume that M has exactly one initial state s_0.

16

Quantitative Analysis

Contents

Will my brake-by-wire system actuate the brakes within one millisecond? Answering this question requires, in part, an **execution-time analysis** of the software that runs on the electronic control unit (ECU) for the brake-by-wire system. Execution time of the software is an example of a **quantitative property** of an embedded system. The constraint that the system actuate the brakes within one millisecond is a **quantitative constraint**. The analysis of quantitative properties for conformance with quantitative constraints is central to the correctness of embedded systems and is the topic of the present chapter.

A quantitative property of an embedded system is any property that can be measured. This includes physical parameters, such as position or velocity of a vehicle controlled by the embedded system, weight of the system, operating temperature, power consumption, or reaction time. Our focus in this chapter is on properties of software-controlled systems, with particular attention to execution time. We present program analysis techniques that can ensure that execution time constraints will be met. We also discuss how similar techniques can be used to analyze other quantitative properties of software, particularly resource usage such as power, energy, and memory.

The analysis of quantitative properties requires adequate models of both the software components of the system and of the environment in which the software executes. The environment includes the processor, operating system, input-output devices, physical components with which the software interacts, and (if applicable) the communication network. The environment is sometimes also referred to as the platform on which the software executes. p. 4 Providing a comprehensive treatment of execution time analysis would require much more than one chapter. The goal of this chapter is more modest. We illustrate key features of programs and their environment that must be considered in quantitative analysis, and we describe qualitatively some analysis techniques that are used. For concreteness, we focus on a single quantity, *execution time*, and only briefly discuss other resource-related quantitative properties.

16.1 Problems of Interest

The typical quantitative analysis problem involves a software task defined by a program P, the environment E in which the program executes, and the quantity of interest q. We assume that q can be given by a function of f_P as follows,

$$q = f_P(x, w)$$

where x denotes the inputs to the program P (such as data read from memory or from sensors, or data received over a network), and w denotes the environment parameters (such as network delays or the contents of the cache when the program begins executing). Defining the function f_P completely is often neither feasible nor necessary; instead, practical quantitative analysis will yield extreme values for q (highest or lowest values), average values for q, or proofs that q satisfies certain threshold constraints. We elaborate on these next.

16.1.1 Extreme-Case Analysis

In extreme-case analysis, we may want to estimate the *largest value* of q for all values of x and w,

$$\max_{x,w} f_P(x,w). \tag{16.1}$$

Alternatively, it can be useful to estimate the *smallest value* of q:

$$\min_{x,w} f_P(x,w). \tag{16.2}$$

If q represents execution time of a program or a program fragment, then the largest value is called the **worst-case execution time** (**WCET**), and the smallest value is called the **best-case execution time** (**BCET**). It may be difficult to determine these numbers exactly, but for many applications, an upper bound on the WCET or a lower bound on the BCET is all that is needed. In each case, when the computed bound equals the actual WCET or BCET, it is said to be a **tight bound**; otherwise, if there is a considerable gap between the actual value and the computed bound, it is said to be a **loose bound**. Computing loose bounds may be much easier than finding tight bounds.

16.1.2 Threshold Analysis

A **threshold property** asks whether the quantity q is always bounded above or below by a threshold T, for any choice of x and w. Formally, the property can be expressed as

$$\forall x, w, \quad f_P(x,w) \leq T \tag{16.3}$$

or

$$\forall x, w, \quad f_P(x,w) \geq T \tag{16.4}$$

Threshold analysis may provide assurances that a quantitative constraint is met, such as the requirement that a brake-by-wire system actuate the brakes within one millisecond.

Threshold analysis may be easier to perform than extreme-case analysis. Unlike extreme-case analysis, threshold analysis does not require us to determine the maximum or minimum value exactly, or even to find a tight bound on these values. Instead, the analysis is provided some guidance in the form of the target value T. Of course, it might be possible to use extreme-case analysis to check a threshold property. Specifically, Constraint 16.3 holds if the WCET does not exceed T, and Constraint 16.4 holds if the BCET is not less than T.

16.1.3 Average-Case Analysis

Often one is interested more in typical resource usage rather than in worst-case scenarios. This is formalized as average-case analysis. Here, the values of input x and environment parameter w are assumed to be drawn randomly from a space of possible values X and W according to probability distributions $\mathcal{D}_x$ and $\mathcal{D}_w$ respectively. Formally, we seek to estimate the value

$$\mathbb{E}_{\mathcal{D}_x,\mathcal{D}_w} f_P(x,w) \tag{16.5}$$

where $\mathbb{E}_{\mathcal{D}_x,\mathcal{D}_w}$ denotes the expected value of $f_P(x,w)$ over the distributions $\mathcal{D}_x$ and $\mathcal{D}_w$.

One difficulty in average-case analysis is to define realistic distributions $\mathcal{D}_x$ and $\mathcal{D}_w$ that capture the true distribution of inputs and environment parameters with which a program will execute.

p.409 In the rest of this chapter, we will focus on a single representative problem, namely, WCET estimation.

16.2 Programs as Graphs

A fundamental abstraction used often in program analysis is to represent a program as a graph indicating the flow of control from one code segment to another. We will illustrate this abstraction and other concepts in this chapter using the following running example:

Example 16.1: Consider the function `modexp` that performs **modular exponentiation**, a key step in many cryptographic algorithms. In modular exponentiation, given a base b, an exponent e, and a modulus m, one must compute $b^e \bmod m$. In the program below, `base`, `exponent` and `mod` represent b, e and m respectively. `EXP_BITS` denotes the number of bits in the exponent. The function uses a standard shift-square-accumulate algorithm, where the base is repeatedly squared, once for each bit position of the exponent, and the base is accumulated into the result only if the corresponding bit is set.

```
1  #define EXP_BITS 32
2
3  typedef unsigned int UI;
4
5  UI modexp(UI base, UI exponent, UI mod) {
6    int i;
7    UI result = 1;
8
9    i = EXP_BITS;
10   while(i > 0) {
11     if ((exponent & 1) == 1) {
12       result = (result * base) % mod;
13     }
14     exponent >>= 1;
15     base = (base * base) % mod;
16     i--;
17   }
18   return result;
19 }
```

16.2.1 Basic Blocks

A **basic block** is a sequence of consecutive program statements in which the flow of control enters only at the beginning of this sequence and leaves only at the end, without halting or the possibility of branching except at the end.

Example 16.2: The following three statements from the `modexp` function in Example 16.1 form a basic block:

```
14    exponent >>= 1;
15    base = (base * base) % mod;
16    i--;
```

Another example of a basic block includes the initializations at the top of the function, comprising lines 7 and 9:

```
7  result = 1;
8
9  i = EXP_BITS;
```

16.2.2 Control-Flow Graphs

A **control-flow graph** (**CFG**) of a program P is a directed graph $G = (V,E)$, where the set of vertices V comprises basic blocks of P, and the set of edges E indicates the flow of control between basic blocks. Figure 16.1 depicts the CFG for the `modexp` program of Example 16.1. Each node of the CFG is labeled with its corresponding basic block. In most cases, this is simply the code as it appears in Example 16.1. The only exception is for conditional statements, such as the conditions in `while` loops and `if` statements; in these cases, we follow the convention of labeling the node with the condition followed by a question mark to indicate the conditional branch.

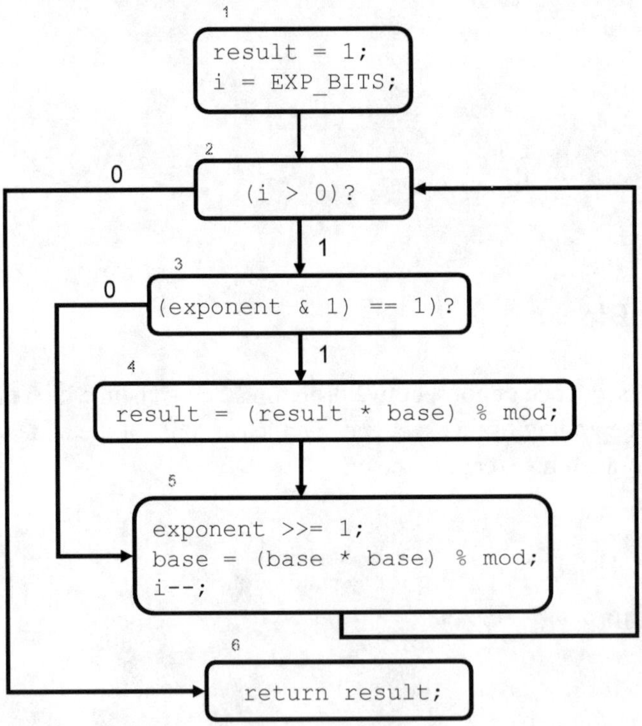

Figure 16.1: Control-flow graph for the `modexp` function of Example 16.1. All incoming edges at a node indicate transfer of control to the start of the basic block for that node, and all outgoing edges from a node indicate an exit from the end of the basic block for that node. For clarity, we label the outgoing edges from a branch statement with 0 or 1 indicating the flow of control in case the branch evaluates to false or true, respectively. An ID number for each basic block is noted above the node for that block; IDs range from 1 to 6 for this example.

Although our illustrative example of a control-flow graph is at the level of C source code, it is possible to use the CFG representation at other levels of program representation as well, including a high-level model as well as low-level assembly code. The level of representation employed depends on the level of detail required by the context. To make them easier to follow, our control-flow graphs will be at the level of source code.

16.2.3 Function Calls

Programs are typically decomposed into several functions in order to systematically organize the code and promote reuse and readability. The control-flow graph (CFG) representation can be extended to reason about code with function calls by introducing special **call** and **return** edges. These edges connect the CFG of the **caller function** – the one making the function call – to that of the **callee function** – the one being called. A **call edge** indicates a transfer of control from the caller to the callee. A **return edge** indicates a transfer of control from the callee back to the caller.

Example 16.3: A slight variant shown below of the modular exponentation program of Example 16.1 uses function calls and can be represented by the CFG with call and return edges in Figure 16.2.

```
#define EXP_BITS 32
typedef unsigned int UI;
UI exponent, base, mod;

UI update(UI r) {
  UI res = r;
  if ((exponent & 1) == 1) {
    res = (res * base) % mod;
  }
  exponent >>= 1;
  base = (base * base) % mod;
  return res;
}

UI modexp_call() {
  UI result = 1; int i;
  i = EXP_BITS;
  while(i > 0) {
    result = update(result);
    i--;
  }
```

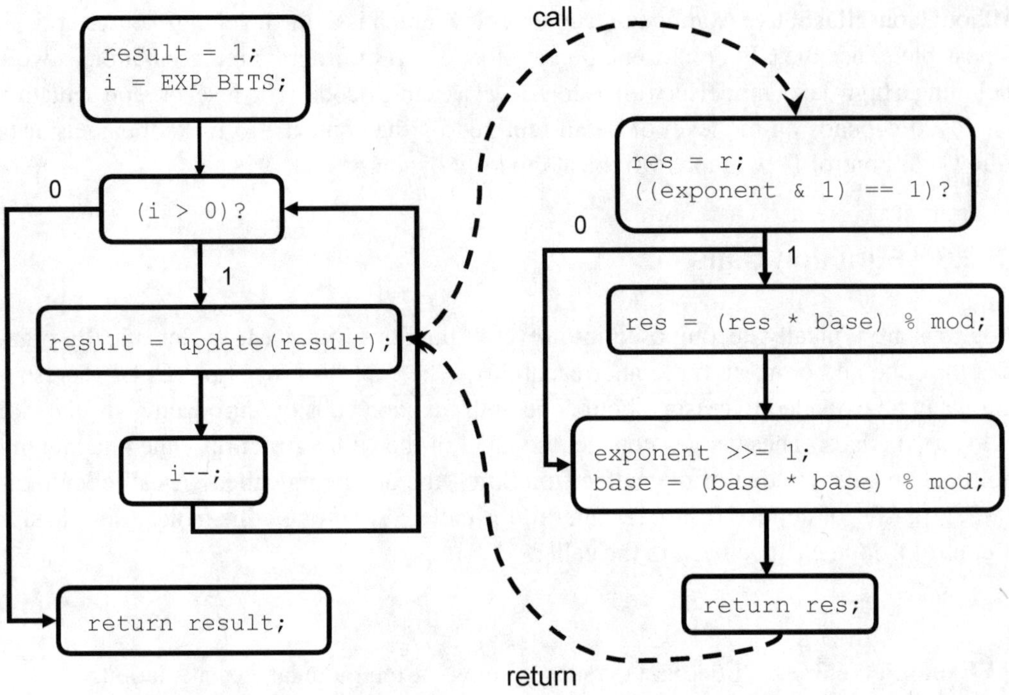

Figure 16.2: Control-flow graphs for the `modexp_call` and `update` functions in Example 16.3. Call/return edges are indicated with dashed lines.

```
    return result;
}
```

In this modified example, the variables `base`, `exponent`, and `mod` are global variables. The update to `base` and `exponent` in the body of the `while` loop, along with the computation of `result` is now performed in a separate function named `update`.

p.244

Non-recursive function calls can also be handled by **inlining**, which is the process of copying the code for the callee into that of the caller. If inlining is performed transitively for all functions called by the code that must be analyzed, the analysis can be performed on the CFG of the code resulting from inlining, without using call and return edges.

16.3 Factors Determining Execution Time

There are several issues one must consider in order to estimate the worst-case execution time of a program. This section outlines some of the main issues and illustrates them with examples. In describing these issues, we take a programmer's viewpoint, starting with the program structure and then considering how the environment can impact the program's execution time.

16.3.1 Loop Bounds

The first point one must consider when bounding the execution time of a program is whether the program terminates. Non-termination of a sequential program can arise from non-terminating loops or from an unbounded sequence of function calls. Therefore, while writing real-time embedded software, the programmer must ensure that all loops are guaranteed p.310 to terminate. In order to guarantee this, one must determine for each loop a bound on the number of times that loop will execute in the worst case. Similarly, all function calls must have bounded recursion depth. The problems of determining bounds on loop iterations or recursion depth are undecidable in general, since the halting problem for Turing machines p.489 can be reduced to either problem. (See Appendix B for an introduction to Turing machines and decidability.)

In this section, we limit ourselves to reasoning about loops. In spite of the undeciable nature of the problem, progress has been made on automatically determining loop bounds for several patterns that arise in practice. Techniques for determining loop bounds are a current research topic and a full survey of these methods is out of the scope of this chapter. We will limit ourselves to presenting illustrative examples for loop bound inference.

The simplest case is that of `for` loops that have a specified constant bound, as in Example 16.4 below. This case occurs often in embedded software, in part due to a discipline of programming enforced by designers who must program for real-time constraints and limited resources.

Example 16.4: Consider the function `modexp1` below. It is a slight variant of the function `modexp` introduced in Example 16.1 that performs modular exponentiation, in which the `while` loop has been expressed as an equivalent `for` loop.

```
#define EXP_BITS 32

typedef unsigned int UI;

UI modexp1(UI base, UI exponent, UI mod) {
  UI result = 1; int i;

  for(i=EXP_BITS; i > 0; i--) {
    if ((exponent & 1) == 1) {
      result = (result * base) % mod;
    }
    exponent >>= 1;
    base = (base * base) % mod;
  }
  return result;
}
```

In the case of this function, it is easy to see that the `for` loop will take exactly `EXP_BITS` iterations, where `EXP_BITS` is defined as the constant 32.

In many cases, the loop bound is not immediately obvious (as it was for the above example). To make this point, here is a variation on Example 16.4.

Example 16.5: The function listed below also performs modular exponentiation, as in Example 16.4. However, in this case, the `for` loop is replaced by a `while` loop with a different loop condition – the loop exits when the value of `exponent` reaches 0. Take a moment to check whether the `while` loop will terminate (and if so, why).

```
typedef unsigned int UI;

UI modexp2(UI base, UI exponent, UI mod) {
  UI result = 1;

  while (exponent != 0) {
    if ((exponent & 1) == 1) {
      result = (result * base) % mod;
    }
    exponent >>= 1;
    base = (base * base) % mod;
```

```
12     }
13     return result;
14 }
```

Now let us analyze the reason that this loop terminates. Notice that `exponent` is an unsigned int, which we will assume to be 32 bits wide. If it starts out equal to 0, the loop terminates right away and the function returns `result = 1`. If not, in each iteration of the loop, notice that line 10 shifts `exponent` one bit to the right. Since `exponent` is an unsigned int, after the right shift, its most significant bit will be 0. Reasoning thus, after at most 32 right shifts, all bits of `exponent` must be set to 0, thus causing the loop to terminate. Therefore, we can conclude that the loop bound is 32.

Let us reflect on the reasoning employed in the above example. The key component of our "proof of termination" was the observation that the number of bits of `exponent` decreases by 1 each time the loop executes. This is a standard argument for proving termination – by defining a **progress measure** or **ranking function** that maps each state of the program to a mathematical structure called a **well order**. Intuitively, a well order is like a program that counts down to zero from some initial value in the natural numbers.

16.3.2 Exponential Path Space

Execution time is a path property. In other words, the amount of time taken by the program is a function of how conditional statements in the program evaluate to true or false. A major source of complexity in execution time analysis (and other program analysis problems as well) is that the number of program paths can be very large — exponential in the size of the program. We illustrate this point with the example below.

Example 16.6: Consider the function `count` listed below, which runs over a two-dimensional array, counting and accumulating non-negative and negative elements of the array separately.

```
#define MAXSIZE 100

int Array[MAXSIZE][MAXSIZE];
```

```
4  int Ptotal, Pcnt, Ntotal, Ncnt;
5  ...
6  void count() {
7    int Outer, Inner;
8    for (Outer = 0; Outer < MAXSIZE; Outer++) {
9      for (Inner = 0; Inner < MAXSIZE; Inner++)  {
10       if (Array[Outer][Inner] >= 0) {
11         Ptotal += Array[Outer][Inner];
12         Pcnt++;
13       } else {
14         Ntotal += Array[Outer][Inner];
15         Ncnt++;
16       }
17     }
18   }
19 }
```

The function includes a nested loop. Each loop executes `MAXSIZE` (100) times. Thus, the inner body of the loop (comprising lines 10–16) will execute 10,000 times – as many times as the number of elements of `Array`. In each iteration of the inner body of the loop, the conditional on line 10 can either evaluate to true or false, thus resulting in 2^{10000} possible ways the loop can execute. In other words, this program has 2^{10000} paths.

Fortunately, as we will see in Section 16.4.1, one does not need to explicitly enumerate all possible program paths in order to perform execution time analysis.

16.3.3 Path Feasibility

Another source of complexity in program analysis is that all program paths may not be executable. A computationally expensive function is irrelevant for execution time analysis if that function is never executed.

A path p in program P is said to be **feasible** if there exists an input x to P such that P executes p on x. In general, even if P is known to terminate, determining whether a path p is feasible is a computationally intractable problem. One can encode the canonical NP-complete p.492 problem, the Boolean satisfiability problem (see Appendix B), as a problem of checking path feasibility in a specially-constructed program. In practice, however, in many cases, it is possible to determine path feasibility.

Example 16.7: Recall Example 13.3 of a software task from the open source Paparazzi unmanned aerial vehicle (UAV) project (Nemer et al., 2006):

```
1  #define PPRZ_MODE_AUTO2 2
2  #define PPRZ_MODE_HOME 3
3  #define VERTICAL_MODE_AUTO_ALT 3
4  #define CLIMB_MAX 1.0
5  ...
6  void altitude_control_task(void) {
7    if (pprz_mode == PPRZ_MODE_AUTO2
8        || pprz_mode == PPRZ_MODE_HOME) {
9      if (vertical_mode == VERTICAL_MODE_AUTO_ALT) {
10       float err = estimator_z - desired_altitude;
11       desired_climb
12              = pre_climb + altitude_pgain * err;
13       if (desired_climb < -CLIMB_MAX) {
14         desired_climb = -CLIMB_MAX;
15       }
16       if (desired_climb > CLIMB_MAX) {
17         desired_climb = CLIMB_MAX;
18       }
19     }
20   }
21 }
```

This program has 11 paths in all. However, the number of *feasible* program paths is only 9. To see this, note that the two conditionals `desired_climb < -CLIMB_MAX` on line 13 and `desired_climb > CLIMB_MAX` on line 16 cannot both be true. Thus, only three out of the four paths through the two innermost conditional statements are feasible. This infeasible inner path can be taken for two possible evaluations of the outermost conditional on lines 7 and 8: either if `pprz_mode == PPRZ_MODE_AUTO2` is true, or if that condition is false, but `pprz_mode == PPRZ_MODE_HOME` is true.

16.3.4 Memory Hierarchy

The preceding sections have focused on properties of programs that affect execution time. We now discuss how properties of the execution platform, specifically of cache memories, can significantly impact execution time. We illustrate this point using Example 16.8.[1] The material on caches introduced in Sec. 9.2.3 is pertinent to this discussion.

Example 16.8: Consider the function `dot_product` listed below, which computes the dot product of two vectors of floating point numbers. Each vector is of dimension `n`, where `n` is an input to the function. The number of iterations of the loop depends on the value of `n`. However, even if we know an upper bound on `n`, hardware effects can still cause execution time to vary widely for similar values of `n`.

```
float dot_product(float *x, float *y, int n) {
  float result = 0.0;
  int i;
  for(i=0; i < n; i++) {
    result += x[i] * y[i];
  }
  return result;
}
```

Suppose this program is executing on a 32-bit processor with a direct-mapped cache. Suppose also that the cache can hold two sets, each of which can hold 4 `floats`. Finally, let us suppose that `x` and `y` are stored contiguously in memory starting with address 0.

Let us first consider what happens if `n` = 2. In this case, the entire arrays `x` and `y` will be in the same block and thus in the same cache set. Thus, in the very first iteration of the loop, the first access to read `x[0]` will be a cache miss, but thereafter every read to `x[i]` and `y[i]` will be a cache hit, yielding best case performance for loads.

Consider next what happens when `n` = 8. In this case, each `x[i]` and `y[i]` map to the same cache set. Thus, not only will the first access to `x[0]` be a miss, the first access to `y[0]` will also be a miss. Moreover, the latter access will evict the block containing `x[0]`-`x[3]`, leading to a cache miss on `x[1]`, `x[2]`, and `x[3]` as well. The reader can see that every access to an `x[i]` or `y[i]` will lead to a cache miss.

Thus, a seemingly small change in the value of `n` from 2 to 8 can lead to a drastic change in execution time of this function.

[1] This example is based on a similar example in Bryant and O'Hallaron (2003).

16.4 Basics of Execution Time Analysis

Execution time analysis is a current research topic, with many problems still to be solved. There have been over two decades of research, resulting in a vast literature. We cannot provide a comprehensive survey of the methods in this chapter. Instead, we will present some of the basic concepts that find widespread use in current techniques and tools for WCET analysis. Readers interested in a more detailed treatment may find an overview in a recent survey paper (Wilhelm et al., 2008) and further details in books (e.g., Li and Malik (1999)) and book chapters (e.g., Wilhelm (2005)). p.409

16.4.1 Optimization Formulation

An intuitive formulation of the WCET problem can be constructed using the view of programs as graphs. Given a program P, let $G = (V, E)$ denote its control-flow graph (CFG). Let $n = |V|$ be the number of nodes (basic blocks) in G, and $m = |E|$ denote the number of edges. We refer to the basic blocks by their index i, where i ranges from 1 to n.

We assume that the CFG has a unique *start* or *source* node s and a unique *sink* or *end* node t. This assumption is not restrictive: If there are multiple start or end nodes, one can add a dummy start/end node to achieve this condition. Usually we will set $s = 1$ and $t = n$.

Let x_i denote the number of times basic block i is executed. We call x_i the **execution count** of basic block i. Let $\mathbf{x} = (x_1, x_2, \ldots, x_n)$ be a vector of variables recording execution counts. Not all valuations of $\mathbf{x}$ correspond to valid program executions. We say that $\mathbf{x}$ is **valid** if the elements of $\mathbf{x}$ correspond to a (valid) execution of the program. The following example illustrates this point.

Example 16.9: Consider the CFG for the modular exponentiation function `modexp` introduced in Example 16.1. There are six basic blocks in this function, labeled 1 to 6 in Figure 16.1. Thus, $\mathbf{x} = (x_1, x_2, \ldots, x_6)$. Basic blocks 1 and 6, the start and end, are each executed only once. Thus, $x_1 = x_6 = 1$; any other valuation cannot correspond to any program execution.

Next consider basic blocks 2 and 3, corresponding to the conditional branches `i > 0` and `(exponent & 1) == 1`. One can observe that x_2 must equal $x_3 + 1$, since the block 3 is executed every time block 2 is executed, except when the loop exits to block 6.

Along similar lines, one can see that basic blocks 3 and 5 must be executed an equal number of times.

Flow Constraints

The intuition expressed in Example 16.9 can be formalized using the theory of **network flow**, which finds use in many contexts including modeling traffic, fluid flow, and the flow of current in an electrical circuit. In particular, in our problem context, the flow must satisfy the following two properties:

1. *Unit Flow at Source:* The control flow from source node $s = 1$ to sink node $t = n$ is a single execution and hence corresponds to unit flow from source to sink. This property is captured by the following two constraints:

$$x_1 = 1 \tag{16.6}$$

$$x_n = 1 \tag{16.7}$$

2. *Conservation of Flow:* For each node (basic block) i, the incoming flow to i from its predecessor nodes equals the outgoing flow from i to its successor nodes.

   To capture this property, we introduce additional variables to record the number of times that each edge in the CFG is executed. Following the notation of Li and Malik (1999), let d_{ij} denote the number of times the edge from node i to node j in the CFG is executed. Then we require that for each node i, $1 \leq i \leq n$,

$$x_i = \sum_{j \in P_i} d_{ji} = \sum_{j \in S_i} d_{ij}, \tag{16.8}$$

   where P_i is the set of predecessors to node i and S_i is the set of successors. For the source node, $P_1 = \emptyset$, so the sum over predecessor nodes is omitted. Similarly, for the sink node, $S_n = \emptyset$, so the sum over successor nodes is omitted.

Taken together, the two sets of constraints presented above suffice to implicitly define all source-to-sink execution paths of the program. Since this constraint-based representation is an *implicit* representation of program paths, this approach is also referred to in the literature as **implicit path enumeration** or **IPET**.

We illustrate the generation of the above constraints with an example.

Example 16.10: Consider again the function `modexp` of Example 16.1, with CFG depicted in Figure 16.1.

The constraints for this CFG are as follows:

$$\begin{aligned} x_1 &= 1 \\ x_6 &= 1 \\ x_1 &= d_{12} \\ x_2 &= d_{12} + d_{52} = d_{23} + d_{26} \\ x_3 &= d_{23} = d_{34} + d_{35} \\ x_4 &= d_{34} = d_{45} \\ x_5 &= d_{35} + d_{45} = d_{52} \\ x_6 &= d_{26} \end{aligned}$$

Any solution to the above system of equations will result in integer values for the x_i and d_{ij} variables. Furthermore, this solution will generate valid execution counts for basic blocks. For example, one such valid solution is

$$x_1 = 1,\ d_{12} = 1,\ x_2 = 2,\ d_{23} = 1,\ x_3 = 1,\ d_{34} = 0,\ d_{35} = 1,$$
$$x_4 = 0,\ d_{45} = 0,\ x_5 = 1,\ d_{52} = 1,\ x_6 = 1,\ d_{26} = 1.$$

Readers are invited to find and examine additional solutions for themselves.

Overall Optimization Problem

We are now in a position to formulate the overall optimization problem to determine worst-case execution time. The key assumption we make in this section is that we know an upper bound w_i on the execution time of the basic block i. (We will later see in Section 16.4.3 how the execution time of a single basic block can be bounded.) Then the WCET is given by the maximum $\sum_{i=1}^{n} w_i x_i$ over valid execution counts x_i.

Putting this together with the constraint formulation of the preceding section, our goal is to find values for x_i that give

$$\max_{x_i,\, 1 \leq i \leq n} \sum_{i=1}^{n} w_i x_i$$

subject to

$$x_1 = x_n = 1$$
$$x_i = \sum_{j \in P_i} d_{ji} = \sum_{j \in S_i} d_{ij}$$

This optimization problem is a form of a **linear programming** (**LP**) problem (also called a p.482 **linear program**), and it is solvable in polynomial time.

However, two major challenges remain:

- This formulation assumes that all source to sink paths in the CFG are feasible and does not bound loops in paths. As we have already seen in Section 16.3, this is not the case in general, so solving the above maximization problem may yield a pessimistic p.409 loose bound on the WCET. We will consider this challenge in Section 16.4.2.
- The upper bounds w_i on execution time of basic blocks i are still to be determined. We will briefly review this topic in Section 16.4.3.

16.4.2 Logical Flow Constraints

In order to ensure that the WCET optimization is not too pessimistic by including paths that cannot be executed, we must add so-called **logical flow constraints**. These constraints rule out infeasible paths and incorporate bounds on the number of loop iterations. We illustrate the use of such constraints with two examples.

Loop Bounds

For programs with loops, it is necessary to use bounds on loop iterations to bound execution counts of basic blocks.

Example 16.11: Consider the modular exponentiation program of Example 16.1 for which we wrote down flow constraints in Example 16.10.

Notice that those constraints impose no upper bound on x_2 or x_3. As argued in Examples 16.4 and 16.5, the bound on the number of loop iterations in this example is 32. However, without imposing this additional constraint, since there is no upper bound on x_2 or x_3, the solution to our WCET optimization will be infinite, implying that there is no upper bound on the WCET. The following single constraint suffices:

$$x_3 \leq 32$$

From this constraint on x_3, we derive the constraint that $x_2 \leq 33$, and also upper bounds on x_4 and x_5. The resulting optimization problem will then return a finite solution, for finite values of w_i.

Adding such bounds on values of x_i does not change the complexity of the optimization problem. It is still a linear programming problem.

Infeasible Paths

Some logical flow constraints rule out combinations of basic blocks that cannot appear together on a single path.

Example 16.12: Consider a snippet of code from Example 16.7 describing a software task from the open source Paparazzi unmanned aerial vehicle (UAV) project (Nemer et al., 2006):

```
#define CLIMB_MAX 1.0
...
void altitude_control_task(void) {
  ...
    err = estimator_z - desired_altitude;
    desired_climb
          = pre_climb + altitude_pgain * err;
    if (desired_climb < -CLIMB_MAX) {
      desired_climb = -CLIMB_MAX;
    }
    if (desired_climb > CLIMB_MAX) {
```

```
            desired_climb = CLIMB_MAX;
        }
        return;
}
```

The CFG for the snippet of code shown above is given in Figure 16.3. The system of flow constraints for this CFG according to the rules in Section 16.4.1 is as follows:

$$
\begin{aligned}
x_1 &= 1 \\
x_5 &= 1 \\
x_1 &= d_{12} + d_{13} \\
x_2 &= d_{12} = d_{23} \\
x_3 &= d_{13} + d_{23} = d_{34} + d_{35} \\
x_4 &= d_{34} = d_{45} \\
x_5 &= d_{35} + d_{45}
\end{aligned}
$$

A solution for the above system of equations is

$$x_1 = x_2 = x_3 = x_4 = x_5 = 1,$$

implying that each basic block gets executed exactly once, and that both conditionals evaluate to *true*. However, as we discussed in Example 16.7, it is impossible for both conditionals to evaluate to *true*. Since `CLIMB_MAX` = 1.0, if `desired_climb` is less than −1.0 in basic block 1, then at the start of basic block 3 it will be set to −1.0.

The following constraint rules out the infeasible path:

$$d_{12} + d_{34} \leq 1 \tag{16.9}$$

This constraint specifies that both conditional statements cannot be *true* together. It is of course possible for both conditionals to be *false*. We can check that this constraint excludes the infeasible path when added to the original system.

More formally, for a program *without loops*, if a set of k edges

$$(i_1, j_1), (i_2, j_2), \ldots, (i_k, j_k)$$

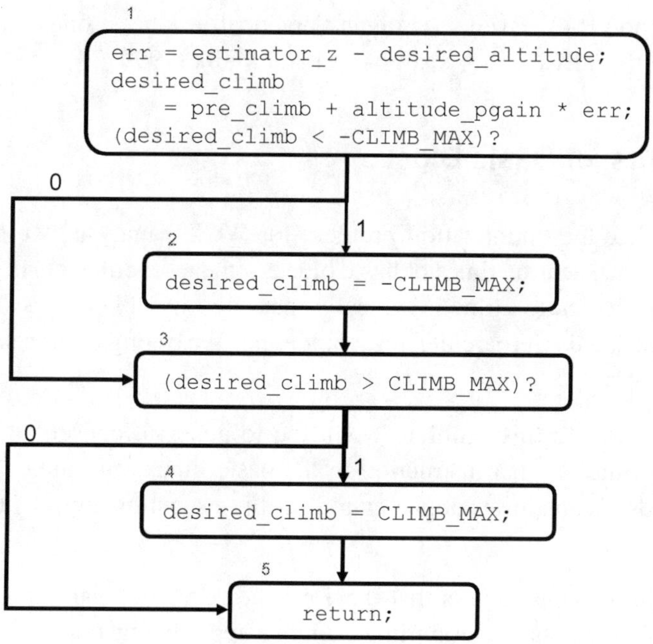

Figure 16.3: Control-flow graph for Example 16.12.

in the CFG cannot be taken together in a program execution, the following constraint is added to the optimization problem:

$$d_{i_1 j_1} + d_{i_2 j_2} + \ldots + d_{i_k j_k} \leq k - 1 \tag{16.10}$$

For programs with loops, the constraint is more complicated since an edge can be traversed multiple times, so the value of a d_{ij} variable can exceed 1. We omit the details in this case; the reader can consult Li and Malik (1999) for a more elaborate discussion of this topic.

In general, the constraints added above to exclude infeasible combinations of edges can change the complexity of the optimization problem, since one must also add the following **integrality** constraints:

$$x_i \in \mathbb{N}, \quad \text{for all } i = 1, 2, \ldots, n \tag{16.11}$$

$$d_{ij} \in \mathbb{N}, \quad \text{for all } i, j = 1, 2, \ldots, n \tag{16.12}$$

In the absence of such integrality constraints, the optimization solver can return fractional values for the x_i and d_{ij} variables. However, adding these constraints results in an integer linear programming (ILP) problem. The ILP problem is known to be NP-hard (see p.493

Appendix B, Section B.4). Even so, in many practical instances, one can solve these ILP problems fairly efficiently (see for example Li and Malik (1999)).

16.4.3 Bounds for Basic Blocks

In order to complete the optimization problem for WCET analysis, we need to compute upper bounds on the execution times of basic blocks – the w_i coefficients in the cost function of Section 16.4.1. Execution time is typically measured in CPU cycles. Generating such bounds requires detailed microarchitectural modeling. We briefly outline some of the issues in this section.

A simplistic approach to this problem would be to generate conservative upper bounds on the execution time of each instruction in the basic block, and then add up these per-instruction bounds to obtain an upper bound on the execution time of the overall basic block.

The problem with this approach is that there can be very wide variation in the execution times for some instructions, resulting in very loose upper bounds on the execution time of a basic block. For instance, consider the latency of memory instructions (loads and stores) for a system with a data cache. The difference between the latency when there is a cache miss versus a hit can be a factor of 100 on some platforms. In these cases, if the analysis does not differentiate between cache hits and misses, it is possible for the computed bound to be a hundred times larger than the execution time actually exhibited.

Several techniques have been proposed to better use program context to predict execution time of instructions more precisely. These techniques involve detailed microarchitectural modeling. We mention two main approaches below:

- *Integer linear programming (ILP) methods:* In this approach, pioneered by Li and Malik (1999), one adds **cache constraints** to the ILP formulation of Section 16.4.1. Cache constraints are linear expressions used to bound the number of cache hits and misses within basic blocks. The approach tracks the memory locations that cause *cache conflicts* – those that map onto the same cache set, but have different tags – and adds linear constraints to record the impact of such conflicts on the number of cache hits and misses. Measurement through simulation or execution on the actual platform must be performed to obtain the cycle count for hits and misses. The cost constraint of the ILP is modified to compute the program path along which the overall number of cycles, including cache hits and misses, is the largest. Further details about this approach are available in Li and Malik (1999).

- *Abstract interpretation methods:* **Abstract interpretation** is a theory of approximation of mathematical structures, in particular those that arise in defining the semantic models of computer systems (Cousot and Cousot (1977)). In particular, in abstract interpretation, one performs **sound approximation**, where the set of behaviors of the system is a subset of that of the model generated by abstract interpretation. In the context of WCET analysis, abstract interpretation has been used to infer invariants at program points, in order to generate loop bounds, and constraints on the state of processor pipelines or caches at the entry and exit locations of basic blocks. For example, such a constraint could specify the conditions under which variables will be available in the data cache (and hence a cache hit will result). Once such constraints are generated, one can run measurements from states satisfying those constraints in order to generate execution time estimates. Further details about this approach can be found in Wilhelm (2005).

In addition to techniques such as those described above, accurate measurement of execution time is critical for finding tight WCET bounds. Some of the measurement techniques are as follows:

1. *Sampling CPU cycle counter:* Certain processors include a register that records the number of CPU cycles elapsed since reset. For example, the **time stamp counter register** on x86 architectures performs this function, and is accessible through a `rdtsc` p.206 ("read time stamp counter") instruction. However, with the advent of multi-core designs and power management features, care must be taken to use such CPU cycle counters to accurately measure timing. For example, it may be necessary to lock the process to a particular CPU.

2. *Using a logic analyzer:* A **logic analyzer** is an electronic instrument used to measure signals and track events in a digital system. In the current context, the events of interest are the entry and exit points of the code to be timed, definable, for example, as valuations of the program counter. Logic analyzers are less intrusive than using cycle counters, since they do not require instrumenting the code, and they can be more accurate. However, the measurement setup is more complicated.

3. *Using a cycle-accurate simulator:* In many cases, timing analysis must be performed when the actual hardware is not yet available. In this situation, a cycle-accurate simulator of the platform provides a good alternative.

Tools for Execution-Time Analysis

Current techniques for execution-time analysis are broadly classified into those primarily based on **static analysis** and those that are **measurement-based**.

Static tools rely on abstract interpretation and **dataflow analysis** to compute facts about the program at selected program locations. These facts are used to identify dependencies between code fragments, generate loop bounds, and identify facts about the platform state, such as the state of the cache. These facts are used to guide timing measurements of basic blocks and combined into an optimization problem as presented in this chapter. Static tools aim to find conservative bounds on extreme-case execution time; however, they are not easy to port to new platforms, often requiring several man-months of effort.

Measurement-based tools are primarily based on testing the program on multiple inputs and then estimating the quantity of interest (e.g., WCET) from those measurements. Static analysis is often employed in performing a guided exploration of the space of program paths and for test generation. Measurement-based tools are easy to port to new platforms and apply broadly to both extreme-case and average-case analysis; however, not all techniques provide guarantees for finding extreme-case execution times.

Further details about many of these tools are available in Wilhelm et al. (2008); Seshia and Rakhlin (2012). Here is a partial list of tools with links to papers and websites:

Name	Primary Type	Institution & Website/References
aiT	Static	AbsInt Angewandte Informatik GmbH (Wilhelm, 2005) http://www.absint.com/ait/
Bound-T	Static	Tidorum Ltd. http://www.bound-t.com/
Chronos	Static	National University of Singapore (Li et al., 2005) http://www.comp.nus.edu.sg/~rpembed/chronos/
Heptane	Static	IRISA Rennes http://www.irisa.fr/aces/work/heptane-demo/heptane.html
SWEET	Static	Mälardalen University http://www.mrtc.mdh.se/projects/wcet/
GameTime	Measurement	UC Berkeley Seshia and Rakhlin (2008)
RapiTime	Measurement	Rapita Systems Ltd. http://www.rapitasystems.com/
SymTA/P	Measurement	Technical University Braunschweig http://www.ida.ing.tu-bs.de/research/projects/symtap/
Vienna M./P.	Measurement	Technical University of Vienna http://www.wcet.at/

16.5 Other Quantitative Analysis Problems

Although we have focused mainly on execution time in this chapter, several other quantitative analysis problems are relevant for embedded systems. We briefly describe two of these in this section.

16.5.1 Memory-bound Analysis

Embedded computing platforms have very limited memory as compared to general-purpose computers. For example, as mentioned in Chapter 9, the Luminary Micro LM3S8962 controller has only 64 KB of RAM. It is therefore essential to structure the program so that it uses memory efficiently. Tools that analyze memory consumption and compute bounds on memory usage can be very useful.

There are two kinds of memory bound analysis that are relevant for embedded systems. In **stack size analysis** (or simply **stack analysis**), one needs to compute an upper bound on the amount of stack-allocated memory used by a program. Recall from Section 9.3.2 that stack memory is allocated whenever a function is called or an interrupt is handled. If the program exceeds the memory allocated for the stack, a stack overflow is said to occur. p.242

If the program does not contain recursive functions and runs uninterrupted, one can bound stack usage by traversing the **call graph** of the program – the graph that tracks which functions call which others. If the space for each stack frame is known, then one can track p.242 the sequence of calls and returns along paths in the call graph in order to compute the worst-case stack size.

Performing stack size analysis for interrupt-driven software is significantly more complicated. We point the interested reader to Brylow et al. (2001).

Heap analysis is the other memory bound analysis problem that is relevant for embedded systems. This problem is harder than stack bound analysis since the amount of heap space used by a function might depend on the values of input data and may not be known prior to run-time. Moreover, the exact amount of heap space used by a program can depend on the implementation of dynamic memory allocation and the garbage collector. p.243

16.5.2 Power and Energy Analysis

Power and energy consumption are increasingly important factors in embedded system design. Many embedded systems are autonomous and limited by battery power, so a designer must ensure that the task can be completed within a limited energy budget. Also, the increasing ubiquity of embedded computing is increasing its energy footprint, which must be reduced for sustainable development.

To first order, the energy consumed by a program running on an embedded device depends on its execution time. However, estimating execution time alone is not sufficient. For example, energy consumption depends on circuit switching activity, which can depend more strongly on the data values with which instructions are executed.

For this reason, most techniques for energy and power estimation of embedded software focus on estimating the average-case consumption. The average case is typically estimated by profiling instructions for several different data values, guided by software benchmarks. For an introduction to this topic, see Tiwari et al. (1994).

16.6 Summary

Quantitative properties, involving physical parameters or specifying resource constraints, are central to embedded systems. This chapter gave an introduction to basic concepts in quantitative analysis. First, we considered various types of quantitative analysis problems, including extreme-case analysis, average-case analysis, and verifying threshold properties. As a representative example, this chapter focused on execution time analysis. Several examples were presented to illustrate the main issues, including loop bounds, path feasibility, path explosion, and cache effects. An optimization formulation that forms the backbone of execution time analysis was presented. Finally, we briefly discussed two other quantitative analysis problems, including computing bounds on memory usage and on power or energy consumption.

Quantitative analysis remains an active field of research – exemplifying the challenges in bridging the cyber and physical aspects of embedded systems.

Exercises

1. This problem studies execution time analysis. Consider the C program listed below:

```
int arr[100];

int foo(int flag) {
  int i;
  int sum = 0;

  if (flag) {
    for(i=0;i<100;i++)
      arr[i] = i;
  }

  for(i=0;i<100;i++)
    sum += arr[i];

  return sum;
}
```

   Assume that this program is run on a processor with data cache of size big enough that the entire array `arr` can fit in the cache.

   (a) How many paths does the function `foo` of this program have? Describe what they are.

   (b) Let T denote the execution time of the second `for` loop in the program. How does executing the first `for` loop affect the value of T? Justify your answer.

2. Consider the program given below:

```
void testFn(int *x, int flag) {
  while (flag != 1) {
    flag = 1;
    *x = flag;
  }
  if (*x > 0)
    *x += 2;
}
```

   In answering the questions below, assume that `x` is not `NULL`.

   (a) Draw the control-flow graph of this program. Identify the basic blocks with unique IDs starting with 1.

Note that we have added a dummy source node, numbered 0, to represent the entry to the function. For convenience, we have also introduced a dummy sink node, although this is not strictly required.

(b) Is there a bound on the number of iterations of the while loop? Justify your answer.

(c) How many total paths does this program have? How many of them are feasible, and why?

(d) Write down the system of flow constraints, including any logical flow constraints, for the control-flow graph of this program.

(e) Consider running this program uninterrupted on a platform with a data cache. Assume that the data pointed to by `x` is not present in the cache at the start of this function.

For each read/write access to `*x`, argue whether it will be a cache hit or miss.

Now, assume that `*x` is present in the cache at the start of this function. Identify the basic blocks whose execution time will be impacted by this modified assumption.

3. Consider the function `check_password` given below that takes two arguments: a user ID `uid` and candidate password `pwd` (both modeled as `int`s for simplicity). This function checks that password against a list of user IDs and passwords stored in an array, returning 1 if the password matches and 0 otherwise.

```
struct entry {
  int user;
  int pass;
};
typedef struct entry entry_t;

entry_t all_pwds[1000];

int check_password(int uid, int pwd) {
  int i = 0;
  int retval = 0;

  while(i < 1000) {
    if (all_pwds[i].user == uid && all_pwds[i].pass == pwd) {
      retval = 1;
      break;
    }
    i++;
  }

  return retval;
}
```

(a) Draw the control-flow graph of the function check_password. State the number of nodes (basic blocks) in the CFG. (Remember that each conditional statement is considered a single basic block by itself.)

Also state the number of paths from entry point to exit point (ignore path feasibility).

(b) Suppose the array all_pwds is sorted based on passwords (either increasing or decreasing order). In this question, we explore if an external client that calls check_password can *infer anything about the passwords* stored in all_pwds by repeatedly calling it and *recording the execution time* of check_password. Figuring out secret data from "physical" information, such as running time, is known as a *side-channel attack*.

In each of the following two cases, what, if anything, can the client infer about the passwords in all_pwds?

(i) The client has exactly one (uid, password) pair present in all_pwds

(ii) The client has NO (uid, password) pairs present in in all_pwds

Assume that the client knows the program but not the contents of the array all_pwds.

4. Consider the code below that implements the logic of a highly simplified vehicle automatic transmission system. The code aims to set the value of current_gear based on a sensor input rpm. LO_VAL and HI_VAL are constants whose exact values are irrelevant to this problem (you can assume that LO_VAL is strictly smaller than HI_VAL).

```
volatile float rpm;

int current_gear; // values range from 1 to 6

void change_gear() {
  if (rpm < LO_VAL)
    set_gear(-1);
  else {
    if (rpm > HI_VAL)
      set_gear(1);
  }

  return;
}

void set_gear(int update) {
  int new_gear = current_gear + update;
  if (new_gear > 6)
    new_gear = 6;
```

```
20    if (new_gear < 1)
21      new_gear = 1;
22
23    current_gear = new_gear;
24
25    return;
26  }
```

This is a 6-speed automatic transmission system, and thus the value of `current_gear` ranges between 1 and 6.

Answer the following questions based on the above code:

(a) Draw the control-flow graph (CFG) of the program starting from `change_gear`, *without inlining* function `set_gear`. In other words, you should draw the CFG using call and return edges.

For brevity, you need not write the code for the basic blocks inside the nodes of the CFG. Just indicate which statements go in which node by using the line numbers in the code listing above.

(b) Count the number of execution paths from the entry point in `set_gear` to its exit point (the return statement). Ignore feasibility issues for this question. Also count the number of paths from the entry point in `change_gear` to its exit point (the return statement), including the paths through `set_gear`. State the number of paths in each case.

(c) Now consider path feasibility. Recalling that `current_gear` ranges between 1 and 6, how many feasible paths does `change_gear` have? Justify your answer.

(d) Give an example of a feasible path and of an infeasible path through the function `change_gear`. Describe each path as a sequence of line numbers, ignoring the line numbers corresponding to function definitions and return statements.

17

Security and Privacy

Contents

Security and **privacy** are two of the foremost design concerns for cyber-physical systems today. Security, broadly speaking, is the state of being protected from harm. Privacy is the state of being kept away from observation. With embedded and cyber-physical systems being increasingly networked with each other and with the Internet, security and privacy concerns are now front and center for system designers.

In a formal sense, there are two primary aspects that differentiate security and privacy from other design criteria. First, the operating environment is considered to be significantly more adversarial in nature than in typical system design. Second, the kinds of properties, specifying desired and undesired behavior, are also different from traditional system specifications (and often impose additional requirements on top of the traditional ones). Let us consider each of these aspects in turn.

The notion of an **attacker** or an **adversary** is central to the theory and practice of security and privacy. An attacker is a malicious agent whose goal is to subvert the operation of the system in some way. The exact subversion depends on characteristics of the system, its goals and requirements, and the capabilities of the attacker. Typically, these characteristics are grouped together into an entity called the **threat model** or **attacker model**. For example, while designing an automobile with no wireless network connectivity, the designers may assume that the only threats arise from people who have physical access to the automobile and knowledge of its components, such as a well-trained mechanic. Transforming an informal threat model (such as the preceding sentence) into a precise, mathematical statement of an attacker's objectives is a challenging task, but one that is essential for principled model-based design of secure embedded systems.

The second defining characteristic of the field of security and privacy are the distinctive properties it is concerned with. Broadly speaking, these properties can be categorized into the following types: confidentiality, integrity, authenticity, and availability. **Confidentiality** is the state of being secret from the attacker. A good example of confidential data is a password or PIN that one uses to access one's bank account. **Integrity** is the state of being unmodified by an attacker. An example of integrity is the property that an attacker, who does not have authorized access to your bank account, cannot modify its contents. **Authenticity** is the state of knowing with a level of assurance the identity of agents you are communicating or interacting with. For instance, when you connect to a website that purports to be your bank, you want to be sure that it indeed is your bank's website and not that of some malicious entity. The process of demonstrating authenticity is known as **authentication**. Finally, **availability** is the property that a system provides a sufficient quality of service to its users. For example, you might expect your bank's website to be available to you 99% of the time.

It is important to note that security and privacy are *not* absolute properties. Do not believe anyone who claims that their system is "completely secure"! Security and privacy can only be guaranteed for a specific threat model and a specific set of properties. As a system designer, if security and privacy are important concerns for you, you must first define your threat model and formalize your properties. Otherwise, any solution you adopt is essentially meaningless.

In this chapter, we seek to give you a basic understanding of security and privacy with a special focus on concepts relevant for cyber-physical system design. The field of security and privacy is too broad for us to cover it in any comprehensive manner in a single chapter; we refer the interested reader to some excellent textbooks on the topic, e.g., (Goodrich and Tamassia, 2011; Smith and Marchesini, 2007). Instead, our goal is more modest: to introduce you to the important basic concepts, and highlight the topics that are specific to, or especially relevant for, embedded, cyber-physical systems.

17.1 Cryptographic Primitives

The field of cryptography is one of the cornerstones of security and privacy. The word "cryptography" is derived from the Latin roots "crypt" meaning *secret* and "graphia" meaning *write*, and thus literally means "the study of secret writing."

We begin with a review of the basic **cryptographic primitives** for encryption and decryption, secure hashing, and authentication. Issues that are particularly relevant for the design and analysis of embedded and cyber-physical systems are highlighted. The reader is warned that examples given in this chapter, particularly those listing code, are given at a high level of abstraction, and omit details that are critical to developing secure cryptographic implementations. See books such as (Menezes et al., 1996; Ferguson et al., 2010) for a more in-depth treatment of the subject.

17.1.1 Encryption and Decryption

Encryption is the process of translating a message into an encoded form with the intent that an adversary cannot recover the former from the latter. The original message is typically referred to as the **plaintext**, and its encoded form as the **ciphertext**. **Decryption** is the process of recovering the plaintext from the ciphertext.

The typical approach to encryption relies on the presence of a secret called the **key**. An encryption algorithm uses the key and the plaintext in a prescribed manner to obtain the ciphertext. The key is shared between the parties that intend to exchange a message securely. Depending on the mode of sharing, there are two broad categories of cryptography. In **symmetric-key cryptography**, the key is a single entity that is known to both sender and receiver. In **public-key cryptography**, also termed **asymmetric cryptography**, the key is split into two portions, a public part and a private part, where the **public key** is known to everyone (including the adversary) and the **private key** is known only to the receiver. In the rest of this section, we present a brief introduction to these two approaches to cryptography.

One of the fundamental tenets of cryptography, known as **Kerckhoff's principle**, states that a cryptographic system (algorithm) should be secure even if everything about the system, except the key, is public knowledge. In practical terms, this means that even if the adversary knows all details about the design and implementation of a cryptographic algorithm, as long as he does not know the key, he should be unable to recover the plaintext from the ciphertext.

Symmetric-Key Cryptography

Let Alice and Bob be two parties that seek to communicate with each other securely. In symmetric-key cryptography, they accomplish this using a shared secret key K. Suppose Alice wishes to encrypt and send Bob an n-bit plaintext message, $M = m_1 m_2 m_3 \ldots m_n \in \{0,1\}^n$. We wish to have an encryption scheme that, given the shared key K, should encode M into ciphertext C with the following two properites. First, the intended recipient Bob should be able to easily recover M from C. Second, any adversary who does not know K should not, by observing C, be able to gain any more information about M.

We present intuition for the operation of symmetric-key cryptography using a simple, idealized scheme known as the **one-time pad**. In this scheme, the two parties Alice and Bob share an n-bit secret key $K = k_1 k_2 k_3 \ldots k_n \in \{0,1\}^n$, where the n bits are chosen independently at random. K is known as the one-time pad. Given K, encryption works by taking the bit-wise XOR of M and K; i.e., $C = M \oplus K$ where $\oplus$ denotes XOR. Alice then sends C over to Bob, who decrypts C by taking the bit-wise XOR of C with K; using properties of the XOR operation, $C \oplus K = M$.

Suppose an adversary Eve observes C. We claim that Eve has no more information about M or K than she had without C. To see this, fix a plaintext message M. Then, every unique ciphertext $C \in \{0,1\}^n$ can be obtained from M with a corresponding unique choice of key K — simply set $K = C \oplus M$ where C is the desired ciphertext. Put another way, a uniformly random bit-string $K \in \{0,1\}^n$ generates a uniformly random ciphertext $C \in \{0,1\}^n$. Thus, looking at this ciphertext, Eve can do no better than guessing at the value of K uniformly at random.

Notice that Alice cannot use the key K more than once! Consider what happens if she does so twice. Then Eve has access to two ciphertexts $C_1 = M_1 \oplus K$ and $C_2 = M_2 \oplus K$. If Eve computes $C_1 \oplus C_2$, using properties of XOR, she learns $M_1 \oplus M_2$. Thus, Eve receives partial information about the messages. In particular, if Eve happens to learn one of the messages, she learns the other one, and can also recover the key K. Thus, this scheme is not secure if the same key is used for multiple communications, hence the name "one-time pad." Fortunately, other, stronger symmetric key cryptography schemes exist.

Most common symmetric key encryption methods use a building block termed a **block cipher**. A **block cipher** is an encryption algorithm based on using a k-bit key K to convert an n-bit plaintext message M into an n-bit ciphertext C. It can be mathematically captured in terms of an encryption function $E : \{0,1\}^k \times \{0,1\}^n \to \{0,1\}^n$; i.e., $E(K,M) = C$. For a fixed key K, the function E_K defined by $E_K(M) = E(K,M)$ must be a permutation from $\{0,1\}^n$ to $\{0,1\}^n$. Decryption is the inverse function of encryption; note that the inverse exists for each K as E_K is invertible. We denote the decryption function corresponding to E_K by $D_K = E_K^{-1}$. Note that for simplicity this model of encryption abstracts it to be (only) a function of the message and the key; in practice, care must be taken to use a suitable block cipher mode that, for example, does not always encrypt the same plaintext to the same ciphertext.

One of the classic block ciphers is the **data encryption standard** (**DES**). Introduced in the mid-1970s, DES was the first block cipher based on modern cryptographic techniques, considered to be "commercial-grade," with an open specification. While the details of DES are beyond the scope of this chapter, we note that versions of DES are still used in certain embedded devices. For instance, a version of DES, known as **3DES**, which involves applying the DES algorithm to a block three times, is used in certain "chip and PIN" payment cards around in the world. The basic version of the DES involves use of a 56-bit key and operates on 64-bit message blocks. While DES initially provided an acceptable level of security, by the mid-1990s, it was getting easier to break it using "brute-force" methods. Therefore, in 2001, the U.S. National Institute of Standards and Technology (NIST) established a new cryptographic standard known as the **advanced encryption standard** or **AES**. This standard is based on a cryptographic scheme called *Rijndael* proposed by Joan Daemen and Vincent Rijmen, two researchers from Belgium. AES uses a message block length of 128 bits and three different key lengths of 128, 192, and 256 bits. Since its adoption, AES has proved to be a strong cryptographic algorithm amenable to efficient implementation both in hardware and in software. It is estimated that the current fastest supercomputers could not successfully mount a brute-force attack on AES within the estimated lifetime of the solar system!

Public-Key Cryptography

In order to use symmetric-key cryptography, Alice and Bob need to have already set up a shared key in advance. This is not always easy to arrange. Public-key cryptography is designed to address this limitation.

In a public-key cryptosystem, each principal (Alice, Bob, etc.) has two keys: a *public key*, known to everyone, and a *private key*, known only to that principal. When Alice wants to send Bob a secret message, she obtains his public key K_B and encrypts her message with it. When Bob receives the message, he decrypts it with his private key k_B. In other words, the encryption function based on K_B must be invertible using the decryption function based on k_B, and moreover, must not be invertible without k_B.

Let us consider the last point. Encryption using a public key must effectively be a **one-way function**: a publicly-known function F such that it is easy to compute $F(M)$ but (virtually) impossible to invert without knowledge of a suitable secret (the private key). Remarkably, there is more than one public-key cryptographic scheme available today, each based on a clever combination of mathematics, algorithm design, and implementation tricks. We focus our attention here on one such scheme proposed in 1978 (Rivest et al.), known as **RSA** for the initials of its creators: Rivest, Shamir and Adleman. For brevity, we mostly limit our discussion to the basic mathematical concepts behind RSA.

Consider the setting where Bob wishes to create his public-private key pair so that others can send him secret messages. The RSA scheme involves three main steps, as detailed below:

1. *Key Generation:* Select two large prime numbers p and q, and compute their product $n = pq$. Then compute $\varphi(n)$ where φ denotes **Euler's totient function**. This function maps an integer n to the number of positive integers less than or equal to n that are relatively prime to n. For the special case of the product of primes, $\varphi(n) = (p-1)(q-1)$. Select a random integer e such that $1 < e < \varphi(n)$ such that the greatest common divisor of e and $\varphi(n)$, denoted $GCD(e, \varphi(n))$ equals 1. Then compute d, $1 < d < \varphi(n)$, such that $ed \equiv 1 \pmod{\varphi(n)}$. The key generation process is now complete. Bob's public key K_B is the pair (n, e). The private key is $k_B = d$.

2. *Encryption:* When Alice wishes to send a message to Bob, she first obtains his public key $K_B = (n, e)$. Given the message M to transmit, she computes the ciphertext C as $C = M^e \pmod{n}$. C is then transmitted to Bob.

3. *Decryption:* Upon receipt of C, Bob computes $C^d \pmod{n}$. By properties of Euler's totient function and modular arithmetic, this quantity equals M, allowing Bob to recover the plaintext message from Alice.

We make two observations on the above scheme. First, the operations in the various steps of RSA make heavy use of non-linear arithmetic on large numbers, especially modular multiplication of large integers and modular exponentiation. These operations must be implemented carefully, especially on embedded platforms, in order to operate efficiently. For instance, modular exponentiation is typically implemented using a version of the square-and-multiply technique outlined in Example 16.1 of Chapter 16. Second, the effectiveness of the approach depends critically on infrastructure to store and maintain public keys. Without such a **public-key infrastructure**, an attacker Eve can masquerade as Bob, advertising her own public key as Bob's, and thus fooling Alice into sending her messages intended for Bob. While such a public-key infrastructure has now been established for servers on the Web using so-called "certificate authorities," it is more challenging to extend that approach for networked embedded systems for a variety of reasons: the large scale of devices that may act as servers, the ad-hoc nature of networks, and the resource constraints on many embedded devices.[1] For this reason, public-key cryptography is not as widely used in embedded devices as is symmetric-key cryptography.

17.1.2 Digital Signatures and Secure Hash Functions

The cryptographic primitives for encryption and decryption help in providing confidentiality of data. However, by themselves, they are not designed to provide integrity or authenticity guarantees. Those properties require the use of related, but distinct, primitives: digital signatures, secure hash functions, and message authentication codes (MACs). This section provides a brief overview of these primitives.

Secure Hash Functions

A **secure hash function** (also known as *cryptographic hash function*) is a deterministic and keyless function $H : \{0,1\}^n \rightarrow \{0,1\}^k$ that maps n-bit messages to k-bit hash values. Typically, n can be arbitrarily large (i.e., the message can be arbitrarily long) while k is a relatively small fixed value. For example, SHA-256 is a secure hash function that maps messages to a 256-bit hash, and is being adopted for authenticating some software packages

[1] *Let's Encrypt*, an ongoing effort to develop a free, automated, and open certificate authority might mitigate some of these challenges; see https://letsencrypt.org/ for more details.

and for hashing passwords in some Linux distributions. For a message M, we term $H(M)$ as the "hash of M."

A secure hash function H has certain important properties that distinguishes it from more traditional hash functions used for non-security purposes:

- *Efficient to compute:* Given a message M, it should be efficient to compute $H(M)$.

Implementation Matters for Cryptography

Embedded platforms typically have limited resources such as memory or energy, and must obey physical constraints, for instance on real-time behavior. Cryptography can be computationally quite demanding. Such computations cannot simply be offloaded "to the cloud" as secret data must be secured at the end point before being sent out on the network. Thus, cryptographic primitives must be implemented in a manner that obeys the constraints imposed by embedded platforms.

Consider, for example, public-key cryptography. These primitives involve modular exponentiation and multiplications that can have large timing, memory, and energy requirements. Consequently, researchers have developed efficient algorithms as well as software and hardware implementations. A good example is **Montgomery multiplication** (Montgomery, 1985), a clever way of performing modular multiplication by replacing the division (implicit in computing the modulus) with addition. Another promising technology is **elliptic curve cryptography** which allows for a level of security comparable with the RSA approach with keys of smaller bit-width, reducing the memory footprint of implementations; see, for example, Paar and Pelzl (2009) for details. Due to the costs of public-key cryptography, it is often used to exchange a shared symmetric key that is then used for all subsequent communication.

Another important aspect of implementing cryptographic schemes is the design and use of **random number generation (RNG)**. A high-quality RNG requires as input a source of randomness with high entropy. Many recommended high entropy sources involve physical processes, such atmospheric noise or atomic decay, but these may not be easy to use in embedded systems, especially in hardware implementations. Alternatively, one can use on-chip thermal noise and the timing of input events from the physical world and network events. Polling the physical world for many such sources of randomness may consume power, leading to implementation trade-offs; see Perrig et al. (2002) for an example.

- *Pre-image resistance:* Given a hash (value) h, it should be computationally infeasible to find a message M such that $h = H(M)$.
- *Second pre-image resistance:* Given a message M_1, it should be computationally infeasible to find another message M_2 such that $H(M_1) = H(M_2)$. This property prevents attackers from taking a known message M_1 and modifying it to match a desired hash value (and hence find the corresponding message M_2).
- *Collision resistance:* It should be computationally infeasible to find two different messages M_1 and M_2 where $H(M_1) = H(M_2)$. This property, a stronger version of second pre-image resistance, prevents an attacker from taking *any* starting message and modifying it to match a given hash value.

A common use of secure hash functions is to verify the integrity of a message or a piece of data. For example, before you install a software update on your computer you might want to verify that the update you downloaded is a valid copy of the package you are updating and not something modified by an attacker. Providing a secure hash of the software update on a separate channel than the one on which the update file is distributed can provide such assurance. After separately downloading this hash value, you can verify, by computing the hash yourself on the software, that the provided hash indeed matches the one you compute. With the growing trend for embedded systems to be networked and to receive software patches over the network, the use of secure hash functions is expected to be a central component of any solution that maintains system integrity.

Digital Signatures

A **digital signature** is a cryptographic mechanism, based on public-key cryptography, for the author of a digital document (message) to authenticate himself/herself to a third party and to provide assurance about the integrity of the document. The use of digital signatures involves the following three stages:

1. *Key generation:* This step is identical to the key generation phase of public-key cryptosystems, resulting in a public key-private key pair.

2. *Signing:* In this step, the author/sender digitally signs the document to be sent to the receiving party.

3. *Verification:* The recipient of a signed document verifies, using the digital signature, that the sender of the document is indeed who he/she purports to be.

We now illustrate how the basic scheme works for the RSA cryptosystem. Suppose Bob wishes to digitally sign a message M before sending it to Alice. Recall from Section 17.1.1 that, using RSA, Bob can generate his public key as $K_B = (n, e)$ and private key $k_B = d$. One simple authentication scheme is for Bob to use his key-pair in reverse: to sign M, he simply encrypts it with his private key, computing $S = M^d \pmod{n}$, and sends S to Alice along with M. When Alice receives this signed message, she uses K_B to recover M from the signature S and verifies it by comparing it with the received message. If they match, the message has been authenticated; otherwise, Alice has detected tampering either in the message or its signature.

The above scheme is simple, but flawed. In particular, given the signatures S_1 and S_2 for two messages M_1 and M_2, notice that an attacker can construct the signature for message $M_1 \cdot M_2$ by simply multiplying together S_1 and $S_2 \pmod{n}$. In order to guard against this attack, one can first compute a secure hash of M_1 and M_2 before computing the signatures on the resultant hashes rather than on the messages.

Message Authentication Codes

A **message authentication code** (**MAC**) is a cryptographic mechanism, based on symmetric-key cryptography, for providing integrity and authenticity to a document. It is thus the symmetric-key analog of a digital signature. As in other symmetric-key schemes, the use of a MAC requires first setting up a shared key between sender and receiver that each of them can use to authenticate messages sent by the other. For this reason, MACs are best suited to settings where such a shared key can be easily set up. For instance, modern automobiles comprise multiple electronic control units (ECUs) communicating with each other on an on-board network such as a **controller-area network** (**CAN**) bus. In this case, if each ECU is pre-programmed with a common key, then each ECU can authenticate messages sent from other ECUs on the CAN bus using MACs.

17.2 Protocol and Network Security

An attacker typically gains access to a system via its connection to a network or some other medium used to connect it with other components. Additionally, it is increasingly the case that a single embedded system comprises many distributed components connected over a network. Thus, one must address fundamental questions about security over a network and using various communication protocols, a field termed **protocol security** or **network**

security. In this section, we review basic ideas related to two topics of particular relevance for embedded systems: *key exchange* and *cryptographic protocol design*.

17.2.1 Key Exchange

We have already seen how secret keys are critical components of both symmetric and asymmetric cryptosystems. How are these keys established in the first place? In a symmetric cryptosystem, we need a mechanism for two communicating parties to agree on a single shared key. In an asymmetric cryptosystem, we need infrastructure for establishing and maintaining public keys so that any party on a network can look up the public key of any other party it wishes to communicate with. In this section, we discuss a classic method for **key exchange** and some alternative schemes customized to specific embedded system design problems, with a focus on symmetric cryptography.

Diffie-Hellman Key Exchange

In 1976, Whitfield Diffie and Martin Hellman introduced what is widely considered the first public-key cryptosystem (Diffie and Hellman, 1976). The crux of their proposal was a clever scheme for two parties to agree on a shared secret key using a communication medium observable by anyone. This scheme is commonly referred to as **Diffie-Hellman** key exchange.

Suppose there are two parties Alice and Bob that wish to agree on a secret key. Everything Alice sends to Bob, and vice-versa, can be viewed by an attacker Eve. Alice and Bob wish to agree on a key while ensuring that it is computationally infeasible for Eve to compute that key by observing their communication. Here is how Alice and Bob can use Diffie-Hellman key exchange to achieve this objective.

To begin with, Alice and Bob need to agree on two parameters. In doing so, they can use various methods including a hard-coded scheme (e.g., a parameter hard-coded into the same program they use) or one of them can announce the parameters to the other in some deterministic fashion (e.g., based on a fixed ordering between their network addresses). The first parameter is a very large prime number p. The second is a number z such that $1 < z < p-1$. Note that an attacker can observe z and p.

After these parameters p and z are agreed upon, Alice randomly selects a number a from the set $\{0,1,\ldots,p-2\}$ and keeps it secret. Similarly, Bob randomly selects a number $b \in \{0,1,\ldots,p-2\}$ and keeps it secret. Alice computes the quantity $A = z^a \pmod{p}$ and sends it to Bob. Likewise, Bob computes $B = z^b \pmod{p}$ and sends it to Alice. In addition to z and p, the attacker Eve can observe A and B, but not a and b.

On receiving B, Alice uses her secret number a to compute $B^a \pmod{p}$. Bob performs the analogous step, computing $A^b \pmod{p}$. Now, observe that

$$A^b \pmod{p} = z^{ab} \pmod{p} = B^a \pmod{p}$$

Thus, amazingly, Alice and Bob have agreed on a shared key $K = z^{ab} \pmod{p}$ simply by communicating on a public channel. Note that they have revealed neither of the secret numbers a or b to the attacker Eve. Without knowledge of a or b, Eve cannot reliably compute K. Moreover, it is computationally very difficult for Eve to compute a or b simply by knowing p, z, A or B, since the underlying problem is one known as the **discrete logarithm** problem, for which no efficient (polynomial-time) algorithm is known. Put another way, the function $f(x) = z^x \pmod{p}$ is a one-way function.

Diffie-Hellman is simple, elegant, and effective: unfortunately, it is typically not practical for resource-constrained embedded systems. Computing it involves modular exponentiation for large primes (typical key sizes range to 2048-bits), and thus impractical for energy-constrained platforms that must obey real-time constraints. We therefore discuss next a few alternative schemes.

Timed Release of Keys

Many networked embedded systems use a **broadcast** medium for communication — one where senders send data over a public channel where all receivers, intended and unintended, can read that data. Examples include on-board automotive networks using the CAN bus and wireless sensor networks. A broadcast medium has many advantages: it is simple, requires little infrastructure, and a sender can quickly reach a large number of receivers. However, it also has certain disadvantages, notably that malicious parties can eavesdrop and inject malicious packets with ease. Reliability of the broadcast medium can also be concern.

How can one achieve secure key exchange in such a broadcast medium, under constraints on timing and energy consumption? This question was studied in the early 2000s, when the first work was done on deploying wireless sensor networks (e.g., see Perrig et al. (2004); Perrig and Tygar (2012)). One of the novel ideas is to leverage *timing properties* of these networks, an idea whose roots go back to a paper by Anderson et al. (1998).

The first property is that of **clock synchronization**, where different nodes on the network have clocks such that the difference between the values of any two clocks is bounded by a small constant. Many sensor nodes can have synchronized clocks via GPS or protocols such as the **precision time protocol** (**PTP**) (Eidson, 2006).

The second property is that of *scheduled transmission*. Each node in the network transmits packets at pre-determined times, following a schedule.

The combination of the above two properties permits the use of a secure broadcast protocol called μTESLA (Perrig et al., 2002). This protocol is designed for broadcast networks comprising two types of nodes, *base stations* and *sensor nodes*. A key property of μTESLA is *secure authentication* — a node receiving a message should be able to determine who the authentic sender is, discarding spoofed messages. It performs this by using a message authentication code (MAC) in each message, which is computed using a secret key. This secret key is broadcast some time after the original message was sent, allowing the receiver to compute the MAC for itself, and determine the authenticity of the received message. Since each message is timestamped, upon receiving a message the receiver verifies that the key used to compute the MAC has not yet been disclosed based on this timestamp, its local clock, the maximum clock skew, and the time schedule at which keys are disclosed. If the check fails, the message is discarded; otherwise, when the key is received, the authenticity of the message is checked by computing the MAC and checking against the received MAC. In this way, the *timed release* of keys is used to share keys between a broadcast sender and its receivers, for the purpose of authenticating messages.

A similar approach has been adopted for CAN-based automotive networks. Lin et al. (2013) have shown how to extend the μTESLA approach for the CAN bus where the nodes do not necessarily share a common global notion of time. Timed delayed release of keys has also been implemented for Time Triggered Ethernet (TTE) (Wasicek et al., 2011).

Other Schemes

In certain application domains, customized schemes have been developed for key exchange using special characteristics of those domains. For instance, Halperin et al. (2008) discuss a key exchange scheme for **implantable medical devices** (**IMD**s) such as pacemakers and defibrillators. They propose a "zero-power" security mechanism on the implanted device based on radio-frequency (RF) power harvesting. The novel key exchange scheme involves a "programmer" device initiating communication with the IMD (and supplying an RF signal to power it). The IMD generates a random value that is communicated as a weak modulated sound wave that is perceptible by the patient. The security of this scheme relies on a threat model based on physical security around the patient's immediate vicinity, one that is plausible for this particular application.

17.2.2 Cryptographic Protocol Design

Some of the most-publicized vulnerabilities in networked embedded systems, at the time of writing, arise from the fact that the communication protocols were designed with little or no security in mind (see the box on page 450). However, even protocols designed for security using cryptographic primitives can have vulnerabilities. Poor protocol design can compromise confidentiality, integrity, and authenticity properties even if one assumes perfect cryptography and the presence of key exchange and key management infrastructure. We explain this point with an example.

Example 17.1: A **replay attack** on a protocol is one where an attacker is able to violate a security property by replaying certain messages in the protocol, possibly with modifications to accompanying data.

Reverse Engineering Systems to Improve Security

Several embedded systems were designed at a time when security was a secondary design concern, if one at all. In recent years, researchers have shown how this lack of attention to security can lead to compromised systems in various domains. The chief approach taken is to reverse engineer protocols that provide access to systems and the working of their components.

We present here three representative examples of vulnerabilities demonstrated in real embedded systems. Halperin et al. (2008) show how implantable medical devices (IMDs) can be read and reprogrammed wirelessly. Koscher et al. (2010) demonstrate for automobiles how the on-board diagnostics bus (OBD-II) protocol can be subverted to adversarially control a wide range of functions, including disabling the brakes and stopping the engine. Ghena et al. (2014) study traffic light intersections and show how their functioning can be partially controlled wirelessly to perform, for example, a denial-of-service attack causing traffic congestion.

Such investigations based on reverse engineering "legacy" embedded systems to understand their functioning and uncover potential security flaws form an important part of the process of securing our cyber-physical infrastructure.

We describe here a replay attack on a fictitious wireless sensor network protocol. This network comprises a base station S and several energy-constrained sensor nodes $N_1, N_2, \ldots, N_k$ communicating via broadcast. Each node spends most of its time in a sleep mode, to conserve energy. When S wants to communicate with a particular node N_i to take a sensor reading, it sends it a message M encrypted with a suitable shared key K_i pre-arranged between S and N_i. N_i then responds with an encrypted sensor reading R. In other words, the protocol has a message exchange as follows:

$$\begin{aligned} S &\rightarrow N_i : E(K_i, M) \\ N_i &\rightarrow S : E(K_i, R) \end{aligned}$$

This protocol, on the face of it, seems secure. However, it has a flaw that can hinder operation of the network. Since the nodes use broadcast communication, an attacker can easily record the message $E(K_i, M)$. The attacker can then replay this message multiple times at a low power so that it is detected by N_i, but not by S. N_i will keep responding with encrypted sensor readings, possibly causing it to run down its battery much faster than anticipated, and disabling its operation.

Notice that there was nothing wrong with the individual cryptographic steps used in this protocol. The problem was with the temporal behavior of the system — that N_i had no way of checking if a message is "fresh." A simple countermeasure to this attack is to attach a fresh random value, called a **nonce**, or a timestamp with each message.

This attack is also an example of a so-called **denial-of-service attack**, abbreviated **DoS attack**, indicating that the impact of the attack is a loss of service provided by the sensor node.

Fortunately, techniques and tools exist to formally model cryptographic protocols and verify their correctness with respect to specific properties and threat models before deploying them. More widespread use of these techniques is needed to improve protocol security.

17.3 Software Security

The field of **software security** is concerned with how errors in the software implementation can impact desired security and privacy properties of the system. Bugs in software can and have been used to steal data, crash a system, and worse, allow an attacker to obtain an arbitrary level of control over a system. Vulnerabilities in software implementations are one of the largest categories of security problems in practice. These vulnerabilities can be especially dangerous for embedded systems. Much of embedded software is programmed in low-level languages such as C, where the programmer can write arbitrary code with few language-level checks. Moreover, the software often runs "bare metal" without an underlying operating system or other software layer that can monitor and trap illegal accesses or operations. Finally, we note that in embedded systems, even "crashing" the system can have severe consequences, since it can completely impair the interaction of the system with the physical world. Consider, for example, a medical device such as a pacemaker, where the consequences of a crash may be that the device stops functioning (e.g. pacing), leading to potentially life-threatening consequences for the patient.

In this section, we give the reader a very brief introduction to software security problems using a few illustrative examples. The reader is referred to relevant books on security (e.g., Smith and Marchesini (2007)) for further details.

Our examples illustrate a vulnerability known as a **buffer overflow** or **buffer overrun**. This class of errors is particularly common in low-level languages such as C. It arises from the absence of automatic bounds-checking for accessing arrays or pointers in C programs. More precisely, a buffer overflow is a error arising from the absence of a bounds check resulting in the program writing past the end of an array or memory region. Attackers can use this out-of-bounds write to corrupt trusted locations in a program such as the value of a secret variable or the return address of a function. (Tip: material in Chapter 9 on Memory Architectures might be worth reviewing.) Let us begin with a simple example.

Example 17.2: The sensors in certain embedded systems use communication protocols where data from various on-board sensors is read as a stream of bytes from a designated port or network socket. The code example below illustrates one such scenario. Here, the programmer expects to read at most 16 bytes of sensor data, storing them into the array `sensor_data`.

```
char sensor_data[16];
int secret_key;

void read_sensor_data() {
  int i = 0;

  // more_data returns 1 if there is more data,
  // and 0 otherwise
```

```
    while(more_data()) {
      sensor_data[i] = get_next_byte();
      i++;
    }

    return;
  }
```

The problem with this code is that it implicitly trusts the sensor stream to be no more than 16 bytes long. Suppose an attacker has control of that stream, either through physical access to the sensors or over the network. Then an attacker can provide more than 16 bytes and cause the program to write past the end of the array `sensor_data`. Notice further how the variable `secret_key` is defined right after `sensor_data`, and assume that the compiler allocates them adjacently. In this case, an attacker can exploit the buffer oveflow vulnerability to provide a stream of length 20 bytes and overwrite `secret_key` with a key of his choosing. This exploit can then be used to compromise the system in other ways.

The example above involves an out-of-bounds write in an array stored in global memory. Consider next the case when the array is stored on the stack. In this case, a buffer overrun vulnerability can be exploited to overwrite the return address of the function, and cause it to execute some code of the attacker's choosing. This can lead to the attacker gaining an arbitrary level of control over the embedded system.

Example 17.3: Consider below a variant on the code in Example 17.2 where the `sensor_data` array is stored on the stack. As before, the function reads a stream of bytes and stores them into `sensor_data`. However, in this case, the read sensor data is then processed within this function and used to set certain globally-stored flags (which can then be used to take control decisions).

```
int sensor_flags[4];

void process_sensor_data() {
  int i = 0;
  char sensor_data[16];

  // more_data returns 1 if there is more data,
  // and 0 otherwise
```

```
  while(more_data()) {
    sensor_data[i] = get_next_byte();
    i++;
  }

  // some code here that sets sensor_flags
  // based on the values in sensor_data

  return;
}
```

Recall from Chapter 9 how the stack frame of a function is laid out. It is possible for the return address of function `process_sensor_data` to be stored in memory right after the end of the array `sensor_data`. Thus, an attacker can exploit the buffer overflow vulnerability to overwrite the return address and cause execution to jump to a location of his choice. This version of the buffer overflow exploit is sometimes (and rather aptly) termed as **stack smashing**.

Moreover, the attacker could write a longer sequence of bytes to memory and include arbitrary code in that sequence. The overwritten return address could be tailored to be *within* this overwritten memory region, thus causing the attacker to control the code that gets executed! This attack is often referred to as a **code injection attack**.

How do we avoid buffer overflow attacks? One easy way is to explicitly check that we never write past the end of the buffer, by keeping track of its length. Another approach is to use higher-level languages that enforce **memory safety** — preventing the program from reading or writing to locations in memory that it does not have privileges for or that the programmer did not intend it to. Exercise 1 gives you some practice with writing code so as to avoid a buffer overflow.

17.4 Information Flow

Many security properties specify restrictions on the flow of information between principals. Confidentiality properties restrict the flow of secret data to channels that are readable by an attacker. For example, your bank balance should be viewable only by you and authorized bank personnel and not by an attacker. Similarly, integrity properties restrict the flow of untrusted values, controlled by an attacker, to trusted channels or locations. Your bank balance should be writable only by trusted deposit or withdrawal operations and not arbitrarily by a malicious party. Thus, **secure information flow**, the systematic study of how the flow of information in a system affects its security and privacy, is a central topic in the literature.

Given a component of an embedded system it is important to understand how information flows from secret locations to attacker-readable "public" locations, or from attacker-controlled untrusted channels to trusted locations. We need techniques to detect illegal information flows and to quantify their impact on the overall security of the system. Additionally, given security and privacy policies, we need techniques to enforce those policies on a system by suitably restricting the flow of information. Our goal, in this section, is to equip the reader with the basic principles of secure information flow so as to be able to develop such techniques.

17.4.1 Examples

In order to motivate the secure information flow problem, we present a series of examples. Although these examples focus on confidentiality, the same approaches and concerns apply for integrity properties as well.

Example 17.4: Medical devices are increasingly software-controlled, personalized, and networked. An example of such a device is a blood glucose meter, used by doctors and patients to monitor a patient's blood glucose level. Consider a hypothetical glucose meter that can take a reading of a patient's glucose level, show it on the device's display, and also transmit it over the network to the patient's hospital. Here is a highly-abstracted version of the software that might perform these tasks.

```
int patient_id; // initialized to the
                // patient's unique identifier
void take_reading() {
  float reading = read_from_sensor();

  display(reading);

  send(network_socket, hospital_server,
       reading, patient_id);

  return;
}
```

The function `take_reading` records a single blood glucose reading (e.g., in mg/dL) and stores it in a floating-point variable `reading`. It then writes this value, suitably formatted, to the device's display. Finally, it transmits the value along with the patient's ID without any encryption over the network to the hospital server for analysis by the patient's doctor.

Suppose that the patient wishes to have his glucose level kept private, in the sense that its value is known only to him and his doctor, but no one else. Does this program achieve that objective? It is easy to see that it does not, as the value of `reading` is transmitted "in the clear" over the network. We can formalize this violation of privacy as an illegal information flow from `reading` to `network_socket`, where the former is a secret location whereas the latter is a public channel visible to an attacker.

With the knowledge of cryptographic primitives presented earlier in this chapter, we know that we can do better. In particular, let us assume the presence of a shared key between the patient's device and the hospital server that permits the use of symmetric-key encryption. Consider the following new version of the above example.

Example 17.5: In this version of the program, we employ symmetric-key cryptography, denoted by `AES`, to protect the reading. The encryption process is abstractly represented by the function `enc_AES`, and `send_enc` differs from `send` only in the type of its third argument.

```
int patient_id; // initialized to the
                // patient's unique identifier
long cipher_text;

struct secret_key_s {
  long key_part1; long key_part2;
}; // struct type storing 128-bit AES key

struct secret_key_s secret_key; // shared key

void take_reading() {
  float reading = read_from_sensor();

  display(reading);

```

```
    enc_AES(&secret_key, reading, &cipher_text);

    send_enc(network_socket, hospital_server,
             cipher_text, patient_id);

    return;
}
```

In this case, there is still a flow of information from `reading` to `network_socket`. In fact, there is also a flow from `secret_key` to `network_socket`. However, both these flows passed through an encryption function denoted by `enc_AES`. In the security literature, such a function is termed a **declassifier** since it encodes the secret data in such a way that it becomes acceptable to transmit the encoded, "declassified" result to an attacker-readable channel. In other words, the declassifier serves as a "dam" blocking the flow of secret information and releasing only information that, by the properties of the cryptographic primitives, provides attackers with no more information than they had before the release.

While the above fix to the code blocks the flow of information about the reading of the patient's blood glucose, note that it does not completely protect the program. In particular, notice that the patient's ID is still being sent in the clear over the network! Thus, even though the attacker cannot tell, without the secret key, what the patient's reading is, he/she knows who transmitted a reading to the hospital (and possibly also at what time). Some private data is still being leaked. One solution to this leak is fairly obvious – we can encrypt both variables `reading` and `patient_id` using the secret key before sending them over the network.

So far we have assumed an implicit threat model where the attacker can only read information sent over the network. However, for many embedded systems, attackers may also have physical access to the device. For instance, consider what happens if the patient loses his device and an unauthorized person (attacker) gains access to it. If the patient's readings are stored on the device, the attacker might be able to read the data stored on it if such an illegal information flow exists. One approach to guard against this attack is to have the patient create a password, much like one creates a password for other personal devices, and have the device prompt one for the password when turned on. This idea is discussed in the example below.

Example 17.6: Suppose that a log of the past 100 readings is stored on the device. The code below sketches an implementation of a function `show_readings` that prompts the user for an integer password, and displays the stored readings only if the current password stored in `patient_pwd` is entered, otherwise displaying an error message that has no information about the stored readings.

```
int patient_id; // initialized to the
                // patient's unique identifier
int patient_pwd; // stored patient password

float stored_readings[100];

void show_readings() {
  int input_pwd = read_input(); // prompt user for
                                // password and read it
  if (input_pwd == patient_pwd) // check password
    display(&stored_readings);
  else
    display_error_mesg();

  return;
}
```

Assuming the attacker does not know the value of `patient_pwd`, we can see that the above code does not leak any of the 100 stored readings. However, it does leak one bit of information: whether the input password provided by the attacker equals the correct password or not. In practice, such a leak is deemed acceptable since, for a strong choice of patient password, it would take even the most determined attacker an exponential number of attempts to log in successfully, and this can be thwarted by disallowing more than a small, fixed number of attempts to provide the correct password.

The above example illustrates the concept of **quantitative information flow** (QIF). QIF is typically defined using a function measuring the amount of information flowing from a secret location to a public location. For instance, one might be interested in computing the number of bits leaked, as illustrated by the example above. If this quantity is deemed small enough, the information leak is tolerated, otherwise, the program must be rewritten.

However, the code in Example 17.6 has a potential flaw. Once again, this flaw stems from a corresponding threat model. In particular, consider the case where an attacker not only has physical possession of the device, but also has the knowledge and resources to perform *invasive* attacks by opening up the device, and reading out the contents of main memory as it executes. In this case, the attacker can read the value of `patient_pwd` simply by booting up the device and reading out the contents of memory as the system is initialized.

How can we protect against such invasive attacks? One approach involves the use of secure hardware or firmware coupled with secure hash functions. Suppose the hardware provides for secure storage where a secure hash of the user password can be written out, and any tampering detected. Then, only the secure hash need be stored in main memory, not the actual password. By the properties of a secure hash function, it would be computationally infeasible for an attacker to reverse engineer the password simply by knowing the hash.

A similar flaw exists in Example 17.5, where the secret key is stored in memory in the clear. However, in this case, simply taking a secure hash of the key is insufficient as the actual key is required to perform encryption and decryption to facilitate secure communication with the server. How do we secure this function? In general, this requires an additional component such as a secure key manager, e.g., implemented in hardware or by a trusted operating system, one that uses a master key to encrypt or decrypt the secret key for the application.

Finally, although our examples illustrate the flow of information through values of variables in software, information can also be leaked through other channels. The term **side channel** is used to refer to a channel that involves a non-traditional way of communicating information, typically using a physical quantity such as time, power consumption, or the amplitude and frequency of an audio signal, or a physical modification to the system, such as a fault, that induces a traditional information leak. We cover some basic concepts on side-channel attacks in Section 17.5.2.

17.4.2 Theory

So far we have spoken of confidentiality and integrity properties only informally. How can we state these properties precisely, similar to the notation introduced in Chapter 13? It turns out that specifying security and privacy properties formally can be quite tricky. Various formalisms have been proposed in the literature over the years, and there is no consensus on a common definition of notions such as confidentiality or integrity. However, some foundational concepts and principles have emerged. We introduce these basic concepts in this section.

The first key concept is that of **non-interference**. The term **non-interference** is used to refer to any of a family of (security) properties that specify how actions taken by one or more principals can or cannot affect ("interfere with") actions taken by others. For instance, for integrity properties, we require that actions of an attacker cannot affect the values of certain trusted data or computations. Similarly, for confidentiality properties, we require that the actions taken by an attacker cannot depend on secret values (thus implying that the attacker has no information about those secrets).

There are several variants of non-interference that are useful for expressing different kinds of security or privacy properties. In defining these, it is customary to use the terms *high* and *low* to denote two security levels. For confidentiality, the high level denotes secret data/channels and the low level denotes public data/channels. For integrity, the high level denotes trusted data/channels while the low level denotes untrusted data/channels. Non-p.67 interference is typically defined over traces of a system, where each input, output, and state element is classified as being either low or high.

The first variant we introduce is **observational determinism** (McLean, 1992; Roscoe, 1995) which is defined as follows: if two traces of a system are initialized to be in states in which the low elements are identical, and they receive the same low inputs, then that implies that the low elements of *all* states and outputs in that trace must be identical. Put another way, the low parts of a system's trace are deterministic functions of the low initial state and the low inputs, and *nothing else*. This, in turn, implies that an attacker who only controls or observes the low parts of a system's trace cannot infer anything about the high parts.

Another variant is **generalized non-interference** (McLean, 1996) which requires a system to exhibit a set of traces with a special property: for any two traces τ_1 and τ_2, there must exist a newly-constructed trace τ_3 such that, at each time step in τ_3, the high inputs are the same as in τ_1 and the low inputs, states, and outputs are the same as in τ_2. Intuitively, by observing the low states and outputs, an attacker cannot tell if she is seeing the high inputs as in τ_1 or as they occur in τ_2.

The different variants of non-interference, however, share something in common. Mathematically, they are all **hyperproperties** (Clarkson and Schneider, 2010). Formally, a **hyperproperty** is a set of sets of traces. Contrast this definition with the definition of a property as a set of traces, as introduced in Chapter 13. The latter is more accurately called a **trace property**, as its truth value can be evaluated on a single, standalone trace of a system. The truth value of a hyperproperty, in contrast, typically[2] cannot be determined by observing a single trace. One needs to observe multiple traces, and compute a relation between them, in order to determine if they together satisfy or violate a hyperproperty.

[2] Clearly, every (trace) property has an equivalent hyperproperty, which is why we use the adverb "typically."

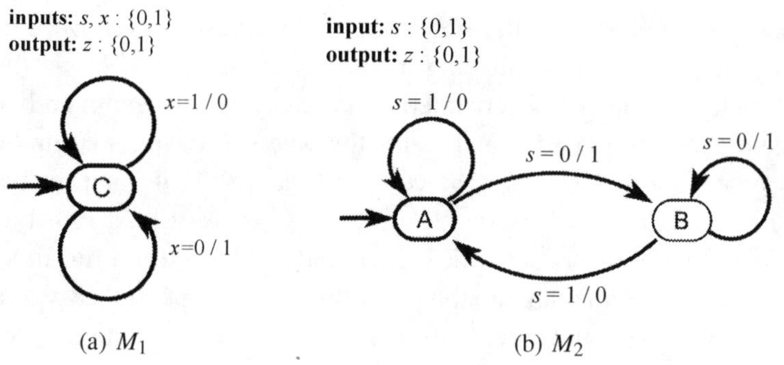

Figure 17.1: An example of a state machine satisfying observational determinism (a) and one that does not (b). Input s is a secret (high) input, and output z is a public (low) output.

Example 17.7: Consider the two state machines in Figure 17.1. In both cases, the state machine has one secret input s and one public output z. Additionally, M_1 has a public input x. An attacker can directly read the value of x and z but not s.

State machine M_1 satisfies observational determinism (OD). For any input sequence, M_1 produces a binary output sequence that depends solely on the values of the public input x and not on s.

However, M_2 does not satisfy OD. For example, consider the input sequence 1010^{ω}. Then, the corresponding output sequence of M_2 will be 0101^{ω}. If the input sequence is switched to 0101^{ω}, then the output will correspondingly change to 1010^{ω}. The adversary can gain information about the input s simply by observing the output z.

17.4.3 Analysis and Enforcement

Several techniques have been developed to analyze and track the flow of information in software systems, and, in some cases, to enforce information flow policies either at design time or run time. We mention some of these methods below, with the caveat that this is a rapidly evolving area and our selection is meant to be representative, not exhaustive.

Taint analysis. In taint analysis, the flow of information is tracked using labels ("taint") attached to data items and by monitoring memory operations. When an illegal information flow is detected, a warning is generated. Taint analysis has been mainly developed for software systems. It can be performed statically, when software is compiled, or dynamically, at run time. For instance, for the code in Example 17.4, taint analysis can detect the flow of secret data from `reading` to `network_socket`. At run time, this can be detected before the write to the network socket, and used to raise a run-time exception or other preventive action. Static taint analysis is a simple concept, and easy to apply, but can generate false alarms. Dynamic taint analysis is just as easy and more precise, but imposes a run-time overhead. System designers must evaluate for themselves the suitability of either mechanism for their contexts.

Formal verification of information flow. Techniques for formal verification, such as the model checking methods covered in Chapter 15, can also be adapted to verify secure information flow. These methods usually operate after the original analysis problem has been reduced to one of checking a safety property on a suitably constructed model (Terauchi and Aiken, 2005). Compared to taint analysis, such methods are much more precise, but they also come at a higher computational cost. As these methods are still rapidly evolving, detailed coverage of these methods is beyond the scope of this chapter at the present time.

Run-time enforcement of policies. The notion of taint allows one to specify simple information flow policies, such as disallowing the flow of secret data to any public location at all times. However, some security and privacy policies are more complex. Consider, for example, a policy that a person's location is public some of the time (e.g., while they are at work or in a public space such as an airport), but is kept secret at selected other times (e.g., when they visit their doctor). Such a policy involves a time-varying secret tag on the variable storing the person's location. The problem of specifying, checking, and enforcing such expressive policies is an active area of research.

17.5 Advanced Topics

Certain problems in security and privacy gain special significance in the context of cyber-physical systems. In this section, we review two of these problems and highlight some of the key issues.

17.5.1 Sensor and Actuator Security

Sensors and actuators form the interface between the cyber and physical worlds. In many cyber-physical systems, these components are easily observed or controlled by an attacker. Detecting attacks on these components and securing them is therefore an important problem. Central to both efforts is to develop realistic threat models and mechanisms to address those threats. We review two representative recent efforts in the area of **sensor security**.

Recent work has focused on attacks on analog sensors, i.e., developing threat models for these attacks as well as countermeasures. The main mode of attack is to employ **electromagnetic interference** (EMI) to modify the sensed signal. Two recent projects have studied EMI attacks in different applications. Foo Kune et al. (2013) investigate EMI attacks at varying power and distances on implantable medical devices and consumer electronics. Shoukry et al. (2013) study the possibility of EMI attacks that spoof sensor values for certain types of automotive sensors. We present here some basic principles from both projects.

Threat Models

In the context of sensor security, EMI can be classified along multiple dimensions. First, such interference can be *invasive*, involving modification to the sensor components, or *non-invasive*, involving observation or remote injection of spurious values. Second, it can be *unintentional* (e.g., arising from lightning strikes or transformers) or *intentional* (injected by an attacker). Third, it can be *high-power*, potentially injecting faults into or disabling the sensor, or *low-power*, which simply injects false values or modifies sensor readings. These dimensions can be used to define informal threat models.

Foo Kune et al. (2013) describe a threat model for *intentional, low-power, non-invasive* attacks. This combination of characteristics is amongst the hardest to defend against, since the low-power and non-intrusive nature of the attacks makes it potentially hard to detect. The researchers design two kinds of EMI attacks. *Baseband attacks* inject signals within the same frequency band in which the generated sensor readings lie. They can be effective on sensors that filter signals outside the operating frequency band. *Amplitude-modulated attacks* start with a carrier signal in the same frequency band of the sensor and modulate it with an attack signal. They can match the resonant frequency of a sensor, thus amplifying the impact of even a low-power attack signal. They demonstrate how these attacks can be performed on implantable cardiac devices, injecting forged signals in leads, inhibiting pacing, causing defibrillation, etc. from 1 to 2 meters away from the device.

The threat model considered by Shoukry et al. (2013) is for *intentional, non-invasive* attacks, with a focus on anti-lock braking systems (ABS) in automobiles. The magnetic wheel speed sensors are attacked by placing a *malicious actuator* in close proximity of the sensor (mountable from outside the vehicle) that modifies the magnetic field measured by the ABS sensor. One form of the attack, in which the actuator disrupts the magnetic field but is not precisely controllable by the attacker, can be "merely" disruptive. Its impact is similar to unintentional, high-power EMI. The trickier form of the attack is a *spoofing* attack, where the ABS system is deceived into measuring an incorrect but precise wheel speed for one or more of the wheels. This is achieved by implementing an active magnetic shield around the actuator. The reader is referred to the paper for additional details. The authors show how their attack can be mounted on real automotive sensors, and demonstrate, in simulation, how the ABS system can be tricked into making an incorrect braking decision in icy road conditions, causing the vehicle to slip off-road.

Countermeasures

Both of the projects mentioned above also discuss potential countermeasures to the attacks.

Foo Kune et al. (2013) consider a range of defenses, both hardware-based and software-based. Hardware or circuit-level approaches include shielding the sensor, filtering the input signals, and common mode noise rejection. However, these methods are reported to have limited effectiveness. Thus, software-based defenses are also introduced, including estimating the EMI level in the environment, adaptive filtering, cardiac probing to see if the sensed signals follow an expected pattern, reverting to safe defaults, and notifying the victim (patient) or physician about the possibility of an attack to allow them to take suitable actions to move away from the EMI source.

Shoukry et al 2015; 2016 take a somewhat different approach that is rooted in control theory and formal methods. The idea is to create a mathematical model of the system and of sensor attacks, and devise algorithms to identify subsets of sensors that are attacked, isolate them, and use the remaining (unattacked) sensors to perform state estimation and control.

17.5.2 Side-channel Attacks

Many embedded systems are accessible to an attacker not just over a network but also physically. These systems are therefore exposed to classes of attacks not possible in other settings. One such class are known as **side-channel attacks**, which involve illegal information flows through side channels. Since the seminal work by Kocher (1996), attacks

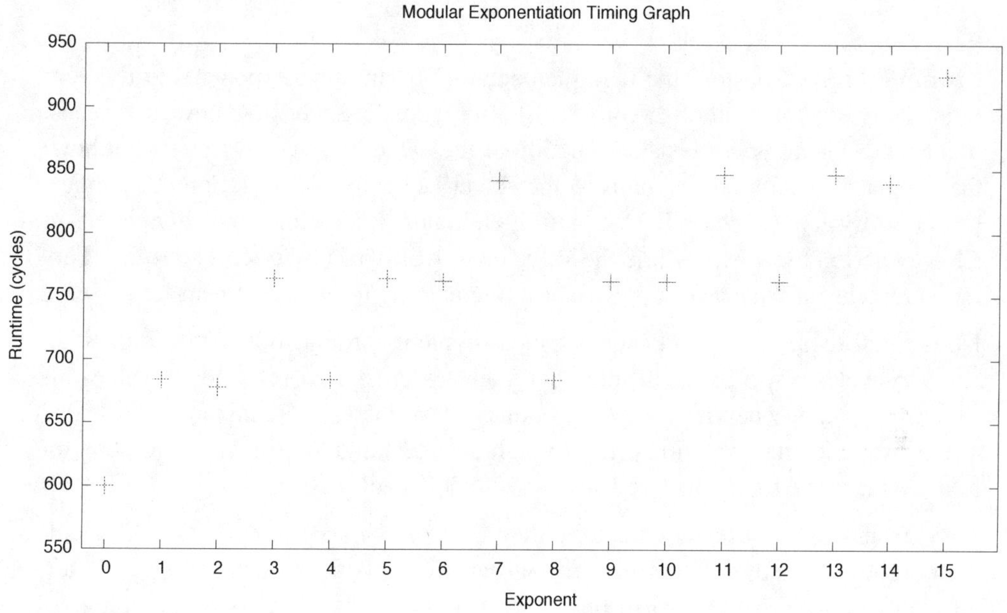

Figure 17.2: Timing data for the `modexp` function of Example 16.1. The graph shows how execution time (in CPU cycles on the y-axis) for `modexp` varies as a function of the value of the 4-bit `exponent` (listed on the x-axis).

have been demonstrated using several types of side channels, including timing (Brumley and Boneh, 2005; Tromer et al., 2010), power (Kocher et al., 1999), faults (Anderson and Kuhn, 1998; Boneh et al., 2001) memory access patterns (Islam et al., 2012), acoustic signals (Zhuang et al., 2009), and data remanence (Anderson and Kuhn, 1998; Halderman et al., 2009). **Timing attacks** involve the observation of the timing behavior of a system rather than the output values it generates. **Power attacks** involve the observation of power consumption; a particularly effective variant, known as **differential power analysis**, are based on a comparative analysis of power consumption for different executions. In **memory access pattern attacks**, observing addresses of memory locations accessed suffices to extract information about encrypted data. **Fault attacks** induce information leaks by modifying normal execution via fault injection.

While an exhaustive survey of side-channel attacks is outside the scope of this book, we illustrate, using the example below, how information can leak through a timing side channel.

Example 17.8: Consider the C implementation of modular exponentiation given as the `modexp` function of Example 16.1. In Figure 17.2 we show how the execution time of `modexp` varies as a function of the value of `exponent`. Recall how the exponent usually corresponds to the secret key in implementations of public-key cryptosystems such as RSA or Diffie-Hellman. Execution time is depicted in CPU cycles on the y axis, while the value of `exponent` ranges from 0 to 15. The measurements are made on a processor implementing the ARMv4 instruction set.

Notice that there are five clusters of measurements around 600, 700, 750, 850, and 925 cycles respectively. It turns out that each cluster corresponds to values of `exponent` that, when represented in binary, have the same number of bits set to 1. For example, the execution times for values 1, 2, 4 and 8, each corresponding to a single bit set to 1, are clustered around 700 CPU cycles.

Thus, from simply observing the execution time of `modexp` and *nothing else*, and with some knowledge of the underlying hardware platform (or access to it), an attacker can infer the number of bits set to 1 in `exponent`, i.e., in the secret key. Such information can significantly narrow down the search space for brute-force enumeration of keys.

In Example 17.8, it is assumed that an attacker can measure the execution time of a fairly small piece of code such as `modexp`, embedded within a larger program. Is this realistic? Probably not, but it does not mean timing attacks cannot be mounted. More sophisticated methods are available for an attacker to measure execution time of a procedure. For instance, Tromer et al. (2010) show how AES can be broken using a timing attack where an attacker simply induces cache hits or misses in another process (by writing to carefully chosen memory locations within its own memory segment), along with an indirect way of measuring whether the victim process suffered a cache hit or a cache miss. This can allow the attacker to know if certain table entries were looked up during the AES computation, thereby allowing him to reconstruct the key.

Side-channel attacks remind us that security compromises are often achieved by breaking assumptions made by system designers. In designing an embedded system, careful consideration must be given to the assumptions one makes and to plausible threat models, in order to achieve a reasonable level of assurance in the system's security.

17.6 Summary

Security and privacy are now amongst the top design concerns for embedded, cyber-physical systems. This chapter gave an introduction to security and privacy with a focus on topics relevant for cyber-physical systems. We covered basic cryptographic primitives covering techniques for encryption and decryption, secure hash functions, and digital signatures. The chapter also gave overviews of protocol security and software security, as well as secure information flow, a fundamental topic that cuts across many sub-areas in security and privacy. We concluded with certain advanced topics, including sensor security and side-channel attacks.

Exercises

1. Consider the buffer overflow vulnerability in Example 17.2. Modify the code so as to prevent the buffer overflow.

2. Suppose a system M has a secret input s, a public input x, and a public output z. Let all three variables be Boolean. Answer the following TRUE/FALSE questions with justification:

   (a) Suppose M satisfies the linear temporal logic (LTL) property $\mathbf{G}\neg z$. Then M must also satisfy observational determinism. p.346 p.460

   (b) Suppose M satisfies the linear temporal logic (LTL) property $\mathbf{G}\,[(s \wedge x) \Rightarrow z]$. Then M must also satisfy observational determinism.

3. Consider the finite-state machine below with one input x and one output z, both taking values in $\{0,1\}$. Both x and z are considered public ("low") signals from a security viewpoint. However, the state of the FSM (i.e., "A" or "B") is considered secret ("high").

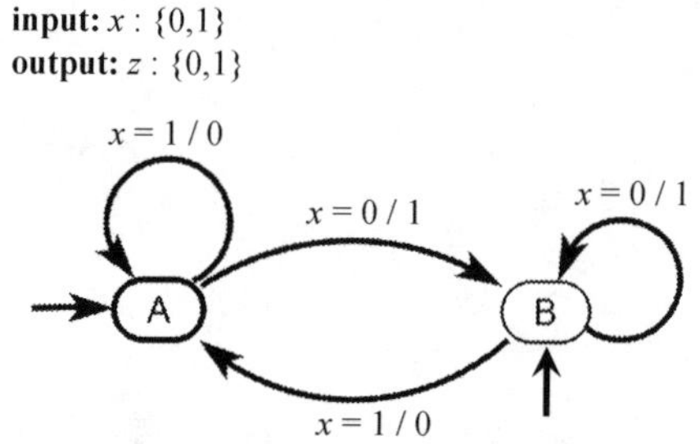

TRUE or FALSE: There is an input sequence an attacker can supply that tells him whether the state machine begins execution in A or in B.

Part IV

Appendices

This part of this text covers some background in mathematics and computer science that is useful to more deeply understand the formal and algorithmic aspects of the main text. Appendix A reviews basic notations in logic, with particular emphasis on sets and functions. Appendix B reviews notions of complexity and computability, which can help a system designer understand the cost of implementing a system and fundamental limits that make certain systems not implementable.

A

Sets and Functions

Contents

This appendix reviews some basic notation for sets and functions.

A.1 Sets

In this section, we review the notation for sets. A **set** is a collection of objects. When object a is in set A, we write $a \in A$. We define the following sets:

- $\mathbb{B} = \{0, 1\}$, the set of **binary digits**.
- $\mathbb{N} = \{0, 1, 2, \cdots\}$, the set of **natural numbers**.
- $\mathbb{Z} = \{\cdots, -1, 0, 1, 2, \cdots\}$, the set of **integers**.
- $\mathbb{R}$, the set of **real numbers**.
- $\mathbb{R}_+$, the set of **non-negative real numbers**.

When set A is entirely contained by set B, we say that A is a **subset** of B and write $A \subseteq B$. For example, $\mathbb{B} \subseteq \mathbb{N} \subseteq \mathbb{Z} \subseteq \mathbb{R}$. The sets may be equal, so the statement $\mathbb{N} \subseteq \mathbb{N}$ is true, for example. The **powerset** of a set A is defined to be the set of all subsets. It is written 2^A. The **empty set**, written $\emptyset$, is always a member of the powerset, $\emptyset \in 2^A$.

We define **set subtraction** as follows,

$$A \setminus B = \{a \in A \ : \ a \notin B\}$$

for all sets A and B. This notation is read "the set of elements a from A such that a is not in B."

A **cartesian product** of sets A and B is a set written $A \times B$ and defined as follows,

$$A \times B = \{(a,b) \ : \ a \in A, b \in B\}.$$

A member of this set (a,b) is called a **tuple**. This notation is read "the set of tuples (a,b) such that a is in A and b is in B." A cartesian product can be formed with three or more sets, in which case the tuples have three or more elements. For example, we might write $(a,b,c) \in A \times B \times C$. A cartesian product of a set A with itself is written $A^2 = A \times A$. A cartesian product of a set A with itself n times, where $n \in \mathbb{N}$ is written A^n. A member of the set A^n is called an n-**tuple**. By convention, A^0 is a **singleton set**, or a set with exactly one element, regardless of the size of A. Specifically, we define $A^0 = \{\emptyset\}$. Note that A^0 is not itself the empty set. It is a singleton set containing the empty set (for insight into the rationale for this definition, see the box on page 477).

A.2 Relations and Functions

A **relation** from set A to set B is a subset of $A \times B$. A **partial function** f from set A to set B is a relation where $(a,b) \in f$ and $(a,b') \in f$ imply that $b = b'$. Such a partial function is written $f \colon A \rightharpoonup B$. A **total function** or simply **function** f from A to B is a partial function where for all $a \in A$, there is a $b \in B$ such that $(a,b) \in f$. Such a function is written $f \colon A \to B$, and the set A is called its **domain** and the set B its **codomain**. Rather than writing $(a,b) \in f$, we can equivalently write $f(a) = b$.

Example A.1: An example of a partial function is $f \colon \mathbb{R} \rightharpoonup \mathbb{R}$ defined by $f(x) = \sqrt{x}$ for all $x \in \mathbb{R}_+$. It is undefined for any $x < 0$ in its domain $\mathbb{R}$.

A partial function $f\colon A \rightharpoonup B$ may be defined by an **assignment rule**, as done in the above example, where an assignment rule simply explains how to obtain the value of $f(a)$ given $a \in A$. Alternatively, the function may be defined by its **graph**, which is a subset of $A \times B$.

Example A.2: The same partial function from the previous example has the graph $f \subseteq \mathbb{R}^2$ given by

$$f = \{(x, y) \in \mathbb{R}^2 \;:\; x \geq 0 \text{ and } y = \sqrt{x}\} \,.$$

Note that we use the same notation f for the function and its graph when it is clear from context which we are talking about.

The **set of all functions** $f\colon A \to B$ is written $(A \to B)$ or B^A. The former notation is used when the exponential notation proves awkward. For a justification of the notation B^A, see the box on page 477.

The **function composition** of $f\colon A \to B$ and $g\colon B \to C$ is written $(g \circ f)\colon A \to C$ and defined by

$$(g \circ f)(a) = g(f(a))$$

for any $a \in A$. Note that in the notation $(g \circ f)$, the function f is applied first. For a function $f\colon A \to A$, the composition with itself can be written $(f \circ f) = f^2$, or more generally

$$\underbrace{(f \circ f \circ \cdots \circ f)}_{n \text{ times}} = f^n$$

for any $n \in \mathbb{N}$. In case $n = 1$, $f^1 = f$. For the special case $n = 0$, the function f^0 is by convention the **identity function**, so $f^0(a) = a$ for all $a \in A$. When the domain and codomain of a function are the same, i.e., $f \in A^A$, then $f^n \in A^A$ for all $n \in \mathbb{N}$.

For every function $f\colon A \to B$, there is an associated **image function** $\hat{f}\colon 2^A \to 2^B$ defined on the powerset of A as follows, p.472

$$\forall\, A' \subseteq A, \quad \hat{f}(A') = \{b \in B \;:\; \exists\, a \in A',\ f(a) = b\}.$$

The image function $\hat{f}$ is applied to *sets* A' of elements in the domain, rather than to single elements. Rather than returning a single value, it returns the set of all values that f would return, given an element of A' as an argument. We call $\hat{f}$ the **lifted** version of f. When there is no ambiguity, we may write the lifted version of f simply as f rather than $\hat{f}$ (see problem 2(c) for an example of a situation where there is ambiguity).

For any $A' \subseteq A$, $\hat{f}(A')$ is called the **image** of A' for the function f. The image $\hat{f}(A)$ of the domain is called the **range** of the function f.

Example A.3: The image $\hat{f}(\mathbb{R})$ of the function $f\colon \mathbb{R} \to \mathbb{R}$ defined by $f(x) = x^2$ is $\mathbb{R}_+$.

A function $f\colon A \to B$ is **onto** (or **surjective**) if $\hat{f}(A) = B$. A function $f\colon A \to B$ is **one-to-one** (or **injective**) if for all $a, a' \in A$,

$$a \neq a' \Rightarrow f(a) \neq f(a'). \tag{A.1}$$

That is, no two distinct values in the domain yield the same values in the codomain. A function that is both one-to-one and onto is **bijective**.

Example A.4: The function $f\colon \mathbb{R} \to \mathbb{R}$ defined by $f(x) = 2x$ is bijective. The function $f\colon \mathbb{Z} \to \mathbb{Z}$ defined by $f(x) = 2x$ is one-to-one, but not onto. The function $f\colon \mathbb{R}^2 \to \mathbb{R}$ defined by $f(x,y) = xy$ is onto but not one-to-one.

The previous example underscores the fact that an essential part of the definition of a function is its domain and codomain.

Proposition A.1. *If $f\colon A \to B$ is onto, then there is a one-to-one function $h\colon B \to A$.*

Proof. Let h be defined by $h(b) = a$ where a is any element in A such that $f(a) = b$. There must always be at least one such element because f is onto. We can now show that h is one-to-one. To do this, consider any two elements $b, b' \in B$ where $b \neq b'$. We need to show that $h(b) \neq h(b')$. Assume to the contrary that $h(b) = h(b') = a$ for some $a \in A$. But then by the definition of h, $f(a) = b$ and $f(a) = b'$, which implies $b = b'$, a contradiction.

□

The converse of this proposition is also easy to prove.

Proposition A.2. *If $h\colon B \to A$ is one-to-one, then there is an onto function $f\colon A \to B$.*

Any bijection $f\colon A \to B$ has an **inverse** $f^{-1}\colon B \to A$ defined as follows,

$$f^{-1}(b) = a \in A \text{ such that } f(a) = b\,, \tag{A.2}$$

for all $b \in B$. This function is defined for all $b \in B$ because f is onto. And for each $b \in B$ there is a single unique $a \in A$ satisfying (A.2) because f is one-to-one. For any bijection f, its inverse is also bijective.

A.2.1 Restriction and Projection

Given a function $f\colon A \to B$ and a subset $C \subseteq A$, we can define a new function $f|_C$ that is the **restriction** of f to C. It is defined so that for all $x \in C$, $f|_C(x) = f(x)$.

Example A.5: The function $f\colon \mathbb{R} \to \mathbb{R}$ defined by $f(x) = x^2$ is not one-to-one. But the function $f|_{\mathbb{R}_+}$ is.

Consider an n-tuple $a = (a_0, a_1, \cdots, a_{n-1}) \in A_0 \times A_1 \times \cdots \times A_{n-1}$. A **projection** of this n-tuple extracts elements of the tuple to create a new tuple. Specifically, let

$$I = (i_0, i_1, \cdots, i_m) \in \{0, 1, \cdots, n-1\}^m$$

for some $m \in \mathbb{N} \setminus \{0\}$. That is, I is an m-tuple of indexes. Then we define the projection of a onto I by

$$\pi_I(a) = (a_{i_0}, a_{i_1}, \cdots, a_{i_m}) \in A_{i_0} \times A_{i_1} \times \cdots \times A_{i_m}\,.$$

The projection may be used to permute elements of a tuple, to discard elements, or to repeat elements.

Projection of a tuple and restriction of a function are related. An n-tuple $a \in A^n$ where $a = (a_0, a_1, \cdots, a_{n-1})$ may be considered a function of the form $a\colon \{0, 1, \cdots, n-1\} \to A$, in which case $a(0) = a_0$, $a(1) = a_1$, etc. Projection is similar to restriction of this function, differing in that restriction, by itself, does not provide the ability to permute, repeat, or renumber elements. But conceptually, the operations are similar, as illustrated by the following example.

Example A.6: Consider a 3-tuple $a = (a_0, a_1, a_2) \in A^3$. This is represented by the function $a\colon \{0,1,2\} \to A$. Let $I = \{1,2\}$. The projection $b = \pi_I(a) = (a_1, a_2)$, which itself can be represented by a function $b\colon \{0,1\} \to A$, where $b(0) = a_1$ and $b(1) = a_2$.

The restriction $a|_I$ is not exactly the same function as b, however. The domain of the first function is $\{1,2\}$, whereas the domain of the second is $\{0,1\}$. In particular, $a|_I(1) = b(0) = a_1$ and $a|_I(2) = b(1) = a_2$.

p.473 A projection may be lifted just like ordinary functions. Given a set of n-tuples $B \subseteq A_0 \times A_1 \times \cdots \times A_{n-1}$ and an m-tuple of indexes $I \in \{0,1,\cdots,n-1\}^m$, the **lifted projection** is

$$\hat{\pi}_I(B) = \{\pi_I(b) \ : \ b \in B\} .$$

A.3 Sequences

A tuple $(a_0, a_1) \in A^2$ can be interpreted as a sequence of length 2. The order of elements in the sequence matters, and is in fact captured by the natural ordering of the natural numbers. The number 0 comes before the number 1. We can generalize this and recognize that a **sequence** of elements from set A of length n is an n-tuple in the set A^n. A^0 represents the
p.472 set of empty sequences, a singleton set (there is only one empty sequence).

The set of all **finite sequences** of elements from the set A is written A^*, where we interpret $*$ as a wildcard that can take on any value in $\mathbb{N}$. A member of this set with length n is an n-tuple, a **finite sequence**.

The set of **infinite sequences** of elements from A is written $A^{\mathbb{N}}$ or A^{ω}. The set of **finite and infinite sequences** is written

$$A^{**} = A^* \cup A^{\mathbb{N}} .$$

p.104 Finite and infinite sequences play an important role in the semantics of concurrent programs. They can be used, for example, to represent streams of messages sent from one part of the program to another. Or they can represent successive assignments of values to a variable. For programs that terminate, finite sequences will be sufficient. For programs that do not terminate, we need infinite sequences.

Exponential Notation for Sets of Functions

The exponential notation B^A for the set of functions of form $f: A \to B$ is worth explaining. Recall that A^2 is the cartesian product of set A with itself, and that 2^A is the powerset of A. These two notations are naturally thought of as sets of functions. A construction attributed to John von Neumann defines the natural numbers as follows, p.472

$$\begin{aligned} \mathbf{0} &= \emptyset \\ \mathbf{1} &= \{\mathbf{0}\} = \{\emptyset\} \\ \mathbf{2} &= \{\mathbf{0},\mathbf{1}\} = \{\emptyset,\{\emptyset\}\} \\ \mathbf{3} &= \{\mathbf{0},\mathbf{1},\mathbf{2}\} = \{\emptyset,\{\emptyset\},\{\emptyset,\{\emptyset\}\}\} \\ &\cdots \end{aligned}$$

With this definition, the powerset $\mathbf{2}^A$ is the set of functions mapping the set A into the set $\mathbf{2}$. Consider one such function, $f \in \mathbf{2}^A$. For each $a \in A$, either $f(a) = \mathbf{0}$ or $f(a) = \mathbf{1}$. If we interpret "**0**" to mean "nonmember" and "**1**" to mean "member," then indeed the set of functions $\mathbf{2}^A$ represents the set of all subsets of A. Each such function defines a subset.

Similarly, the cartesian product $A^{\mathbf{2}}$ can be interpreted as the set of functions of form $f: \mathbf{2} \to A$, or using von Neumann's numbers, $f: \{\mathbf{0},\mathbf{1}\} \to A$. Consider a tuple $a = (a_0, a_1) \in A^{\mathbf{2}}$. It is natural to associate with this tuple a function $a: \{\mathbf{0},\mathbf{1}\} \to A$ where $a(\mathbf{0}) = a_0$ and $a(\mathbf{1}) = a_1$. The argument to the function is the index into the tuple. We can now interpret the set of functions B^A of form $f: A \to B$ as a set of tuples indexed by the set A instead of by the natural numbers.

Let $\omega = \{\emptyset, \{\emptyset\}, \{\emptyset, \{\emptyset\}\}, \cdots\}$ represent the set of **von Neumann numbers**. This set is closely related to the set $\mathbb{N}$ (see problem 2). Given a set A, it is now natural to interpret A^ω as the set of all infinite sequences of elements from A, the same as $A^{\mathbb{N}}$.

The singleton set $A^{\mathbf{0}}$ can now be interpreted as the set of all functions whose domain is the empty set and codomain is A. There is exactly one such function (no two such functions are distinguishable), and that function has an empty graph. Before, we defined $A^{\mathbf{0}} = \{\emptyset\}$. Using von Neumann numbers, $A^{\mathbf{0}} = \mathbf{1}$, corresponding nicely with the definition of a zero exponent on ordinary numbers. Moreover, you can think of $A^{\mathbf{0}} = \{\emptyset\}$ as the set of all functions with an empty graph. p.472 p.473

It is customary in the literature to omit the bold face font for $A^{\mathbf{0}}$, $\mathbf{2}^A$, and $A^{\mathbf{2}}$, writing instead simply A^0, 2^A, and A^2.

Exercises

1. This problem explores properties of onto and one-to-one functions.

   (a) Show that if $f\colon A \to B$ is onto and $g\colon B \to C$ is onto, then $(g \circ f)\colon A \to C$ is onto.

   (b) Show that if $f\colon A \to B$ is one-to-one and $g\colon B \to C$ is one-to-one, then $(g \circ f)\colon A \to C$ is one-to-one.

p.477

2. Let $\omega = \{\emptyset, \{\emptyset\}, \{\emptyset, \{\emptyset\}\}, \cdots\}$ be the von Neumann numbers as defined in the box on page 477. This problem explores the relationship between this set and $\mathbb{N}$, the set of natural numbers.

   (a) Let $f\colon \omega \to \mathbb{N}$ be defined by

$$f(x) = |x|, \quad \forall\, x \in \omega\,.$$

p.474

   That is, $f(x)$ is the size of the set x. Show that f is bijective.

p.473

   (b) The lifted version of the function f in part (a) is written $\hat{f}$. What is the value of $\hat{f}(\{0, \{0\}\})$? What is the value of $f(\{0, \{0\}\})$? Note that on page 473 it is noted that when there is no ambiguity, $\hat{f}$ may be written simply f. For this function, is there such ambiguity?

B

Complexity and Computability

Contents

Complexity theory and **computability theory** are areas of Computer Science that study the *efficiency* and the *limits of computation*. Informally, computability theory studies *which problems can be solved* by computers, while complexity theory studies *how efficiently* a problem can be solved by computers. Both areas are *problem-centric*, meaning that they are more concerned with the intrinsic ease or difficulty of problems and less concerned with specific techniques (algorithms) for solving them.

In this appendix, we very briefly review selected topics from complexity and computability theory that are relevant for this book. There are excellent books that offer a detailed treatment of these topics, including Papadimitriou (1994), Sipser (2005), and Hopcroft et al. (2007). We begin with a discussion of the complexity of algorithms. Algorithms are realized by computer programs, and we show that there are limitations on what computer programs can do. We then describe Turing machines, which can be used to define what we have come to accept as "computation," and show how the limitations of programs manifest themselves as undecidable problems. Finally, we close with a discussion of the complexity of *problems*, as distinct from the complexity of the algorithms that solve the problems.

B.1 Effectiveness and Complexity of Algorithms

An **algorithm** is a step-by-step procedure for solving a problem. To be **effective**, an algorithm must complete in a finite number of steps and use a finite amount of resources (such as memory). To be **useful**, an algorithm must complete in a *reasonable* number of steps and use a reasonable amount of resources. Of course, what is "reasonable" will depend on the problem being solved.

Some problems are known to have no effective algorithm, as we will see below when we discuss undecidability. For other problems, one or more effective algorithms are known, but it is not known whether the best algorithm has been found, by some measure of "best." There are even problems where we know that there exists an effective algorithm, but no effective algorithm is known. The following example describes such a problem.

Example B.1: Consider a function $f\colon \mathbb{N} \to \{\text{YES, NO}\}$ where $f(n) = \text{YES}$ if there is a sequence of n consecutive fives in the decimal representation of π, and $f(n) = \text{NO}$ otherwise. This function has one of two forms. Either

$$f(n) = \text{YES} \quad \forall\, n \in \mathbb{N},$$

or there is a $k \in \mathbb{N}$ such that

$$f(n) = \begin{cases} \text{YES} & \text{if } n < k \\ \text{NO} & \text{otherwise} \end{cases}$$

It is not known which of these two forms is correct, nor, if the second form is correct, what k is. However, no matter what the answer is, there is an effective algorithm for solving this problem. In fact, the algorithm is rather simple. Either the algorithm immediately returns YES, or it compares n to k and returns YES if

$n < k$. We know that one of these is the right algorithm, but we do not know which. Knowing that one of these is correct is sufficient to know that there is an effective algorithm.

For a problem with known effective algorithms, there are typically many algorithms that will solve the problem. Generally, we prefer algorithms with lower complexity. How do we choose among these? This is the topic of the next subsection.

B.1.1 Big O Notation

Many problems have several known algorithms for solving them, as illustrated in the following example.

Example B.2: Suppose we have a list $(a_1, a_2, \cdots, a_n)$ of n integers, arranged in increasing order. We would like to determine whether the list contains a particular integer b. Here are two algorithms that accomplish this:

1. Use a **linear search**. Starting at the beginning of the list, compare the input b against each entry in the list. If it is equal, return YES. Otherwise, proceed to the next entry in the list. In the worst case, this algorithm will require n comparisons before it can give an answer.
2. Use a **binary search**. Start in the middle of the list and compare b to the entry $a_{(n/2)}$ in the middle. If it is equal, return YES. Otherwise, determine whether $b < a_{(n/2)}$. If it is, then repeat the search, but over only the first half of the list. Otherwise, repeat the search over the second half of the list. Although each step of this algorithm is more complicated than the steps of the first algorithm, usually fewer steps will be required. In the worst case, $\log_2(n)$ steps are required.

The difference between these two algorithms can be quite dramatic if n is large. Suppose that $n = 4096$. The first algorithm will require 4096 steps in the worst case, whereas the second algorithm will require only 12 steps in the worst case.

The number of steps required by an algorithm is the **time complexity** of the algorithm. It is customary when comparing algorithms to simplify the measure of time complexity by ignoring some details. In the previous example, we might ignore the complexity of each step of the algorithm and consider only how the complexity grows with the input size n. So if algorithm (1) in Example B.2 takes $K_1 n$ seconds to execute, and algorithm (2) takes $K_2 \log_2(n)$ seconds to execute, we would typically ignore the constant factors K_1 and K_2. For large n, they are not usually very helpful in determining which algorithm is better.

To facilitate such comparisons, it is customary to use **big O notation**. This notation finds the term in a time complexity measure that grows fastest as a function of the size of the input, for large input sizes, and ignores all other terms. In addition, it discards any constant factors in the term. Such a measure is an **asymptotic complexity** measure because it studies only the limiting growth rate as the size of the input gets large.

Example B.3: Suppose that an algorithm has time complexity $5 + 2n + 7n^3$, where n is the size of the input. This algorithm is said to have $O(n^3)$ time complexity, which is read "order n cubed." The term $7n^3$ grows fastest with n, and the number 7 is a relatively unimportant constant factor.

The following complexity measures are commonly used:

1. **constant time**: The time complexity does not depend at all on the size of the input. The complexity is $O(1)$.
2. **logarithmic time**: $O(\log_m(n))$ complexity, for any fixed m.
3. **linear time**: $O(n)$ complexity.
4. **quadratic time**: $O(n^2)$ complexity.
5. **polynomial time**: $O(n^m)$ complexity, for any fixed $m \in \mathbb{N}$.
6. **exponential time**: $O(m^n)$ complexity for any $m > 1$.
7. **factorial time**: $O(n!)$ complexity.

The above list is ordered by costliness. Algorithms later in the list are usually more expensive to realize than algorithms earlier in the list, at least for large input size n.

Example B.4: Algorithm 1 in Example B.2 is a linear-time algorithm, whereas algorithm 2 is a logarithmic-time algorithm. For large n, algorithm (2) is more efficient.

The number of steps required by an algorithm, of course, is not the only measure of its cost. Some algorithms execute in rather few steps but require a great deal of memory. The size of the memory required can be similarly characterized using big O notation, giving a measure of **space complexity**.

B.2 Problems, Algorithms, and Programs

Algorithms are developed to solve some problem. How do we know whether we have found the best algorithm to solve a problem? The time complexity of *known* algorithms can be compared, but what about algorithms we have not thought of? Are there problems for which there is no algorithm that can solve them? These are difficult questions.

Assume that the input to an algorithm is a member of a set W of all possible inputs, and the output is a member of a set Z of all possible outputs. The algorithm computes a function $f\colon W \to Z$. The function f, a mathematical object, is the **problem** to be solved, and the algorithm is the **mechanism** by which the problem is solved.

It is important to understand the distinction between the problem and the mechanism. Many different algorithms may solve the same problem. Some algorithms will be better than others; for example, one algorithm may have lower time complexity than another. We next address two interesting questions:

- Is there a function of the form $f\colon W \to Z$ for which there is no algorithm that can compute the function for all inputs $w \in W$? This is a computability question.
- Given a particular function $f\colon W \to Z$, is there a lower bound on the time complexity of an algorithm to compute the function? This is a complexity question.

If W is a finite set, then the answer to the first question is clearly no. Given a particular function $f\colon W \to Z$, one algorithm that will always work uses a lookup table listing $f(w)$ for all $w \in W$. Given an input $w \in W$, this algorithm simply looks up the answer in the table. This is a constant-time algorithm; it requires only one step, a table lookup. Hence, this algorithm provides the answer to the second question, which is that if W is a finite set, then the lowest time complexity is constant time.

A lookup table algorithm may not be the best choice, even though its time complexity is constant. Suppose that W is the set of all 32-bit integers. This is a finite set with 2^{32} elements, so a table will require more than four billion entries. In addition to time complexity, we must consider the memory required to implement the algorithm.

The above questions become particularly interesting when the set W of possible inputs is infinite. We will focus on **decision problems**, where $Z = \{\text{YES}, \text{NO}\}$, a set with only two elements. A decision problem seeks a yes or no answer for each $w \in W$. The simplest p.471 infinite set of possible inputs is $W = \mathbb{N}$, the natural numbers. Hence, we will next consider fundamental limits on decision problems of the form $f\colon \mathbb{N} \to \{\text{YES}, \text{NO}\}$. We will see next that for such problems, the answer to the first question above is yes. There are functions of this form that are not computable.

B.2.1 Fundamental Limitations of Programs

One way to describe an algorithm is to give a computer program. A computer program is p.476 always representable as a member of the set $\{0,1\}^*$, i.e., the set of finite sequences of bits. A **programming language** is a subset of $\{0,1\}^*$. It turns out that not all decision problems can be solved by computer programs.

Proposition B.1. *No programming language can express a program for each and every function of the form* $f\colon \mathbb{N} \to \{\mathit{YES}, \mathit{NO}\}$.

Proof. To prove this proposition, it is enough to show that there are strictly more functions of the form $f\colon \mathbb{N} \to \{\text{YES}, \text{NO}\}$ than there are programs in a programming language. It is sufficient to show that the set $\{\text{YES}, \text{NO}\}^{\mathbb{N}}$ is strictly larger than the set $\{0,1\}^*$, because a programming language is a subset of $\{0,1\}^*$. This can be done with a variant of **Cantor's diagonal argument**, which goes as follows.

First, note that the members of the set $\{0,1\}^*$ can be listed in order. Specifically, we list them in the order of binary numbers,

$$\lambda, 0, 1, 00, 01, 10, 11, 000, 001, 010, 011, \cdots, \tag{B.1}$$

where λ is the empty sequence. This list is infinite, but it includes all members of the set $\{0,1\}^*$. Because the members of the set can be so listed, the set $\{0,1\}^*$ is said to be **countable** or **countably infinite**.

For any programming language, every program that can be written will appear somewhere in the list (B.1). Assume the first such program in the list realizes the decision function $f_1 \colon \mathbb{N} \to \{\text{YES, NO}\}$, the second one in the list realizes $f_2 \colon \mathbb{N} \to \{\text{YES, NO}\}$, etc. We can now construct a function $g \colon \mathbb{N} \to \{\text{YES, NO}\}$ that is not computed by any program in the list. Specifically, let

$$g(i) = \begin{cases} \text{YES} & \text{if } f_i(i) = \text{NO} \\ \text{NO} & \text{if } f_i(i) = \text{YES} \end{cases}$$

for all $i \in \mathbb{N}$. This function g differs from every function f_i in the list, and hence it is not included in the list. Thus, there is no computer program in the language that computes function g.

□

This theorem tells us that programs, and hence algorithms, are not capable of solving all decision problems. We next explore the class of problems they can solve, known as the **effectively computable** functions. We do this using Turing machines.

B.3 Turing Machines and Undecidability

In 1936, **Alan Turing** proposed a model for computation that is now called the **Turing machine** (Turing, 1936). A Turing machine, depicted in Figure B.1, is similar to a finite-state machine, but with an unlimited amount of memory. This memory has the form p.47 of an infinite tape that the Turing machine can read from and write to. The machine comprises a finite-state machine (FSM) controller, a read/write head, and an infinite tape organized as a sequence of cells. Each cell contains a value drawn from a finite set Σ or the special value □, which indicates an **empty cell**. The FSM acts as a control for the read/write head by producing outputs that move the read/write head over the tape.

In Figure B.1, the symbols on the non-empty cells of the tape are drawn from the set $\Sigma = \{0, 1\}$, the binary digits. The FSM has two output ports. The top output port is *write*, which has type Σ and produces a value to write to the cell at the current position of the read/write p.45 head. The bottom output port is *move*, which has type $\{L, R\}$, where the output symbol L causes the read/write head to move to the left (but not off the beginning of the tape), and R causes it to move to the right. The FSM has one input port, *read*, which has type Σ and receives the current value held by the cell under the read/write head.

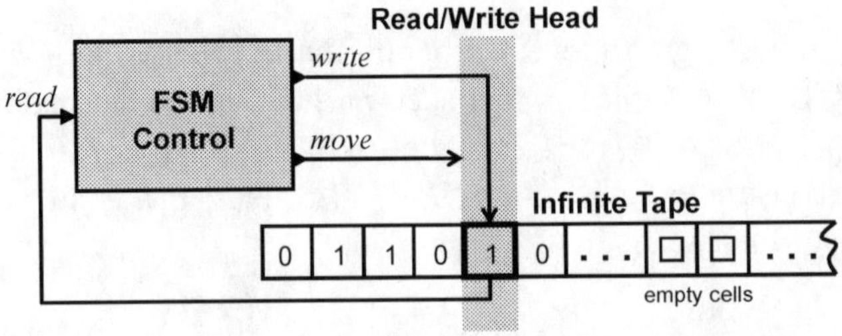

Figure B.1: Illustration of a Turing machine.

The tape is initialized with an **input string**, which is an element of the set Σ^* of p.476 finite sequences of elements of Σ, followed by an infinite sequence of empty cells. The Turing machine starts in the initial state of the FSM with the read/write head on the left p.44 end of the tape. At each reaction, the FSM receives as input the value from the current cell under the read/write head. It produces an output that specifies the new value of that cell (which may be the same as the current value) and a command to move the head left or right.

p.367 The control FSM of the Turing machine has two final states: an **accepting state** accept and a **rejecting state** reject. If the Turing machine reaches accept or reject after a finite number of reactions, then it is said to **terminate**, and the execution is called a **halting computation**. If it terminates in accept, then the execution is called an **accepting computation**. If it terminates in reject, then the execution is called a **rejecting computation**. It is also possible for a Turing machine to reach neither accept nor reject, meaning that it does not halt. When a Turing machine does not halt, we say that it **loops**.

p.56 When the control FSM is deterministic, we say that the Turing machine is also deterministic. Given an input string $w \in \Sigma^*$, a deterministic Turing machine D will exhibit a unique computation. Therefore, given an input string $w \in \Sigma^*$, a deterministic Turing machine D will either halt or not, and if it halts, it will either accept w or reject it. For simplicity, we will limit ourselves in this section to deterministic Turing machines, unless explicitly stated otherwise.

B.3.1 Structure of a Turing Machine

More formally, each **deterministic Turing machine** can be represented by a pair $D = (\Sigma, M)$, where Σ is a finite set of symbols and M is any FSM with the following properties:

- a finite set $States_M$ of states that includes two final states accept and reject;

- an input port *read* of type Σ; p.45
- an output port *write* of type Σ; and
- an output port *move* of type $\{L, R\}$.

As with any FSM, it also must have an initial state s_0 and a transition function $update_M$, as explained in Section 3.3.3. If the read/write head is over a cell containing $\square$, then the input to the *read* port of the FSM will be *absent*. If at a reaction the *write* output of the FSM is *absent*, then the cell under the read/write head will be erased, setting its contents to $\square$. p.55

A Turing machine described by $D = (\Sigma, M)$ is a synchronous composition of two machines, the FSM M and a tape T. The tape T is distinctly not an FSM, since it does not have finite state. Nonetheless, the tape is a (extended) state machine, and can be described using the same five-tuple used in Section 3.3.3 for FSMs, except that the set $States_T$ is now infinite. The data on the tape can be modeled as a function with domain $\mathbb{N}$ and codomain $\Sigma \cup \{\square\}$, and the position of the read/write head can be modeled as a natural number, so p.107 p.47

$$States_T = \mathbb{N} \times (\Sigma \cup \{\square\})^{\mathbb{N}}.$$

The machine T has input port *write* of type Σ, input port *move* of type $\{L, R\}$, and output port *read* of type Σ. The $update_T$ transition function is now easy to define formally (see Exercise 1).

Note that the machine T is the same for all Turing machines, so there is no need to include it in the description $D = (\Sigma, M)$ of a particular Turing machine. The description D can be understood as a program in a rather special programming language. Since all sets in the formal description of a Turing machine are finite, any Turing machine can be encoded as a finite sequence of bits in $\{0, 1\}^*$. p.484 p.476

Note that although the control FSM M and the tape machine T both generate output, the Turing machine itself does not. It only computes by transitioning between states of the control FSM, updating the tape, and moving left (*L*) or right (*R*). On any input string w, we are only concerned with whether the Turing machine halts, and if so, whether it accepts or rejects w. Thus, a Turing machine attempts to map an input string $w \in \Sigma^*$ to $\{accept, reject\}$, but for some input strings, it may be unable to produce an answer.

We can now see that Proposition B.1 applies, and the fact that a Turing machine may not produce an answer on some input strings is not surprising. Let $\Sigma = \{0, 1\}$. Then any input string $w \in \Sigma^*$ can be interpreted as a binary encoding of a natural number in $\mathbb{N}$. Thus, a Turing machine implements a partial function of the form $f \colon \mathbb{N} \to \{accept, reject\}$. The function is partial because for some $n \in \mathbb{N}$, the machine may loop. Since a Turing machine is a program, Proposition B.1 tells that Turing machines are incapable of realizing all functions of the form $f \colon \mathbb{N} \to \{accept, reject\}$. This limitation manifests itself as looping. p.472

A principle that lies at heart of computer science, known as the **Church-Turing thesis**, asserts that every effectively computable function can be realized by a Turing machine. This principle is named for mathematicians Alonzo Church and Alan Turing. Our intuitive notion of computation, as expressed in today's computers, is equivalent to the Turing machine model of computation in this sense. Computers can realize exactly the functions that can be realized by Turing machines: no more, no less. This connection between the informal notion of an algorithm and the precise Turing machine model of computation is not a theorem; it cannot be proved. It is a principle that underlies what we mean by computation.

p.485

B.3.2 Decidable and Undecidable Problems

Turing machines, as described here, are designed to solve decision problems, which only have a YES or NO answer. The input string to a Turing machine represents the encoding of a **problem instance**. If the Turing machine *accepts*, it is viewed as a YES answer, and if it *rejects*, it is viewed as a NO answer. There is the third possibility that the Turing machine might loop.

Example B.5: Consider the problem of determining, given a directed graph G with two nodes s and t in G, whether there is a path from s to t. One can think of writing down the problem as a long string listing all nodes and edges of G, followed by s and t. Thus, an instance of this path problem can be presented to the Turing machine as an input string on its tape. The *instance* of the problem is the particular graph G, and nodes s and t. If there exists a path from s to t in G, then this is a YES problem instance; otherwise, it is a NO problem instance.

Turing machines are typically designed to solve *problems*, rather than specific problem instances. In this example, we would typically design a Turing machine that, for *any* graph G, nodes s and t, determines whether there is a path in G from s to t.

p.484 Recall that a decision problem is a function $f\colon W \to \{\text{YES}, \text{NO}\}$. For a Turing machine, p.476 the domain is a set $W \subseteq \Sigma^*$ of finite sequences of symbols from the set Σ. A problem instance is a particular $w \in W$, for which the "answer" to the problem is either $f(w) = \text{YES}$ or $f(w) = \text{NO}$. Let $Y \subseteq W$ denote the set of all YES instances of problem f. That is,

$$Y = \{w \in W \mid f(w) = \text{YES}\}.$$

Given a decision problem f, a Turing machine $D = (\Sigma, M)$ is called a **decision procedure** for f if D accepts every string $w \in Y$, and D rejects every $w \in W \setminus Y$, where $\setminus$ denotes set subtraction. Note that a decision procedure always halts for any input string $w \in W$. p.472

A problem f is said to be **decidable** (or **solvable**) if there exists a Turing machine that is a decision procedure for f. Otherwise, we say that the problem is **undecidable** (or **unsolvable**). For an undecidable problem f, there is no Turing machine that terminates with the correct answer $f(w)$ for all input strings $w \in W$.

One of the important philosophical results of 20th century mathematics and computer science is the existence of problems that are undecidable. One of the first problems to be proved undecidable is the so-called **halting problem** for Turing machines. This problem can be stated as follows:

> Given a Turing machine $D = (\Sigma, M)$ initialized with input string $w \in \Sigma^*$ on its tape, decide whether or not M will halt.

Proposition B.2. *(Turing, 1936) The halting problem is undecidable.*

Probing Further: Recursive Functions and Sets

Logicians make distinctions between the functions that can be realized by Turing machines. The so-called **total recursive functions** are those where a Turing machine realizing the function terminates for all inputs $w \in \Sigma^*$. The **partial recursive functions** are those where a Turing machine may or may not terminate on a particular input $w \in \Sigma^*$. By these definitions, every total recursive function is also a partial recursive function, but not vice-versa.

Logicians also use Turing machines to make useful distinctions between sets. Consider sets of natural numbers, and consider Turing machines where $\Sigma = \{0, 1\}$ and an input $w \in \Sigma^*$ is the binary encoding of a natural number. Then a set C of natural numbers is a **computable set** (or synonymously a **recursive set** or **decidable set**) if there is a Turing machine that terminates for all inputs $w \in \mathbb{N}$ and yields *accept* if $w \in C$ and *reject* if $w \notin C$. A set $E \subset \mathbb{N}$ is a **computably enumerable set** (or synonymously a **recursively enumerable set** or a **semidecidable set**) if there is a Turing machine that terminates if and only if the input w is in E.

Proof. This is a decision problem $h\colon W' \to \{\text{YES}, \text{NO}\}$, where W' denotes the set of all Turing machines and their inputs. The proposition can be proved using a variant of
p.484 Cantor's diagonal argument.

It is sufficient to prove the theorem for the subset of Turing machines with binary tape symbols, $\Sigma = \{0, 1\}$. Moreover, we can assume without loss of generality that every Turing machine in this set can be represented by a finite sequence of binary digits (bits), so

$$W' = \Sigma^* \times \Sigma^*.$$

Assume further that every finite sequence of bits represents a Turing machine. The form of the decision problem becomes

$$h\colon \Sigma^* \times \Sigma^* \to \{\text{YES}, \text{NO}\}. \tag{B.2}$$

We seek a procedure to determine the value of $h(D, w)$, where D is a finite sequence of bits representing a Turing machine and w is a finite sequence of bits representing an input to the Turing machine. The answer $h(D, w)$ will be YES if the Turing machine D halts with input w and NO if it loops.

p.485 Consider the set of all effectively computable functions of the form

$$f\colon \Sigma^* \times \Sigma^* \to \{\text{YES}, \text{NO}\}.$$

p.488 These functions can be given by a Turing machine (by the Church-Turing thesis), and hence the set of such functions can be enumerated $f_0, f_1, f_2, \cdots$. We will show that the halting problem (B.2) is not on this list. That is, there is no f_i such that $h = f_i$.

Consider a sequence of Turing machines $D_0, D_1, \cdots$ where D_i is the sequence of bits representing the ith Turing machine, and D_i halts if $f_i(D_i, D_i) = \text{NO}$ and loops otherwise. Since f_i is a computable function, we can clearly construct such a Turing machine. Not one of the computable functions in the list $f_0, f_1, f_2, \cdots$ can possibly equal the function h, because every function f_i in the list gives the wrong answer for input (D_i, D_i). If Turing machine D_i halts on input $w = D_i$, function f_i evaluates to $f_i(D_i, D_i) = \text{NO}$, whereas $h(D_i, D_i) = \text{YES}$. Since no function in the list $f_0, f_1, f_2, \cdots$ of computable functions works, the function h is not computable. □

B.4 Intractability: P and NP

Section B.1 above studied asymptotic complexity, a measure of how quickly the cost (in p.482 time or space) of solving a problem with a particular algorithm grows with the size of the p.483 input. In this section, we consider *problems* rather than *algorithms*. We are interested in whether an algorithm with a particular asymptotic complexity *exists* to solve a problem. This is not the same as asking whether an algorithm with a particular complexity class is *known*.

A **complexity class** is a collection of problems for which there exist algorithms with the same asymptotic complexity. In this section, we very briefly introduce the complexity classes **P** and **NP**.

First recall the concept of a deterministic Turing machine from the preceding section. A **nondeterministic Turing machine** $N = (\Sigma, M)$ is identical to its deterministic counterpart, except that the control FSM M can be a nondeterministic FSM. On any input string $w \in \Sigma^*$, p.64 a nondeterministic Turing machine N can exhibit several computations. N is said to **accept** w if *any* computation accepts w, and N **rejects** w if *all* its computations reject w.

A decision problem is a function $f\colon W \to \{\text{YES}, \text{NO}\}$, where $W \subseteq \Sigma^*$. N is said to be a p.484 decision procedure for f if for each input $w \in W$, *all* of its computations halt, no matter what p.489 nondeterministic choices are made. Note that a particular execution of a nondeterministic Turing machine N may give the wrong answer. That is, it could yield NO for input w when the right answer is $f(w) = \text{YES}$. It can still be a decision procedure, however, because we define the final answer to be YES if *any* execution yields YES. We do not require that *all* executions yield YES. This subtle point underlies the expressive power of nondeterministic Turing machines.

An execution that accepts an input w is called a **certificate**. A certificate can be represented by a finite list of choices made by the Turing machine such that it accepts w. We need only one valid certificate to know that $f(w) = \text{YES}$.

Given the above definitions, we are ready to introduce P and NP. **P** is the set of problems decidable by a *deterministic* Turing machine in polynomial time. **NP**, on the other hand, p.482 is the set of problems decidable by a *nondeterministic* Turing machine in polynomial time. That is, a problem f is in NP if there is a nondeterministic Turing machine N that is a decision procedure for f, and for all inputs $w \in W$, *every* execution of the Turing machine has time complexity no greater than $O(n^m)$, for some $m \in \mathbb{N}$. p.482

An equivalent alternative definition of NP is the set of all problems for which one can check the validity of a certificate for a YES answer in polynomial time. Specifically, a problem

f is in NP if there is a nondeterministic Turing machine N that is a decision procedure for f, and given an input w and a certificate, we can check in polynomial time whether the certificate is valid (i.e., whether the choices it lists do indeed result in accepting w). Note that this says nothing about NO answers. This asymmetry is part of the meaning of NP.

An important notion that helps systematize the study of complexity classes is that of **completeness**, in which we identify problems that are "representative" of a complexity class. In the context of NP, we say that a problem A is **NP-hard** if any other problem B in NP can be reduced ("translated") to A in polynomial time. Intuitively, A is "as hard as" any problem in NP — if we had a polynomial-time algorithm for A, we could derive one for B by first translating the instance of B to one of A, and then invoking the algorithm to solve A. A problem A is said to be **NP-complete** if (i) A is in NP, and (ii) A is NP-hard. In other words, an NP-complete problem is a problem in NP that is as hard as any other problem in NP.

Several core problems in the modeling, design, and analysis of embedded systems are NP-complete. One of these is the very first problem to be proved NP-complete, the **Boolean satisfiability (SAT)** problem. The SAT problem is to decide, given a propositional logic formula p.348 ϕ expressed over Boolean variables $x_1, x_2, \ldots, x_n$, whether there exists a valuation of the x_i variables such that $\phi(x_1, x_2, \ldots, x_n) = true$. If there exists such a valuation, we say ϕ is **satisfiable**; otherwise, we say that ϕ is **unsatisfiable**. The SAT problem is a decision problem of the form $f \colon W \to \{\text{YES}, \text{NO}\}$, where each $w \in W$ is an encoding of a propositional logic formula ϕ.

Example B.6: Consider the following propositional logic formula ϕ:

$$(x_1 \vee \neg x_2) \wedge (\neg x_1 \vee x_3 \vee x_2) \wedge (x_1 \vee \neg x_3)$$

We can see that setting $x_1 = x_3 = true$ will make ϕ evaluate to $true$. It is possible to construct a nondeterministic Turing machine that takes as input an encoding of the formula, where the nondeterministic choices correspond to choices of valuations for each variable x_i, and where the machine will accept the input formula if it is satisfiable and reject it otherwise. If the input w encodes the above formula ϕ, then one of the certificates demonstrating that $f(w) = \text{YES}$ is the choices $x_1 = x_2 = x_3 = true$.

Next consider the alternative formula ϕ':

$$(x_1 \vee \neg x_2) \wedge (\neg x_1 \vee x_2) \wedge (x_1 \vee x_2) \wedge (\neg x_1 \vee \neg x_2)$$

In this case, no matter how we assign Boolean values to the x_i variables, we cannot make $\phi' = true$. Thus while ϕ is satisfiable, ϕ' is unsatisfiable. The same nondeterministic Turing machine as above will reject an input w' that is an encoding of ϕ'. Rejecting this input means that *all* choices result in executions that terminate in reject.

Another problem that is very useful, but NP-complete, is checking the feasibility of an **integer linear program (ILP)**. Informally, the feasibility problem for integer linear programs is to find a valuation of integer variables such that each inequality in a collection of linear inequalities over those variables is satisfied.

Given that both SAT and ILP are NP-complete, one can transform an instance of either problem into an instance of the other problem, in polynomial time.

Example B.7: The following integer linear program is equivalent to the SAT problem corresponding to formula ϕ' of Example B.6:

$$\begin{aligned} &\text{find } x_1, x_2 \in \{0,1\} \\ &\text{such that:} \\ x_1 - x_2 &\geq 0 \\ -x_1 + x_2 &\geq 0 \\ x_1 + x_2 &\geq 1 \\ -x_1 - x_2 &\geq -1 \end{aligned}$$

One can observe that there is no valuation of x_1 and x_2 that will make all the above inequalities simultaneously true.

NP-complete problems seem to be harder than those in P; for large enough input sizes, these problems can become **intractable**, meaning that they cannot be practically solved. In general, it appears that to determine that $f(w) = \text{YES}$ for some w without being given a certificate, we might have to explore *all* executions of the nondeterministic Turing machine before finding, on the last possibility, an execution that accepts w. The number of possible executions can be exponential in the size of the input. Indeed, there are no known polynomial-time algorithms that solve NP-complete problems. Surprisingly, as of this writing, there is no proof that no such algorithm exists. It is widely believed that NP is a strictly larger set of problems than P, but without a proof, we cannot be sure. The **P versus NP** question is one of the great unsolved problems in mathematics today.

Despite the lack of polynomial-time algorithms for solving NP-complete problems, many such problems turn out to be solvable in practice. SAT problems, for example, can often be solved rather quickly, and a number of very effective **SAT solvers** are available. These solvers use algorithms that have worst-case exponential complexity, which means that for some inputs they can take a very long time to complete. Yet for most inputs, they complete quickly. Hence, we should not be deterred from tackling a problem just because it is NP-complete.

B.5 Summary

This appendix has very briefly introduced two rather large interrelated topics, the theories of complexity and computability. The chapter began with a discussion of complexity measures for algorithms, and then established a fundamental distinction between a problem to be solved and an algorithm for solving the problem. It then showed that there are problems that cannot be solved. We then explained Turing machines, which are capable of describing solution procedures for all problems that have come to be considered "computable." The chapter then closed with a brief discussion of the complexity classes P and NP, which are classes of problems that can be solved by algorithms with comparable complexity.

Exercises

1. Complete the formal definition of the tape machine T by giving the initial state of T and the mathematical description of its transition function $update_T$.

2. *Directed, acyclic graphs* (DAGs) have several uses in modeling, design, and analysis of embedded systems; e.g., they are used to represent precedence graphs of tasks (see Chapter 12) and control-flow graphs of loop-free programs (see Chapter 16). p.323

   p.494 A common operation on DAGs is to topologically sort the nodes of the graph. Formally, consider a DAG $G = (V, E)$ where V is the set of vertices $\{v_1, v_2, \ldots, v_n\}$ and E is the set of edges. A **topological sort** of G is a linear ordering of vertices $\{v_1, v_2, \ldots, v_n\}$ such that if $(v_i, v_j) \in E$ (i.e., there is a directed edge from v_i to v_j), then vertex v_i appears before vertex v_j in this ordering.

The following algorithm due to Kahn (1962) topologically sorts the vertices of a DAG:

```
input  : A DAG G = (V,E) with n vertices and m edges.
output: A list L of vertices in V in topologically-sorted order.
L ← empty list
S ← {v | v is a vertex with no incoming edges}
while S is non-empty do
    Remove vertex v from S
    Insert v at end of list L
    for each vertex u such that edge (v,u) is in E do
        Mark edge (v,u)
        if all incoming edges to u are marked then
            Add u to set S
        end
    end
end
L contains all vertices of G in topologically sorted order.
```

Algorithm B.1: Topological sorting of vertices in a DAG

State the asymptotic time complexity of Algorithm B.1 using Big O notation. Prove the correctness of your answer.

Bibliography

Abelson, H. and G. J. Sussman, 1996: *Structure and Interpretation of Computer Programs*. MIT Press, 2nd ed.

Adam, T. L., K. M. Chandy, and J. R. Dickson, 1974: A comparison of list schedules for parallel processing systems. *Communications of the ACM*, **17(12)**, 685–690.

Adve, S. V. and K. Gharachorloo, 1996: Shared memory consistency models: A tutorial. *IEEE Computer*, **29(12)**, 66–76.

Allen, J., 1975: Computer architecture for signal processing. *Proceedings of the IEEE*, **63(4)**, 624– 633.

Alpern, B. and F. B. Schneider, 1987: Recognizing safety and liveness. *Distributed Computing*, **2(3)**, 117–126.

Alur, R., 2015: *Principles of Cyber-Physical Systems*. MIT Press.

Alur, R., C. Courcoubetis, and D. Dill, 1991: Model-checking for probabilistic real-time systems. In *Proc. 18th Intl. Colloquium on Automata, Languages and Programming (ICALP)*, pp. 115–126.

Alur, R. and D. L. Dill, 1994: A theory of timed automata. *Theoretical Computer Science*, **126(2)**, 183–235.

Alur, R. and T. A. Henzinger, 1993: Real-time logics: Complexity and expressiveness. *Information and Computation*, **104(1)**, 35–77.

Anderson, R., F. Bergadano, B. Crispo, J.-H. Lee, C. Manifavas, and R. Needham, 1998: A new family of authentication protocols. *ACM SIGOPS Operating Systems Review*, **32(4)**, 9–20.

Anderson, R. and M. Kuhn, 1998: Low cost attacks on tamper resistant devices. In *Security Protocols*, Springer, pp. 125–136.

André, C., 1996: SyncCharts: a visual representation of reactive behaviors. Tech. Rep. RR 95–52, revision: RR (96–56), University of Sophia-Antipolis. Available from: http://www-sop.inria.fr/members/Charles.Andre/CA\%20Publis/SYNCCHARTS/overview.html.

ARM Limited, 2006: Cortex™- M3 technical reference manual. Tech. rep. Available from: http://www.arm.com.

Audsley, N. C., A. Burns, R. I. Davis, K. W. Tindell, and A. J. Wellings, 2005: Fixed priority pre-emptive scheduling: An historical perspective. *Real-Time Systems*, **8(2-3)**, 173–198. Available from: http://www.springerlink.com/content/w602g7305r125702/.

Ball, T., V. Levin, and S. K. Rajamani, 2011: A decade of software model checking with SLAM. *Communications of the ACM*, **54(7)**, 68–76.

Ball, T., R. Majumdar, T. Millstein, and S. K. Rajamani, 2001: Automatic predicate abstraction of c programs. In *ACM SIGPLAN Conference on Programming Language Design and Implementation*, vol. 36 of *ACM SIGPLAN Notices*, pp. 203–213.

Ball, T. and S. K. Rajamani, 2001: The SLAM toolkit. In *13th International Conference on Computer Aided Verification (CAV)*, Springer, vol. 2102 of *Lecture Notes in Computer Science*, pp. 260–264.

Barr, M. and A. Massa, 2006: *Programming Embedded Systems*. O'Reilly, 2nd ed.

Barrett, C., R. Sebastiani, S. A. Seshia, and C. Tinelli, 2009: Satisfiability modulo theories. In Biere, A., H. van Maaren, and T. Walsh, eds., *Handbook of Satisfiability*, IOS Press, vol. 4, chap. 8, pp. 825–885.

Ben-Ari, M., Z. Manna, and A. Pnueli, 1981: The temporal logic of branching time. In *8th Annual ACM Symposium on Principles of Programming Languages*.

Benveniste, A. and G. Berry, 1991: The synchronous approach to reactive and real-time systems. *Proceedings of the IEEE*, **79(9)**, 1270–1282.

Berger, A. S., 2002: *Embedded Systems Design: An Introduction to Processes, Tools, & Techniques*. CMP Books.

Berry, G., 1999: *The Constructive Semantics of Pure Esterel - Draft Version 3*. Book Draft. Available from: http://www-sop.inria.fr/meije/esterel/doc/main-papers.html.

—, 2003: The effectiveness of synchronous languages for the development of safety-critical systems. White paper, Esterel Technologies.

Berry, G. and G. Gonthier, 1992: The Esterel synchronous programming language: Design, semantics, implementation. *Science of Computer Programming*, **19(2)**, 87–152.

Biere, A., A. Cimatti, E. M. Clarke, and Y. Zhu, 1999: Symbolic model checking without BDDs. In *5th International Conference on Tools and Algorithms for Construction and Analysis of Systems (TACAS)*, Springer, vol. 1579 of *Lecture Notes in Computer Science*, pp. 193–207.

Boehm, H.-J., 2005: Threads cannot be implemented as a library. In *Programming Language Design and Implementation (PLDI)*, ACM SIGPLAN Notices, vol. 40(6), pp. 261 – 268.

Boneh, D., R. A. DeMillo, and R. J. Lipton, 2001: On the importance of eliminating errors in cryptographic computations. *Journal of cryptology*, **14(2)**, 101–119.

Booch, G., I. Jacobson, and J. Rumbaugh, 1998: *The Unified Modeling Language User Guide*. Addison-Wesley.

Brumley, D. and D. Boneh, 2005: Remote timing attacks are practical. *Computer Networks*, **48(5)**, 701–716.

Bryant, R. E., 1986: Graph-based algorithms for Boolean function manipulation. *IEEE Transactions on Computers*, **C-35(8)**, 677–691.

Bryant, R. E. and D. R. O'Hallaron, 2003: *Computer Systems: A Programmer's Perspective*. Prentice Hall.

Brylow, D., N. Damgaard, and J. Palsberg, 2001: Static checking of interrupt-driven software. In *Proc. Intl. Conference on Software Engineering (ICSE)*, pp. 47–56.

Buck, J. T., 1993: *Scheduling Dynamic Dataflow Graphs with Bounded Memory Using the Token Flow Model*. Ph.d. thesis, University of California, Berkeley. Available from: http://ptolemy.eecs.berkeley.edu/publications/papers/93/jbuckThesis/.

Burns, A. and S. Baruah, 2008: Sustainability in real-time scheduling. *Journal of Computing Science and Engineering*, **2(1)**, 74–97.

Burns, A. and A. Wellings, 2001: *Real-Time Systems and Programming Languages: Ada 95, Real-Time Java and Real-Time POSIX*. Addison-Wesley, 3rd ed.

Buttazzo, G. C., 2005a: *Hard Real-Time Computing Systems: Predictable Scheduling Algorithms and Applications*. Springer, 2nd ed.

—, 2005b: Rate monotonic vs. EDF: judgment day. *Real-Time Systems*, **29(1)**, 5–26.

Cassandras, C. G., 1993: *Discrete Event Systems, Modeling and Performance Analysis*. Irwin.

Cataldo, A., E. A. Lee, X. Liu, E. Matsikoudis, and H. Zheng, 2006: A constructive fixed-point theorem and the feedback semantics of timed systems. In *Workshop on Discrete Event Systems (WODES)*, Ann Arbor, Michigan. Available from: http://ptolemy.eecs.berkeley.edu/publications/papers/06/constructive/.

Chapman, B., G. Jost, and R. van der Pas, 2007: *Using OpenMP: Portable Shared Memory Parallel Programming*. MIT Press.

Chetto, H., M. Silly, and T. Bouchentouf, 1990: Dynamic scheduling of real-time tasks under precedence constraints. *Real-Time Systems*, **2(3)**, 181–194.

Clarke, E. M. and E. A. Emerson, 1981: Design and synthesis of synchronization skeletons using branching-time temporal logic. In *Logic of Programs*, pp. 52–71.

Clarke, E. M., O. Grumberg, S. Jha, Y. Lu, and H. Veith, 2000: Counterexample-guided abstraction refinement. In *12th International Conference on Computer Aided Verification (CAV)*, Springer, vol. 1855 of *Lecture Notes in Computer Science*, pp. 154–169.

Clarke, E. M., O. Grumberg, and D. Peled, 1999: *Model Checking*. MIT Press.

Clarkson, M. R. and F. B. Schneider, 2010: Hyperproperties. *Journal of Computer Security*, **18(6)**, 1157–1210.

Coffman, E. G., Jr., M. J. Elphick, and A. Shoshani, 1971: System deadlocks. *Computing Surveys*, **3(2)**, 67–78.

Coffman, E. G., Jr. (Ed), 1976: *Computer and Job Scheduling Theory*. Wiley.

Conway, R. W., W. L. Maxwell, and L. W. Miller, 1967: *Theory of Scheduling*. Addison-Wesley.

Cousot, P. and R. Cousot, 1977: Abstract interpretation: A unified lattice model for static analysis of programs by construction or approximation of fixpoints. In *Symposium on Principles of Programming Languages (POPL)*, ACM Press, pp. 238–252.

Dennis, J. B., 1974: First version data flow procedure language. Tech. Rep. MAC TM61, MIT Laboratory for Computer Science.

Derenzo, S. E., 2003: *Practical Interfacing in the Laboratory: Using a PC for Instrumentation, Data Analysis and Control*. Cambridge University Press.

Diffie, W. and M. E. Hellman, 1976: New directions in cryptography. *Information Theory, IEEE Transactions on*, **22(6)**, 644–654.

Dijkstra, E. W., 1968: Go to statement considered harmful (letter to the editor). *Communications of the ACM*, **11(3)**, 147–148.

Eden, M. and M. Kagan, 1997: The Pentium® processor with MMX™ technology. In *IEEE International Conference (COMPCON)*, IEEE, San Jose, CA, USA, pp. 260–262.

Edwards, S. A., 2000: *Languages for Digital Embedded Systems*. Kluwer Academic Publishers.

Edwards, S. A. and E. A. Lee, 2003: The semantics and execution of a synchronous block-diagram language. *Science of Computer Programming*, **48(1)**, 21–42.

Eidson, J. C., 2006: *Measurement, Control, and Communication Using IEEE 1588*. Springer.

Eidson, J. C., E. A. Lee, S. Matic, S. A. Seshia, and J. Zou, 2009: Time-centric models for designing embedded cyber-physical systems. Technical Report UCB/EECS-2009-135, EECS Department, University of California, Berkeley. Available from: http://www.eecs.berkeley.edu/Pubs/TechRpts/2009/EECS-2009-135.html.

Einstein, A., 1907: Uber das relativitatsprinzip und die aus demselben gezogene folgerungen. *Jahrbuch der Radioaktivitat und Elektronik*, **4**, 411–462.

Emerson, E. A. and E. M. Clarke, 1980: Characterizing correctness properties of parallel programs using fixpoints. In *Proc. 7th Intl. Colloquium on Automata, Languages and Programming (ICALP)*, Lecture Notes in Computer Science 85, pp. 169–181.

European Cooperation for Space Standardization, 2002: Space engineering – SpaceWire – links, nodes, routers, and networks (draft ECSS-E-50-12A). Available from: http://spacewire.esa.int/.

Ferguson, N., B. Schneier, and T. Kohno, 2010: *Cryptography Engineering: Design Principles and Practical Applications*. Wiley.

Fielding, R. T. and R. N. Taylor, 2002: Principled design of the modern web architecture. *ACM Transactions on Internet Technology (TOIT)*, **2(2)**, 115–150.

Fishman, G. S., 2001: *Discrete-Event Simulation: Modeling, Programming, and Analysis*. Springer-Verlag.

Foo Kune, D., J. Backes, S. S. Clark, D. B. Kramer, M. R. Reynolds, K. Fu, Y. Kim, and W. Xu, 2013: Ghost talk: Mitigating EMI signal injection attacks against analog sensors. In *Proceedings of the 34th Annual IEEE Symposium on Security and Privacy*.

Fujimoto, R., 2000: *Parallel and Distributed Simulation Systems*. John Wiley and Sons.

Gajski, D. D., S. Abdi, A. Gerstlauer, and G. Schirner, 2009: *Embedded System Design - Modeling, Synthesis, and Verification*. Springer.

Galison, P., 2003: *Einstein's Clocks, Poincaré's Maps*. W. W. Norton & Company, New York.

Galletly, J., 1996: *Occam-2*. University College London Press, 2nd ed.

Gamma, E., R. Helm, R. Johnson, and J. Vlissides, 1994: *Design Patterns: Elements of Reusable Object-Oriented Software*. Addison Wesley.

Geilen, M. and T. Basten, 2003: Requirements on the execution of Kahn process networks. In *European Symposium on Programming Languages and Systems*, Springer, LNCS, pp. 319–334.

Ghena, B., W. Beyer, A. Hillaker, J. Pevarnek, and J. A. Halderman, 2014: Green lights forever: analyzing the security of traffic infrastructure. In *Proceedings of the 8th USENIX conference on Offensive Technologies*, USENIX Association, pp. 7–7.

Ghosal, A., T. A. Henzinger, C. M. Kirsch, and M. A. Sanvido, 2004: Event-driven programming with logical execution times. In *Seventh International Workshop on Hybrid Systems: Computation and Control (HSCC)*, Springer-Verlag, vol. LNCS 2993, pp. 357–371.

Goldstein, H., 1980: *Classical Mechanics*. Addison-Wesley, 2nd ed.

Goodrich, M. T. and R. Tamassia, 2011: *Introduction to Computer Security*. Addison Wesley.

Graham, R. L., 1969: Bounds on multiprocessing timing anomalies. *SIAM Journal on Applied Mathematics*, **17(2)**, 416–429.

Halbwachs, N., P. Caspi, P. Raymond, and D. Pilaud, 1991: The synchronous data flow programming language LUSTRE. *Proceedings of the IEEE*, **79(9)**, 1305–1319.

Halderman, J. A., S. D. Schoen, N. Heninger, W. Clarkson, W. Paul, J. A. Calandrino, A. J. Feldman, J. Appelbaum, and E. W. Felten, 2009: Lest we remember: cold-boot attacks on encryption keys. *Communications of the ACM*, **52(5)**, 91–98.

Halperin, D., T. S. Heydt-Benjamin, B. Ransford, S. S. Clark, B. Defend, W. Morgan, K. Fu, T. Kohno, and W. H. Maisel, 2008: Pacemakers and implantable cardiac defibrillators: Software radio attacks and zero-power defenses. In *Proceedings of the 29th Annual IEEE Symposium on Security and Privacy*, pp. 129–142.

Hansson, H. and B. Jonsson, 1994: A logic for reasoning about time and reliability. *Formal Aspects of Computing*, **6**, 512–535.

Harel, D., 1987: Statecharts: A visual formalism for complex systems. *Science of Computer Programming*, **8**, 231–274.

Harel, D., H. Lachover, A. Naamad, A. Pnueli, M. Politi, R. Sherman, A. Shtull-Trauring, and M. Trakhtenbrot, 1990: STATEMATE: A working environment for the development of complex reactive systems. *IEEE Transactions on Software Engineering*, **16(4)**, 403 – 414.

Harel, D. and A. Pnueli, 1985: On the development of reactive systems. In Apt, K. R., ed., *Logic and Models for Verification and Specification of Concurrent Systems*, Springer-Verlag, vol. F13 of *NATO ASI Series*, pp. 477–498.

Harter, E. K., 1987: Response times in level structured systems. *ACM Transactions on Computer Systems*, **5(3)**, 232–248.

Hayes, B., 2007: Computing in a parallel universe. *American Scientist*, **95**, 476–480.

Henzinger, T. A., B. Horowitz, and C. M. Kirsch, 2003: Giotto: A time-triggered language for embedded programming. *Proceedings of IEEE*, **91(1)**, 84–99.

Hoare, C. A. R., 1978: Communicating sequential processes. *Communications of the ACM*, **21(8)**, 666–677.

Hoffmann, G., D. G. Rajnarqan, S. L. Waslander, D. Dostal, J. S. Jang, and C. J. Tomlin, 2004: The Stanford testbed of autonomous rotorcraft for multi agent control (starmac). In *Digital Avionics Systems Conference (DASC)*.

Holzmann, G. J., 2004: *The SPIN Model Checker – Primer and Reference Manual*. Addison-Wesley, Boston.

Hopcroft, J. and J. Ullman, 1979: *Introduction to Automata Theory, Languages, and Computation*. Addison-Wesley, Reading, MA.

Hopcroft, J. E., R. Motwani, and J. D. Ullman, 2007: *Introduction to Automata Theory, Languages, and Computation*. Addison-Wesley, 3rd ed.

Horn, W., 1974: Some simple scheduling algorithms. *Naval Research Logistics Quarterly*, **21(1)**, 177 – 185.

Islam, M. S., M. Kuzu, and M. Kantarcioglu, 2012: Access pattern disclosure on searchable encryption: Ramification, attack and mitigation. In *19th Annual Network and Distributed System Security Symposium (NDSS)*.

Jackson, J. R., 1955: Scheduling a production line to minimize maximum tardiness. Management Science Research Project 43, University of California Los Angeles.

Jantsch, A., 2003: *Modeling Embedded Systems and SoCs - Concurrency and Time in Models of Computation*. Morgan Kaufmann.

Jensen, E. D., C. D. Locke, and H. Tokuda, 1985: A time-driven scheduling model for real-time operating systems. In *Real-Time Systems Symposium (RTSS)*, IEEE, pp. 112–122.

Jin, X., A. Donzé, J. Deshmukh, and S. A. Seshia, 2015: Mining requirements from closed-loop control models. *IEEE Transactions on Computer-Aided Design of Circuits and Systems*, **34(11)**, 1704–1717.

Joseph, M. and P. Pandya, 1986: Finding response times in a real-time system. *The Computer Journal (British Computer Society)*, **29(5)**, 390–395.

Kahn, A. B., 1962: Topological sorting of large networks. *Communications of the ACM*, **5(11)**, 558–562.

Kahn, G., 1974: The semantics of a simple language for parallel programming. In *Proc. of the IFIP Congress 74*, North-Holland Publishing Co., pp. 471–475.

Kahn, G. and D. B. MacQueen, 1977: Coroutines and networks of parallel processes. In Gilchrist, B., ed., *Information Processing*, North-Holland Publishing Co., pp. 993–998.

Kamal, R., 2008: *Embedded Systems: Architecture, Programming, and Design*. McGraw Hill.

Kamen, E. W., 1999: *Industrial Controls and Manufacturing*. Academic Press.

Klein, M. H., T. Ralya, B. Pollak, R. Obenza, and M. G. Harbour, 1993: *A Practitioner's Guide for Real-Time Analysis*. Kluwer Academic Publishers.

Kocher, P., J. Jaffe, and B. Jun, 1999: Differential power analysis. In *Advances in CryptologyCRYPTO99*, Springer, pp. 388–397.

Kocher, P. C., 1996: Timing attacks on implementations of diffie-hellman, rsa, dss, and other systems. In *Advances in CryptologyCRYPTO96*, Springer, pp. 104–113.

Kodosky, J., J. MacCrisken, and G. Rymar, 1991: Visual programming using structured data flow. In *IEEE Workshop on Visual Languages*, IEEE Computer Society Press, Kobe, Japan, pp. 34–39.

Kohler, W. H., 1975: A preliminary evaluation of the critical path method for scheduling tasks on multiprocessor systems. *IEEE Transactions on Computers*, **24(12)**, 1235–1238.

Koopman, P., 2010: *Better Embedded System Software*. Drumnadrochit Education. Available from: http://www.koopman.us/book.html.

Kopetz, H., 1997: *Real-Time Systems : Design Principles for Distributed Embedded Applications*. Springer.

Kopetz, H. and G. Bauer, 2003: The time-triggered architecture. *Proceedings of the IEEE*, **91(1)**, 112–126.

Kopetz, H. and G. Grunsteidl, 1994: TTP - a protocol for fault-tolerant real-time systems. *Computer*, **27(1)**, 14–23.

Koscher, K., A. Czeskis, F. Roesner, S. Patel, T. Kohno, S. Checkoway, D. McCoy, B. Kantor, D. Anderson, H. Shacham, et al., 2010: Experimental security analysis of a modern automobile. In *IEEE Symposium on Security and Privacy (SP)*, IEEE, pp. 447–462.

Kremen, R., 2008: Operating inside a beating heart. *Technology Review*, October 21, 2008. Available from: http://www.technologyreview.com/biomedicine/21582/.

Kurshan, R., 1994: Automata-theoretic verification of coordinating processes. In Cohen, G. and J.-P. Quadrat, eds., *11th International Conference on Analysis and Optimization of Systems – Discrete Event Systems*, Springer Berlin / Heidelberg, vol. 199 of *Lecture Notes in Control and Information Sciences*, pp. 16–28.

Lamport, L., 1977: Proving the correctness of multiprocess programs. *IEEE Trans. Software Eng.*, **3(2)**, 125–143.

—, 1979: How to make a multiprocessor computer that correctly executes multiprocess programs. *IEEE Transactions on Computers*, **28(9)**, 690–691.

Landau, L. D. and E. M. Lifshitz, 1976: *Mechanics*. Pergamon Press, 3rd ed.

Lapsley, P., J. Bier, A. Shoham, and E. A. Lee, 1997: *DSP Processor Fudamentals – Architectures and Features*. IEEE Press, New York.

Lawler, E. L., 1973: Optimal scheduling of a single machine subject to precedence constraints. *Management Science*, **19(5)**, 544–546.

Le Guernic, P., T. Gauthier, M. Le Borgne, and C. Le Maire, 1991: Programming real-time applications with SIGNAL. *Proceedings of the IEEE*, **79(9)**, 1321 – 1336.

Lea, D., 1997: *Concurrent Programming in Java: Design Principles and Patterns*. Addison-Wesley, Reading MA.

—, 2005: The java.util.concurrent synchronizer framework. *Science of Computer Programming*, **58(3)**, 293–309.

Lee, E. A., 1999: Modeling concurrent real-time processes using discrete events. *Annals of Software Engineering*, **7**, 25–45.

—, 2001: Soft walls - modifying flight control systems to limit the flight space of commercial aircraft. Technical Memorandum UCB/ERL M001/31, UC Berkeley. Available from: http://ptolemy.eecs.berkeley.edu/publications/papers/01/softwalls2/.

—, 2003: Soft walls: Frequently asked questions. Technical Memorandum UCB/ERL M03/31, UC Berkeley. Available from: http://ptolemy.eecs.berkeley.edu/papers/03/softwalls/.

—, 2006: The problem with threads. *Computer*, **39(5)**, 33–42.

—, 2009a: Computing needs time. Tech. Rep. UCB/EECS-2009-30, EECS Department, University of California, Berkeley. Available from: http://www.eecs.berkeley.edu/Pubs/TechRpts/2009/EECS-2009-30.html.

—, 2009b: Disciplined message passing. Technical Report UCB/EECS-2009-7, EECS Department, University of California, Berkeley. Available from: http://www.eecs.berkeley.edu/Pubs/TechRpts/2009/EECS-2009-7.html.

Lee, E. A. and S. Ha, 1989: Scheduling strategies for multiprocessor real-time DSP. In *Global Telecommunications Conference (GLOBECOM)*, vol. 2, pp. 1279 –1283.

Lee, E. A., S. Matic, S. A. Seshia, and J. Zou, 2009: The case for timing-centric distributed software. In *IEEE International Conference on Distributed Computing Systems Workshops: Workshop on Cyber-Physical Systems*, IEEE, Montreal, Canada, pp. 57–64. Available from: http://chess.eecs.berkeley.edu/pubs/607.html.

Lee, E. A. and E. Matsikoudis, 2009: The semantics of dataflow with firing. In Huet, G., G. Plotkin, J.-J. Lévy, and Y. Bertot, eds., *From Semantics to Computer Science: Essays in memory of Gilles Kahn*, Cambridge University Press. Available from: http://ptolemy.eecs.berkeley.edu/publications/papers/08/DataflowWithFiring/.

Lee, E. A. and D. G. Messerschmitt, 1987: Synchronous data flow. *Proceedings of the IEEE*, **75(9)**, 1235–1245.

Lee, E. A. and T. M. Parks, 1995: Dataflow process networks. *Proceedings of the IEEE*, **83(5)**, 773–801.

Lee, E. A. and S. Tripakis, 2010: Modal models in Ptolemy. In *3rd International Workshop on Equation-Based Object-Oriented Modeling Languages and Tools (EOOLT)*, Linköping University Electronic Press, Linköping University, Oslo, Norway, vol. 47, pp. 11–21. Available from: http://chess.eecs.berkeley.edu/pubs/700.html.

Lee, E. A. and P. Varaiya, 2003: *Structure and Interpretation of Signals and Systems*. Addison Wesley.

—, 2011: *Structure and Interpretation of Signals and Systems*. LeeVaraiya.org, 2nd ed. Available from: http://LeeVaraiya.org.

Lee, E. A. and H. Zheng, 2005: Operational semantics of hybrid systems. In Morari, M. and L. Thiele, eds., *Hybrid Systems: Computation and Control (HSCC)*, Springer-Verlag, Zurich, Switzerland, vol. LNCS 3414, pp. 25–53.

—, 2007: Leveraging synchronous language principles for heterogeneous modeling and design of embedded systems. In *EMSOFT*, ACM, Salzburg, Austria, pp. 114 – 123.

Lee, I. and V. Gehlot, 1985: Language constructs for distributed real-time programming. In *Proc. Real-Time Systems Symposium (RTSS)*, San Diego, CA, pp. 57–66.

Lee, R. B., 1996: Subword parallelism with MAX2. *IEEE Micro*, **16(4)**, 51–59.

Lemkin, M. and B. E. Boser, 1999: A three-axis micromachined accelerometer with a cmos position-sense interface and digital offset-trim electronics. *IEEE J. of Solid-State Circuits*, **34(4)**, 456–468.

Leung, J. Y.-T. and J. Whitehead, 1982: On the complexity of fixed priority scheduling of periodic real-time tasks. *Performance Evaluation*, **2(4)**, 237–250.

Li, X., Y. Liang, T. Mitra, and A. Roychoudhury, 2005: Chronos: A timing analyzer for embedded software. Technical report, National University of Singapore.

Li, Y.-T. S. and S. Malik, 1999: *Performance Analysis of Real-Time Embedded Software*. Kluwer Academic Publishers.

Lin, C.-W., Q. Zhu, C. Phung, and A. Sangiovanni-Vincentelli, 2013: Security-aware mapping for can-based real-time distributed automotive systems. In *Computer-Aided Design (ICCAD), 2013 IEEE/ACM International Conference on*, IEEE, pp. 115–121.

Liu, C. L. and J. W. Layland, 1973: Scheduling algorithms for multiprogramming in a hard real time environment. *Journal of the ACM*, **20(1)**, 46–61.

Liu, J. and E. A. Lee, 2003: Timed multitasking for real-time embedded software. *IEEE Control Systems Magazine*, **23(1)**, 65–75.

Liu, J. W. S., 2000: *Real-Time Systems*. Prentice-Hall.

Liu, X. and E. A. Lee, 2008: CPO semantics of timed interactive actor networks. *Theoretical Computer Science*, **409(1)**, 110–125.

Liu, X., E. Matsikoudis, and E. A. Lee, 2006: Modeling timed concurrent systems. In *CONCUR 2006 - Concurrency Theory*, Springer, Bonn, Germany, vol. LNCS 4137, pp. 1–15.

Luminary Micro®, 2008a: Stellaris® LM3S8962 evaluation board user's manual. Tech. rep., Luminary Micro, Inc. Available from: http://www.luminarymicro.com.

—, 2008b: Stellaris® LM3S8962 microcontroller data sheet. Tech. rep., Luminary Micro, Inc. Available from: http://www.luminarymicro.com.

—, 2008c: Stellaris® peripheral driver library - user's guide. Tech. rep., Luminary Micro, Inc. Available from: http://www.luminarymicro.com.

Maler, O., Z. Manna, and A. Pnueli, 1992: From timed to hybrid systems. In *Real-Time: Theory and Practice, REX Workshop*, Springer-Verlag, pp. 447–484.

Maler, O. and D. Nickovic, 2004: Monitoring temporal properties of continuous signals. In *Proc. International Conference on Formal Modelling and Analysis of Timed Systems (FORMATS)*, Springer, vol. 3253 of *Lecture Notes in Computer Science*, pp. 152–166.

Malik, S. and L. Zhang, 2009: Boolean satisfiability: From theoretical hardness to practical success. *Communications of the ACM*, **52(8)**, 76–82.

Manna, Z. and A. Pnueli, 1992: *The Temporal Logic of Reactive and Concurrent Systems*. Springer, Berlin.

—, 1993: Verifying hybrid systems. In *Hybrid Systems*, vol. LNCS 736, pp. 4–35.

Marion, J. B. and S. Thornton, 1995: *Classical Dynamics of Systems and Particles*. Thomson, 4th ed.

Marwedel, P., 2011: *Embedded System Design - Embedded Systems Foundations of Cyber-Physical Systems*. Springer, 2nd ed. Available from: http://springer.com/978-94-007-0256-1.

McLean, J., 1992: Proving noninterference and functional correctness using traces. *Journal of Computer security*, **1(1)**, 37–57.

—, 1996: A general theory of composition for a class of possibilistic properties. *Software Engineering, IEEE Transactions on*, **22(1)**, 53–67.

Mealy, G. H., 1955: A method for synthesizing sequential circuits. *Bell System Technical Journal*, **34**, 1045–1079.

Menezes, A. J., P. C. van Oorschot, and S. A. Vanstone, 1996: *Handbook of Applied Cryptography*. CRC Press.

Milner, R., 1980: *A Calculus of Communicating Systems*, vol. 92 of *Lecture Notes in Computer Science*. Springer.

Mishra, P. and N. D. Dutt, 2005: *Functional Verification of Programmable Embedded Processors - A Top-down Approach*. Springer.

Misra, J., 1986: Distributed discrete event simulation. *ACM Computing Surveys*, **18(1)**, 39–65.

Montgomery, P. L., 1985: Modular multiplication without trial division. *Mathematics of Computation*, **44(170)**, 519–521.

Moore, E. F., 1956: Gedanken-experiments on sequential machines. *Annals of Mathematical Studies*, **34(Automata Studies, C. E. Shannon and J. McCarthy (Eds.))**, 129–153.

Murata, T., 1989: Petri nets: Properties, analysis and applications. *Proceedings of IEEE*, **77(4)**, 541–580.

Nemer, F., H. Cass, P. Sainrat, J.-P. Bahsoun, and M. D. Michiel, 2006: Papabench: A free real-time benchmark. In *6th Intl. Workshop on Worst-Case Execution Time (WCET) Analysis*. Available from: http://www.irit.fr/recherches/ARCHI/MARCH/rubrique.php3?id_rubrique=97.

Noergaard, T., 2005: *Embedded Systems Architecture: A Comprehensive Guide for Engineers and Programmers*. Elsevier.

Oshana, R., 2006: *DSP Software Development Techniques for Embedded and Real-Time Systems*. Embedded Technology Series, Elsevier.

Ousterhout, J. K., 1996: Why threads are a bad idea (for most purposes) (invited presentation). In *Usenix Annual Technical Conference*.

Paar, C. and J. Pelzl, 2009: *Understanding cryptography: a textbook for students and practitioners*. Springer Science & Business Media.

Papadimitriou, C., 1994: *Computational Complexity*. Addison-Wesley.

Parab, J. S., V. G. Shelake, R. K. Kamat, and G. M. Naik, 2007: *Exploring C for Microcontrollers*. Springer.

Parks, T. M., 1995: Bounded scheduling of process networks. Ph.D. Thesis Tech. Report UCB/ERL M95/105, UC Berkeley. Available from: http://ptolemy.eecs.berkeley.edu/papers/95/parksThesis.

Patterson, D. A. and D. R. Ditzel, 1980: The case for the reduced instruction set computer. *ACM SIGARCH Computer Architecture News*, **8(6)**, 25–33.

Patterson, D. A. and J. L. Hennessy, 1996: *Computer Architecture: A Quantitative Approach*. Morgan Kaufmann, 2nd ed.

Perrig, A., J. Stankovic, and D. Wagner, 2004: Security in wireless sensor networks. *Communications of the ACM*, **47(6)**, 53–57.

Perrig, A., R. Szewczyk, J. D. Tygar, V. Wen, and D. E. Culler, 2002: SPINS: Security protocols for sensor networks. *Wireless networks*, **8(5)**, 521–534.

Perrig, A. and J. D. Tygar, 2012: *Secure Broadcast Communication in Wired and Wireless Networks*. Springer Science & Business Media.

Plotkin, G., 1981: *A Structural Approach to Operational Semantics*.

Pnueli, A., 1977: The temporal logic of programs. In *18th Annual Symposium on Foundations of Computer Science (FOCS)*, pp. 46–57.

Pottie, G. and W. Kaiser, 2005: *Principles of Embedded Networked Systems Design*. Cambridge University Press.

Price, H. and R. Corry, eds., 2007: *Causation, Physics, and the Constitution of Reality*. Clarendon Press, Oxford.

Queille, J.-P. and J. Sifakis, 1981: Iterative methods for the analysis of Petri nets. In *Selected Papers from the First and the Second European Workshop on Application and Theory of Petri Nets*, pp. 161–167.

Ravindran, B., J. Anderson, and E. D. Jensen, 2007: On distributed real-time scheduling in networked embedded systems in the presence of crash failures. In *IFIFP Workshop on Software Technologies for Future Embedded and Ubiquitous Systems (SEUS)*, IEEE ISORC.

Rice, J., 2008: Heart surgeons as video gamers. *Technology Review*, June 10, 2008. Available from: http://www.technologyreview.com/biomedicine/20873/.

Rivest, R. L., A. Shamir, and L. Adleman, 1978: A method for obtaining digital signatures and public-key cryptosystems. *Communications of the ACM*, **21(2)**, 120–126.

Roscoe, A. W., 1995: Csp and determinism in security modelling. In *Security and Privacy, 1995. Proceedings., 1995 IEEE Symposium on*, IEEE, pp. 114–127.

Sander, I. and A. Jantsch, 2004: System modeling and transformational design refinement in forsyde. *IEEE Transactions on Computer-Aided Design of Circuits and Systems*, **23(1)**, 17–32.

Schaumont, P. R., 2010: *A Practical Introduction to Hardware/Software Codesign*. Springer. Available from: http://www.springerlink.com/content/978-1-4419-5999-7.

Scott, D. and C. Strachey, 1971: Toward a mathematical semantics for computer languages. In *Symposium on Computers and Automata*, Polytechnic Institute of Brooklyn, pp. 19–46.

Seshia, S. A., 2015: Combining induction, deduction, and structure for verification and synthesis. *Proceedings of the IEEE*, **103(11)**, 2036–2051.

Seshia, S. A. and A. Rakhlin, 2008: Game-theoretic timing analysis. In *Proc. IEEE/ACM International Conference on Computer-Aided Design (ICCAD)*, pp. 575–582.

—, 2012: Quantitative analysis of systems using game-theoretic learning. *ACM Transactions on Embedded Computing Systems (TECS)*, **11(S2)**, 55:1–55:27.

Sha, L., T. Abdelzaher, K.-E. Årzén, A. Cervin, T. Baker, A. Burns, G. Buttazzo, M. Caccamo, J. Lehoczky, and A. K. Mok, 2004: Real time scheduling theory: A historical perspective. *Real-Time Systems*, **28(2)**, 101–155.

Sha, L., R. Rajkumar, and J. P. Hehoczky, 1990: Priority inheritance protocols: An approach to real-time synchronization. *IEEE Transactions on Computers*, **39(9)**, 1175–1185.

Shoukry, Y., M. Chong, M. Wakiaki, P. Nuzzo, A. Sangiovanni-Vincentelli, S. A. Seshia, J. P. Hespanha, and P. Tabuada, 2016: SMT-based observer design for cyber physical systems under sensor attacks. In *Proceedings of the International Conference on Cyber-Physical Systems (ICCPS)*.

Shoukry, Y., P. D. Martin, P. Tabuada, and M. B. Srivastava, 2013: Non-invasive spoofing attacks for anti-lock braking systems. In *15th International Workshop on Cryptographic Hardware and Embedded Systems (CHES)*, pp. 55–72.

Shoukry, Y., P. Nuzzo, A. Puggelli, A. L. Sangiovanni-Vincentelli, S. A. Seshia, and P. Tabuada, 2015: Secure state estimation under sensor attacks: A satisfiability modulo theory approach. In *Proceedings of the American Control Conference (ACC)*.

Simon, D. E., 2006: *An Embedded Software Primer*. Addison-Wesley.

Sipser, M., 2005: *Introduction to the Theory of Computation*. Course Technology (Thomson), 2nd ed.

Smith, S. and J. Marchesini, 2007: *The Craft of System Security*. Addison-Wesley.

Sriram, S. and S. S. Bhattacharyya, 2009: *Embedded Multiprocessors: Scheduling and Synchronization*. CRC press, 2nd ed.

Stankovic, J. A., I. Lee, A. Mok, and R. Rajkumar, 2005: Opportunities and obligations for physical computing systems. *Computer*, 23–31.

Stankovic, J. A. and K. Ramamritham, 1987: The design of the Spring kernel. In *Real-Time Systems Symposium (RTSS)*, IEEE, pp. 146–157.

—, 1988: *Tutorial on Hard Real-Time Systems*. IEEE Computer Society Press.

Sutter, H. and J. Larus, 2005: Software and the concurrency revolution. *ACM Queue*, **3(7)**, 54–62.

Terauchi, T. and A. Aiken, 2005: Secure information flow as a safety problem. In *In Proc. of Static Analysis Symposium (SAS)*, pp. 352–367.

Tiwari, V., S. Malik, and A. Wolfe, 1994: Power analysis of embedded software: a first step towards software power minimization. *IEEE Transactions on VLSI*, **2(4)**, 437–445.

Tremblay, M., J. M. O'Connor, V. Narayannan, and H. Liang, 1996: VIS speeds new media processing. *IEEE Micro*, **16(4)**, 10–20.

Tromer, E., D. A. Osvik, and A. Shamir, 2010: Efficient cache attacks on aes, and countermeasures. *Journal of Cryptology*, **23(1)**, 37–71.

Turing, A. M., 1936: On computable numbers with an application to the entscheidungsproblem. *Proceedings of the London Mathematical Society*, **42**, 230–265.

Vahid, F. and T. Givargis, 2010: *Programming Embedded Systems - An Introduction to Time-Oriented Programming*. UniWorld Publishing, 2nd ed. Available from: http://www.programmingembeddedsystems.com/.

Valvano, J. W., 2007: *Embedded Microcomputer Systems - Real Time Interfacing*. Thomson, 2nd ed.

Vardi, M. Y. and P. Wolper, 1986: Automata-theoretic techniques for modal logics of programs. *Journal of Computer and System Sciences*, **32(2)**, 183–221.

von der Beeck, M., 1994: A comparison of Statecharts variants. In Langmaack, H., W. P. de Roever, and J. Vytopil, eds., *Third International Symposium on Formal Techniques in Real-Time and Fault-Tolerant Systems*, Springer-Verlag, Lübeck, Germany, vol. 863 of *Lecture Notes in Computer Science*, pp. 128–148.

Wang, Y., S. Lafortune, T. Kelly, M. Kudlur, and S. Mahlke, 2009: The theory of deadlock avoidance via discrete control. In *Principles of Programming Languages (POPL)*, ACM SIGPLAN Notices, Savannah, Georgia, USA, vol. 44, pp. 252–263.

Wasicek, A., C. E. Salloum, and H. Kopetz, 2011: Authentication in time-triggered systems using time-delayed release of keys. In *14th IEEE International Symposium on Object/Component/Service-Oriented Real-Time Distributed Computing (ISORC)*, pp. 31–39.

Wiener, N., 1948: *Cybernetics: Or Control and Communication in the Animal and the Machine*. Librairie Hermann & Cie, Paris, and MIT Press.Cambridge, MA.

Wilhelm, R., 2005: Determining Bounds on Execution Times. In Zurawski, R., ed., *Handbook on Embedded Systems*, CRC Press.

Wilhelm, R., J. Engblom, A. Ermedahl, N. Holsti, S. Thesing, D. Whalley, G. Bernat, C. Ferdinand, R. Heckmann, T. Mitra, F. Mueller, I. Puaut, P. Puschner, J. Staschulat, and P. Stenstr, 2008: The worst-case execution-time problem - overview of methods and survey of tools. *ACM Transactions on Embedded Computing Systems (TECS)*, **7(3)**, 1–53.

Wolf, W., 2000: *Computers as Components: Principles of Embedded Computer Systems Design*. Morgan Kaufman.

Wolfe, V., S. Davidson, and I. Lee, 1993: RTC: Language support for real-time concurrency. *Real-Time Systems*, **5(1)**, 63–87.

Wolper, P., M. Y. Vardi, and A. P. Sistla, 1983: Reasoning about infinite computation paths. In *24th Annual Symposium on Foundations of Computer Science (FOCS)*, pp. 185–194.

Young, W., W. Boebert, and R. Kain, 1985: Proving a computer system secure. *Scientific Honeyweller*, **6(2)**, 18–27.

Zeigler, B., 1976: *Theory of Modeling and Simulation*. Wiley Interscience, New York.

Zeigler, B. P., H. Praehofer, and T. G. Kim, 2000: *Theory of Modeling and Simulation*. Academic Press, 2nd ed.

Zhao, Y., E. A. Lee, and J. Liu, 2007: A programming model for time-synchronized distributed real-time systems. In *Real-Time and Embedded Technology and Applications Symposium (RTAS)*, IEEE, Bellevue, WA, USA, pp. 259 – 268.

Zhuang, L., F. Zhou, and J. D. Tygar, 2009: Keyboard acoustic emanations revisited. *ACM Transactions on Information and System Security (TISSEC)*, **13(1)**, 3.

Notation Index

Index

Printed in the United States
by Baker & Taylor Publisher Services